The Nature of
Mathematical Knowledge

The Nature of
Mathematical Knowledge

PHILIP KITCHER

But how I caught it, found it, or came by it,
What stuff 'tis made of, whereof it is born,
I am to learn. . . .
The Merchant of Venice, Act I, Scene I

New York   Oxford
OXFORD UNIVERSITY PRESS
1984

Copyright © 1984 by Oxford University Press, Inc.

Library of Congress Cataloging in Publication Data

Kitcher, Philip, 1947–
   The nature of mathematical knowledge.

   Bibliography: p.
   Includes index.
   1. Mathematics—Philosophy.   I. Title.
QA8.4.K53   511   81-22378
ISBN 0-19-503541-0   AACR2

Printing (last digit):   9 8 7 6 5 4 3 2 1

Printed in the United States of America

Preface

This book has evolved over a number of years. I owe much to many people. I would like to begin by recording my gratitude, both to individuals and to institutions.

As a graduate student on the Philosophy side of the Program in History and Philosophy of Science at Princeton University I received an education whose quality I have appreciated ever since. I learned much from Paul Benacerraf and Michael Mahoney who jointly directed my doctoral dissertation. The former gave me whatever ability I have to criticize and refine my own pet ideas; the latter inspired my interest in the history of mathematics and showed me how to practice it. In addition, I am very grateful to Peter Hempel and to Thomas Kuhn. Their teaching has had a pervasive influence on my thinking and writing.

An older debt is to the staff of the Royal Mathematical School at Christ's Hospital. I was fortunate to receive, while still in my teens, an extraordinarily rich education in mathematics. Since some of the talented mathematicians and dedicated teachers who kindled my love of mathematics are now dead, I cannot thank them in person. But I want to record my indebtedness to Messrs. R. Rae, J. Bullard, J. I. Gowers, N. T. Fryer, I. McConnell, and, especially, W. Armistead.

Many colleagues and friends have commented on parts of this book or on papers which worked toward it. Michael Resnik and Ivor Grattan-Guinness have been extremely generous with criticisms and suggestions. I am grateful to Leslie Tharp, Penelope Maddy, and Emily Grosholz for helpful correspondence. Discussions with Roger Cooke, David Fair, Allan Gibbard, Alvin Goldman, Jaegwon Kim, Hilary Kornblith, Timothy McCarthy, and George Sher have been extremely valuable. I would also like to thank two chairmen of the Department of Philosophy of the University of Vermont, Steven Cahn and Bill Mann, for their encouragement and support.

My research on this book has been aided by three grants. In the summer of 1975 and again in the summer of 1979 I received research grants from the Graduate College of the University of Vermont that enabled me to write papers which advanced my thinking about mathematical knowledge. In 1980 I held a summer fellowship from the National Endowment for the Humanities. During this period I was able to complete most of the first draft of this book. I appreciate very much the support of these institutions.

Like my colleagues in the Department of Philosophy at the University of Vermont, I have been lucky to be able to exploit the secretarial skills of Leslie Weiger. Her exceptional proficiency at the keyboard, together with her intelligence and understanding, made the production of my manuscript far easier than I had any reason to expect it to be. I am very grateful to her, and also to Sandra Gavett of the Bailey-Howe Library at the University of Vermont, whose expertise enabled me to consult rare primary sources in the history of mathematics.

Although some of the debts I have mentioned are substantial, none can compare with that I owe Patricia Kitcher. This is no place to detail what she has given to me in personal terms. It is enough to say that she has been both my greatest source of encouragement and my best critic. Always patient, constructive, intelligent, sensitive, and thorough, she has influenced virtually every page of this book. I dedicate it to her, in love and gratitude.

November 1982   P.K.

Acknowledgments

I am grateful to a number of people and institutions who have granted me permission to quote from previously published works. I would like to thank Paul Benacerraf and Hilary Putnam for allowing me to quote from the postscript to Kurt Gödel's "What Is Cantor's Continuum Problem?," which originally appeared in their *Philosophy of Mathematics: Selected Readings;* Prentice-Hall for permission to quote from *Philosophy of Logic* by W. V. Quine; Yale University Press for permission to quote from G. Frege *On the Foundations of Geometry and Formal Theories of Arithmetic,* edited and translated by Eike-Hennert W. Kluge; and Dover Publications for permission to quote from the translation of Descartes's *Géométrie* by D. E. Smith and M. Latham, and also from the translation of R. Dedekind's *Essays on the Theory of Numbers* by W. Beman.

  I would also like to thank several journals and their editors who have given me permission to reproduce pages from some of my articles: *The Philosophical Review* for large portions of "A Priori Knowledge" (*Philosophical Review,* 89 (1980): 3–23), in Chapter 1; *The Australasian Journal of Philosophy* for a part of "Apriority and Necessity" (*Australasian Journal of Philosophy,* 58 (1980): 89–101), in the final section of Chapter 1; *Philosophical Studies* for parts of "Arithmetic for the Millian" (*Philosophical Studies,* 37 (1980): 215–36), in Chapters 4 and 6; *Philosophical Topics* for two pages from "How Kant Almost Wrote 'Two Dogmas of Empiricism' (And Why He Didn't)" (*Philosophical Topics,* 12 (1981): 217–49), in Chapter 4; *Isis* for a few paragraphs from "Fluxions, Limits, and Infinite Littlenesse" (*Isis,* 64 (1973): 33–49), in Chapter 10; and *Noûs* for two pages from "Mathematical Rigor—Who Needs It?" (*Noûs,* 15 (1981): 469–93), in Chapter 10.

                                                                            P.K.

Contents

The Nature of
Mathematical Knowledge

Introduction

In this book I shall provide a theory about mathematical knowledge. I take as my starting point the obvious and uncontroversial thesis that most people know some mathematics and some people know a large amount of mathematics. My goal is to understand how the mathematical knowledge of the ordinary person and of the expert mathematician is obtained.

The theory that I shall elaborate breaks with traditional thinking about mathematical knowledge in a number of different respects. Virtually every philosopher who has discussed mathematics has claimed that our knowledge of mathematical truths is different in kind from our knowledge of the propositions of the natural sciences. This almost unanimous judgment reflects two obvious features of mathematics. For the ordinary person, as for the philosopher, mathematics is a shining example of human knowledge, a subject which can be used as a standard against which claims to knowledge in other areas can be measured. However, this knowledge does not seem to grow in the same way as other bodies of human knowledge. Mathematicians do not seem to perform experiments or to await the results of observations. Thus there arises the conviction that mathematical knowledge must be obtained from a source different from perceptual experience. To put the point in a familiar philosophical idiom, mathematical knowledge is a priori.

The doctrine that mathematical knowledge is a priori—*mathematical apriorism* for short—has been articulated in many different ways during the course of reflection about mathematics. To name only the most prominent defenders of mathematical apriorism since the seventeenth century, Descartes, Locke, Berkeley, Kant, Frege, Hilbert, Brouwer, and Carnap all developed the central apriorist thesis in different ways. Most of the disputes in philosophy of mathematics conducted in our century represent internal differences of opinion among apriorists. The theory which I shall advance in this book abandons the common presupposition of these debates. I shall offer a picture of mathematical knowledge which rejects mathematical apriorism.

Although mathematical apriorism has been—and continues to be—an extremely popular doctrine, it has not gone completely unquestioned. J. S. Mill attempted to argue that mathematics is an empirical science, thereby making himself the subject of Frege's biting criticism. More recently, W. V. Quine, Hilary Putnam, and Imre Lakatos have, in different ways, challenged the apriorist thesis.[1] However, none of these writers has offered any systematic account of our mathematical knowledge. Quine insists that mathematical statements, like all other statements, are vulnerable to empirical disconfirmation, but he does not explain how we have come to know the parts of mathematics that we do. Putnam suggests that mathematics involves what he calls "quasi-empirical" inferences, but he does not provide any extended discussion of the notion of a quasi-empirical inference nor does he tell us how we reach the starting points for such inferences. Lakatos attempts to apply some of Popper's ideas about the methods of natural science to episodes from the history of mathematics, but it is very hard to glean from his treatment a clear picture of how our mathematical knowledge has been acquired. Finally, if we return to Mill, we face the problem that many of his formulations are imprecise (almost inviting the well-known Fregean ironies) and, in addition, Mill only considers the most rudimentary parts of mathematics.

Hence I think that it is fair to conclude that the alternative to mathematical apriorism—*mathematical empiricism*—has never been given a detailed articulation. I shall try to provide the missing account. In doing so, I do not pretend to be repudiating completely the ideas of the authors mentioned in the last paragraph. It should be evident in what follows that I have gained much from insights of Quine and Putnam. I have also learned from Mill's derided account of arithmetic. My quarrel with earlier empiricists is, for the most part, that their accounts have been incomplete rather than mistaken.

The theory of mathematical knowledge which I shall propose breaks with tradition not only by rejecting mathematical apriorism. I shall also abandon a tacit assumption which pervades much thinking about knowledge in general and mathematical knowledge in particular. We are inclined to forget that knowers form a community, painting a picture of a person as having built up by herself the entire body of her knowledge of (for example) mathematics. Yet it is a commonplace that we learn, and that we learn mathematics, from others. Traditional views of mathematical knowledge would probably not deny the commonplace, but would question its epistemological relevance. I shall give it a central place in my account of mathematical knowledge.

A third break with the usual approaches to mathematical knowledge consists in my emphasis on the historical development of mathematics. I suggest that the knowledge of one generation of mathematicians is obtained by extending

---

1. See W. V. Quine, "Two Dogmas of Empiricism," *Philosophy of Logic*, chapter 7; Hilary Putnam, "What is Mathematical Truth?"; Imre Lakatos, *Proofs and Refutations*. For Mill's views, see *A System of Logic*, especially book two, chapters 5 and 6, and book three, chapter 24.

the knowledge of the previous generation. To understand the epistemological order of mathematics one must understand the historical order. (As will become clear later in this introduction, this does not quite mean that the epistemological order *is* the historical order.) Most philosophers of mathematics have regarded the history of mathematics as epistemologically irrelevant. (Lakatos's principal insight, it seems to me, was to recognize that this is a mistake.)[2] They have supposed that, independently of the historical process through which mathematics has been elaborated, the individual mathematician of the present day can reconstruct the body of knowledge bequeathed to us by our predecessors, achieving systematic knowledge which does not reflect the patterns of inference instantiated in the painful historical process.

At this point, I can sketch the theory of mathematical knowledge which I shall present, and thereby bring into focus the ways in which it differs from previous approaches. I shall explain the knowledge of individuals by tracing it to the knowledge of their communities. More exactly, I shall suppose that the knowledge of an individual is grounded in the knowledge of community authorities. The knowledge of the authorities of later communities is grounded in the knowledge of the authorities of earlier communities. Putting these two points together, we can envisage the mathematical knowledge of someone at the present day to be explained by reference to a chain of prior knowers. At the most recent end of the chain stand the authorities of our present community—the teachers and textbooks of today. Behind them is a sequence of earlier authorities. However, if this explanation is to be ultimately satisfactory, we must understand how the chain of knowers is itself initiated. Here I appeal to ordinary perception. Mathematical knowledge arises from rudimentary knowledge acquired by perception. Several millennia ago, our ancestors, probably somewhere in Mesopotamia, set the enterprise in motion by learning through practical experience some elementary truths of arithmetic and geometry. From these humble beginnings mathematics has flowered into the impressive body of knowledge which we have been fortunate to inherit.

2. Although I agree with Lakatos that the history of mathematics is epistemologically relevant, my development of this theme will be somewhat different from his treatment of it. In particular, my study of the growth of mathematical knowledge will be placed in the context of a prior discussion of epistemological issues which diverges from Lakatos's epistemological assumptions at many points. I have decided not to include any explicit comparison of my own views with those of Lakatos for several reasons. First, as I have just noted, his epistemological framework differs greatly from my own, so that direct comparison would require much reformulation and stage-setting. Second, I have already advanced my main criticisms of Lakatos in a review of *Proofs and Refutations*. Third, given the development of Lakatos's own methodological ideas, it does not seem appropriate to spend a great deal of time on the Popperian brand of falsificationism which he later rejected.

In an interesting recent paper ("Towards a Theory of Mathematical Research Programmes"), Michael Hallett has begun to develop an account of the growth of mathematical knowledge along the lines of Lakatos's later philosophy of science. This work has some points of contact with my own approach, but I think that it is less well articulated than the theory offered in this book and that it retains some faulty epistemological assumptions from the Popper-Lakatos tradition.

I anticipate charges that this account is plainly too crude or that it is absurd. These charges are likely to stem from two major problems that I shall attempt to overcome. First, is it possible to claim that the humble experiences of Babylonian chandlers or Egyptian bricklayers could give us genuinely *mathematical* knowledge? Doesn't mathematics describe a reality which is far too refined to be penetrated by perception? I shall try to explain how perceptual origins for mathematical knowledge are possible by giving an account of what mathematics is about, a picture of mathematical reality, if you like. This will block the first major line of objection to my theory, and also supply a basis for my response to the second. The second criticism springs from recognition of the extent to which contemporary mathematics differs from the subject begun by our Mesopotamian ancestors. Is it really credible that, from such primitive beginnings, we should have been led to our present corpus of abstract knowledge? My aim will be to show that it is indeed credible to suppose that our knowledge should grow in this way. I shall try to disclose patterns of rational inference which can lead the creative mathematician to extend the knowledge which his authorities have passed on to him. As the sum of these rational endeavors, the knowledge of the community increases from generation to generation, and I shall describe how, in one important case, a sequence of rational transitions transformed the character of mathematics.

Thus my presentation of my theory centers on answering two main questions: What is mathematics about? How does mathematical knowledge grow? Once these questions have been answered, I suggest that the approach to mathematical knowledge briefly presented above will yield an adequate explanation of mathematical knowledge. In answering them, I depart further from some popular views about mathematics.

Currently, the most widely accepted thesis about the nature of mathematical reality is Platonism. Platonists regard mathematical statements as descriptive of a realm of mind-independent abstract objects—such as numbers and sets. This position gains credibility *faute de mieux*. It is widely assumed that Platonism is forced on us if we want to accommodate the results and methods of classical mathematics. Certainly, the views of traditional opponents of Platonism—nominalists and constructivists—have usually involved restrictions of mathematics, restrictions which mathematicians have often seen as mutilations of their discipline. However, I shall try to show that these sacrifices are not inevitable. I shall assemble the elements of a non-Platonistic picture of mathematical reality from various sources, developing a version of constructivism which answers to a range of desiderata, most notably to the demands that the objectivity and utility of mathematics should be explicable and that the methods of classical mathematics should not be curtailed.

My answer to the question of how mathematical knowledge evolves can hardly be said to challenge previous *philosophical* conceptions, since, as I have already noted, philosophers have had very little to say about the history of mathematics. However, my treatment will differ from the usual discussions of historians of mathematics and will challenge philosophical assumptions which those discussions presuppose. Although the history of mathematics is an undeveloped part of the history of science—it is thus an immature subfield of a discipline which has only recently come of age—recent years have seen the appearance of a number of interesting studies of mathematical figures, problems, and concepts. Yet, although these studies have replaced casual anecdote with sophisticated analysis of texts and proofs, they are confined by the apriorist perspective which has dominated philosophy of mathematics. In analyzing the history of mathematics with the aim of showing how mathematical knowledge grows, I shall be asking questions which standardly do not occur in historical discussions. The contrast can best be understood by considering the parallel situation in the history and philosophy of natural science. Philosophers of science have spent considerable effort in discussing the types of inferences which occur in natural science and the desiderata which play a role in theory choice. Their conclusions form the backdrop for historical discussion, so that one can draw on philosophical views about theory and evidence to investigate the work of a particular figure and, conversely, historical investigations can prompt revisions of one's philosophical models. Since virtually no philosophical attention has been given to the question of how mathematical knowledge evolves, of what kinds of inferences and desiderata function in the growth of mathematics, this interplay between history and philosophy is absent in the mathematical case. As a result, I think that the historiography of mathematics has been stunted: some of the most fascinating questions have rarely found their way into historical discussion. In attempting to show how the historical development of mathematics can be seen as a sequence of rational transitions, I shall sometimes be doing history with a different emphasis from that which is usual. Instead of focussing on the question of how to reconstruct the proofs of the great mathematicians of the past, I shall attend to a wider range of issues. How and why does mathematical language change? Why do some mathematical questions come to assume an overriding importance? How are standards of proof modified? By raising these questions and suggesting how they should be answered, I intend not only to complete my epistemological project but also to outline a novel approach to the history of mathematics.

To summarize, my theory of mathematical knowledge traces the knowledge of the contemporary individual, through the knowledge of her authorities, through a chain of prior authorities, to perceptual knowledge acquired by our remote ancestors. This theory rejects mathematical apriorism, and ascribes to the present mathematical community and to previous communities an epistemological significance with which they are not usually credited. I intend to elaborate this theory by giving an account of mathematical reality, an account which will forestall worries about how perceptual experience could have initiated the tradition, and by identifying those patterns of rational transition which have led from primitive beginnings to the mathematics of today. Both of these endeavors involve me in further heterodoxy. My account of mathematical reality rejects the Platonist view of mathematics, diverging also from previous versions of nominalism and constructivism. And my account of the growth of mathematical knowledge is intended to point toward a new historiography of mathematics.

Let me now explain how the chapters that follow carry out the projects which I have announced. The first part of the book is devoted to a critique of apriorism. I begin with some general epistemological points which are needed if we are to understand the apriorist thesis. Once the thesis has been clearly stated, it is then possible to begin systematic evaluation of the versions of mathematical apriorism which have been proposed. I divide these into three major groups, corresponding to three positions about the nature of mathematical truth, and I argue that none of the three ways of articulating apriorism will succeed. At this point, it is possible for me to explain more precisely my own positive theory. Chapter 5 uses the previous critique of apriorism to elaborate my own picture, to suggest how the genuine insights of some apriorists may be accommodated, and to set the stage for the subsequent development of my own theory.

Chapter 6 undertakes the task of providing an account of mathematical reality. The second major enterprise, that of explaining how mathematical knowledge evolves, occupies the remaining four chapters. In Chapter 7, I compare mathematical change with scientific change, attempting to show that the growth of mathematical knowledge is far more similar to the growth of scientific knowledge than is usually appreciated and using the comparison to pose my problem more precisely. The next two chapters are devoted to assembling the elements of my account. Chapter 8 surveys the types of changes in mathematics which are of epistemological interest, and Chapter 9 describes some types of inference and principles of theory choice which are involved in the growth of mathematics. Finally, in Chapter 10, I try to show that the elements I have assembled can be fitted together to yield a coherent account of the development of analysis from the middle of the seventeenth century to the end of the nineteenth century. This case study is intended to rebut the charge that no empiricist account can do credit to our knowledge of advanced mathematics and to show how the history of mathematics looks from the perspective of my theory of mathematical knowledge.

Although I believe that the best way to reveal the advantages of my theory is to give a detailed presentation of it and a critique of previous approaches, there are a few concerns and objections which I want to address before I launch my main exposition and argument. Consider first the worry that the type of empiricism which I favor will be limited to a utilitarian view of mathematics. In a deservedly popular book, G. H. Hardy argues eloquently against the thesis that the activity of mathematics can be defended on the grounds that it is practically important: ''The 'real' mathematics of the 'real' mathematicians, the mathematics of Fermat and Euler and Gauss and Riemann, is almost wholly 'useless' (and this is as true of 'applied' as of 'pure' mathematics.''[3] The claim is an overstatement but it has a sound core. One would be hard pressed to explain the utility of the great theorems of number theory (one of Hardy's favorite fields). Yet despite the fact that the roots of mathematical knowledge lie in simple perceptual experiences, and although those experiences give rise to items of knowledge which have obvious practical value, we should not assume that the development of mathematics preserves the pragmatic significance which accrues to the rudiments. There is no obvious reason to rule out the possibility of patterns of mathematical inference and principles of mathematical theory choice which would lead us to develop the "useless" parts of mathematics. Indeed, brief reflection on the natural sciences will remind us that enterprises which begin with practical problems may end in theories which have little practical utility. (Inquiries which begin with everyday concerns about the structure of matter may terminate in investigations of the elusive properties of short-lived particles.) Hence we should not convict mathematical empiricism in advance for overemphasizing the usefulness of mathematics. As I shall try to show in Chapters 9 and 10, my evolutionary epistemology can account for the "real" mathematics of the "real" mathematicians whom Hardy mentions—as well as for the "real" mathematics which is of practical significance, some of the mathematics of Archimedes, Newton, Laplace, Fourier, and von Neumann.

A second natural concern about my theory arises from my remarks about the historical order and the epistemological order. Do I intend to claim that our knowledge of some part of mathematics is based on the actual historical process through which that part of mathematics was originally introduced? The historicism which I advocate is not so crude. I allow for the possibility that new principles (and concepts) are originally adopted on inadequate grounds, and that it is only later that they obtain their justification through the exhibition of an entirely different relation to previous mathematics. Nor would I deny that, as time goes on, new ways of justifying old extensions of mathematics are discovered so that, when we trace the epistemological order of mathematics, it may diverge at some points from the order of historical development. What matters is that we should be able to describe a sequence of transitions leading from perceptually justified mathematical knowledge to current mathematics. In giving this description, we shall follow the historical order *grosso modo,* in that we shall appeal to antecedently justified principles to justify further extensions.

My point is easily illustrated with an example. Imagine that statements about complex numbers were first adopted for relatively poor reasons, but that, after their introduction, it was found that the newly accepted statements could be used to solve a variety of traditional mathematical problems. Then it is appro-

3. *A Mathematician's Apology,* p. 119.

priate to regard the knowledge of later communities (including our own community) as based on the recognition of the success of the theory of complex numbers, rather than on the inadequate grounds which originally inspired the theory. We can continue to uphold the general historicist claim that later modifications of mathematical practice are justified in virtue of their relation to elements of prior practices, without committing ourselves to the crude historicist thesis that the relation must be exhibited in the historical genesis of the modification.

Another natural response to my account would be that, as so far presented, it contains no mention of proof. Although I shall have plenty to say about the concept of proof—and about the apriorist construal of proof—in what follows, I want to forestall a misinterpretation of my theory. I do not intend to deny that much mathematical knowledge is gained by constructing or following the sequences of statements contained in mathematics books and labelled "proofs." Nor am I suggesting that the kinds of inferences involved in these proofs are anything other than what logicians and philosophers of mathematics have traditionally taken them to be. My point is that if we are to understand how the activity of following a proof generates knowledge we must be able to understand how the person who follows the proof knows the principles from which the proof begins. If we are to give an account of how someone reaches the starting points for her deductive inferences then, I suggest, we shall have to use the picture of mathematical knowledge which I have outlined. To be explicit, I envisage the complete explanation of a mathematician's knowledge of the theorem she has just proved to run as follows. We begin by showing how the mathematician's knowledge of the principles from which the proof begins, together with the activity of following the proof, engenders knowledge of the theorem. Then we turn our attention to the knowledge of the first principles, either tracing this back through a chain of previous proofs to knowledge gleaned from authorities or, perhaps, recognizing that this knowledge was obtained directly from authorities. We then account for the knowledge of the authorities by appeal to prior authorities, exhibiting how it evolved through a sequence of rational transitions from perceptually based, rudimentary mathematical knowledge. In emphasizing the role of nondeductive inferences in mathematics, I am not opposing the thesis that much of our mathematical knowledge is gained by following proofs but rather exposing the conditions which make it possible for proofs to give us knowledge.

One final point to which I wish to respond is the charge that, on the account I have sketched, it is hard to understand how mathematical creativity is possible. Someone may worry that I have depicted the individual mathematician as subservient to the authority of the community and that I have portrayed the community as dominated by tradition. However, despite the fact that the young mathematician begins his career by acquiring knowledge from authorities, and although the knowledge that is transmitted is shaped by the previous development of mathematics, two kinds of creative accomplishment are possible. The

first consists in adding to the store of mathematical results without amending
the basic framework within which mathematics is done: it sometimes takes
great creativity—even genius—to show that the existing resources (concepts,
principles) suffice for the proof of an important conjecture. The second type of
creativity is more dramatic. Moved by considerations which govern the development of mathematics at all times, a mathematician may modify, even transform, the elements of the practice which he inherited from his teachers, introducing new concepts, principles, questions, or methods of reasoning. How this
is possible and what kinds of considerations can induce revision of the authoritative doctrine of a community are topics which I shall investigate in some
detail in the later chapters of this book.

Let me conclude this Introduction by acknowledging some constraints on an
adequate theory of mathematical knowledge. An answer to the question "How
do we know the mathematics we do?" ought to fit within the general account
of human knowledge offered by epistemologists and psychologists. Since the
details of epistemology and of the psychology of cognition are both matters of
controversy, I have tried to remain neutral wherever the development of my
theory permitted. Nevertheless, it is true that the theory I propose can easily
be recast in the favored terminology of a currently popular psychological theory, the approach of "ecological realism" which stems from the work of J.
J. Gibson and his students.[4] Some of the central ideas of ecological realism
can be used to add further detail to my account of mathemetical knowledge.
From a different perspective, my account may be seen as resolving a problem
for ecological realism, the problem of how to fit mathematical knowledge into
the ecological approach.

Ecological realism offers a theory of perception according to which perception is *direct*. What this means is that the idea of perception as a process in
which the mind engages in complicated inferences and computations to construct a perception from scanty data is abandoned. Instead, ecological realists
emphasize the richness of sensory information, claiming that questions about
how we compute or construct to achieve awareness of intricate features of our
environment only arise because the perceptual data have been misrepresented
as impoverished. The view that perceptual data are rich is encouraging to any
theory which claims, as mine does, that we can identify a perceptual basis for
mathematical knowledge. Even more promising for my particular account, is
the doctrine that what an organism primarily perceives are the *affordances* of
things in its environment. Gibson and his followers introduce the technical term
"affordance" to mark out what the environment *"offers* animals, what it *pro-*

---

4. I am grateful to an anonymous reader for bringing to my attention a recent book, *Direct Perception,* by Claire F. Michaels and Claudia Carello, which provides a succinct account of this work
for the nonpsychologist. I should note that, while I find the psychological claims of ecological
realism interesting, I do not endorse many of the philosophical points put forward by Michaels and
Carello.

vides or *furnishes,* either for good or ill."[5] Examples are easily found: lettuce affords eating to rabbits; a tree affords refuge to a squirrel pursued by a dog. What is distinctive about ecological realism is the use it makes of this concept: ". . . for Gibson, *it is the affordance that is perceived.*"[6] If this doctrine is correct, then the picture of mathematical reality proposed below in Chapter 6 will lend itself to a simple psychological story of the basis of mathematical knowledge. The constructivist position I defend claims that mathematics is an idealized science of operations which we can perform on objects in our environment. Specifically, mathematics offers an idealized description of operations of collecting and ordering which we are able to perform with respect to any objects. If we say that a *universal affordance* is an affordance which any environment offers to any human, then we may state my theory as the claim that mathematics is an idealized science of particular universal affordances. In this form, the theory expresses clearly the widespread utility of mathematics and, given the ecological realist claim that affordances are the objects of perception, it is also easy to see how mathematical knowledge is possible.

I offer this thumbnail sketch of ecological realism only to indicate how psychological theory might develop further my account of mathematical knowledge, and how, by the same token, my theory is constrained by the requirement that such development ought to be forthcoming. If ecological realism is correct, then the fact that my view of mathematical knowledge can so easily be integrated with it should provide further support for my view. However, I want to stress that, to the best of my knowledge, no psychological theory which currently enjoys wide support is incompatible with the picture of mathematics I propose.[7] Hence, although I would be pleased if ecological realism should receive further confirmation, I believe that the ideas advanced in this book could survive its demise.

I have explicitly tried to construct a theory of mathematical knowledge which will honor certain general considerations about knowledge and will also do justice to the historical development of mathematics. The account offered below is the direct result of applying those constraints. But any correct picture of mathematics ought to meet further requirements, requirements which I have not explicitly used in working out my theory. Among these are the demands made by psychology. In these last paragraphs I have tried to suggest briefly why I think those demands can also be met.

5. J. J. Gibson, *The Ecological Approach to Visual Perception.* Cited in Michaels and Carello, *Direct Perception,* p. 42.

6. *Direct Perception,* p. 42.

7. In particular, even a psychologist who holds that we have tacit mathematical knowledge, which we employ in unconscious computational processes, need not reject my theory as an account of our explicit mathematical knowledge. This combination of positions is the analog of the idea that a successful linguist may have both tacit knowledge of the grammar of a language and explicit knowledge of grammar which is based on empirical investigations.

1
Epistemological Preliminaries

I

What is knowledge? What conditions must be met if someone is to know something? These questions are, not surprisingly, central to epistemology. Fortunately, we shall not need detailed answers to them in order to undertake an investigation of mathematical knowledge. However, we shall need to decide between two major *kinds* of answer.

Most philosophers before our century assumed, for the most part implicitly, that the correct account of knowledge should be *psychologistic*. They supposed that states of knowledge are states of true belief—for someone to know that *p*, it must be true that *p* and the person must believe that *p*—but they recognized that not any state of true belief is a state of knowledge. Adopting a psychologistic account, they tacitly assumed that the question of whether a person's true belief counts as knowledge depends on whether the presence of the state of true belief can be explained in an appropriate fashion. The difference between an item of knowledge and mere true belief turns on the factors which produced the belief—thus the issue revolves around the way in which a particular mental state was generated. In many cases, of course, the explanation of the presence of belief must describe a process which includes events extrinsic to the believer if it is to settle the issue of whether the belief counts as an item of knowledge, but the approach is appropriately called 'psychologistic' in that it focuses on processes which produce belief, processes which will always contain, at their latter end, psychological events. Pursuing the psychologistic approach to knowledge, many of the great philosophers of the seventeenth, eighteenth, and nineteenth centuries took an important part of their epistemological task to be one of describing psychological processes which can engender various kinds of knowledge.

Twentieth-century epistemology has frequently been characterized by an attitude of explicit distaste for theories of knowledge which describe the psychological capacities and activities of the subject. This attitude has fostered an *apsychologistic* approach to knowledge, an approach which proposes that knowledge is differentiated from true belief in ways which are independent of the causal antecedents of a subject's states.[1] What is crucial is the character of the subject's belief system, the nature of the propositions believed, and their logical interconnections. Abstracting from differences among rival versions of apsychologistic epistemology, we can present the heart of the approach by considering the way in which it would tackle the question of whether a person's true belief that *p* counts as knowledge that *p*. The idea would be to disregard the psychological life of the subject, looking just at the various propositions she believes. If *p* is "connected in the right way" to other propositions which are believed, then we count the subject as knowing that *p*. Of course, apsychologistic epistemology will have to supply a criterion for propositions to be "connected in the right way"—it is here that we shall encounter different versions of the apsychologistic approach—but proponents of this view of knowledge will emphasize that the criterion is to be given in *logical* terms. We are concerned with logical relations among propositions, not with psychological relations among mental states.

It is important for me to resolve the issue between these two approaches to knowledge because apsychologistic epistemology, at least in some of its popular versions, allows no place to the questions about mathematical knowledge which concern me most. Frequently, an apsychologistic epistemology is developed by attributing to some propositions a special status. These propositions, labelled as "self-justifying," "self-evident," and so forth, are supposed to have the property of counting automatically as items of knowledge if they occur on the list of propositions which the subject believes. Now the major candidates for this special status, apart from minimal propositions about "perceptual appearances," have been so-called "a priori truths," including at least some propositions of logic and mathematics. Once apsychologistic epistemology has granted to some axioms of mathematics this privileged epistemic role, the question of how we know these axioms disappears and one who raises that question (as I intend to do) can be accused of dabbling in those psychological mysteries from which twentieth-century epistemology has liberated itself. Since I think that some of the support for mathematical apriorism stems from acceptance of an apsychologistic epistemology with consequent dismissal of the question of how we might come to a priori mathematical knowledge, it is important for me to undermine the apsychologistic approach.

Luckily, I am not alone in rejecting apsychologistic epistemology. Recently a number of writers have made a persuasive case for the claim that if a subject

---

1. This apsychologistic approach is present in the writings of Russell, Moore, Ayer, C. I. Lewis, R. Chisholm, R. Firth, W. Sellars, and K. Lehrer, and is presupposed by the discussions of science offered by Carnap, Hempel, and Nagel.

is to know that p then his belief must have the right kind of causal (including psychological) antecedents.[2] There is a general method of finding counterexamples to the proposals of apsychologistic epistemologists. Given a set of "logical constraints" on belief systems which purportedly distinguish knowledge from true belief, we can describe cases in which subjects meet those constraints fortuitously, so that, intuitively, they fail to know because the logical interconnections required by the proposal are not reflected in psychological connections made by the subject. Even though we have at our disposal excellent reasons for believing a proposition, we may still come to believe it in some epistemically defective way. There are numerous possible cases ranging from the banal to the exotic. We may believe the proposition because we like the sound of a poetic formulation of it. Or we may credulously accept the testimony of some disreputable source. Or—conceivably—we may be the victims of the recreational gropings of some deranged neurophysiologists. These possible explanations of the formation of belief enable us to challenge any apsychologistic account of knowledge with scenarios in which the putative conditions for knowledge are met but in which, by our intuitive standards, the subject fails to know.

Let me illustrate the point by considering that part of our knowledge with which I am chiefly concerned. The logical positivists hoped to understand the notion of a priori knowledge and to defend the apriority of mathematics without venturing into psychology. The simplest of their suggestions for analyzing a priori knowledge was to propose that X knows a priori that p if and only if X believes that p and p is analytically true.[3] After some hard work to show that mathematics is analytic, they could then draw the conclusion that we know a priori the mathematical truths that we believe. Irrespective of its merits as a proposal about *a priori knowledge,* the positivist analysis fails to identify a sufficient condition for *knowledge.* Grant, for the sake of argument, the positivist thesis that mathematics is analytic, and imagine a mathematician who comes to believe that some unobvious theorem is true. Her belief is exhibited in continued efforts to prove the theorem. Finally, she succeeds. It seems eminently possible that the original reasons which led to belief are not good enough for it initially to count as an item of knowledge, so that we would naturally claim that the mathematician has come to know something which she only believed (or conjectured) before. The positivistic proposal forces us to attribute knowledge from the beginning. Worse still, we can imagine that the mathematician has many colleagues who believe the theorem because of dreams,

2. This point has been clearly made by Gilbert Harman (*Thought,* chapter 2) and by Alvin Goldman ("What Is Justified Belief?"). My own thinking about basic epistemological issues owes much to Goldman's work, not only in "What Is Justified Belief?" but also in his other papers on epistemology.

3. A. J. Ayer, *Language, Truth and Logic,* chapter 4; M. Schlick, "The Foundation of Knowledge," especially p. 224.

trances, fits of Pythagorean ecstasy, and so forth. Not only does the positivistic approach fail to separate the mathematician after she has found the proof from her younger self, but it also gives her the same status as her colleagues.

Is this a function of the fact that the positivistic proposal is too crude? Let us try a natural modification. Distinguish among the class of analytic truths those which are *elementary* (basic laws of logic, immediate consequences of definitions, and, perhaps, a few others). Restrict the proposal so that it claims only that elementary analytic truths can be known (a priori) merely by being believed. Even this more cautious version of the original claim is vulnerable. If you believe the basic laws of logic because you have faith in the testimony of a maverick mathematician, who has deluded himself into believing that the principles of some inconsistent system (the system of Frege's *Grundgesetze,* for example) are true, and if you ignore readily available evidence which exposes your favored teacher as a misguided fanatic, then you do not know those laws.

This is not the *coup de grâce* for apsychologistic epistemology. Further epicycles can be added to circumvent the counterexamples so far adduced. So, for example, someone concerned to defend the apsychologistic program may demand that the subject should have certain other beliefs (such as beliefs about his beliefs) if belief in an analytic truth is to constitute knowledge. Readers of the recent literature in epistemology will be familiar with some of these manoeuvres. I suggest that they are *merely* epicycles; although they patch up one local problem for an apsychologistic epistemology, they leave the root difficulty untouched. Given any apsychologistic condition, we can always tell a new version of our story and thereby defeat the proposal. Our success results from the fact that the mere presence in the subject of a particular belief or of a set of beliefs is always compatible with peculiar stories about causal antecedents. The predicament of the apsychologistic epistemologist is comparable to that of a seventeenth-century Ptolemaic astronomer, struggling desperately to reduce Kepler's elliptical orbits to a complex of circular, earth-centered motions.

We can strengthen the case for psychologistic epistemology by disarming an objection to it, an objection which may originally have prompted the apsychologistic approach and which may represent the continued struggles to defend the approach as worthwhile.[4] Does not psychologistic epistemology commit the genetic fallacy, confusing the context of discovery with the context of justification? The question arises from the correct observation that there are cases in which we are originally led to belief in an epistemically defective way but later acquire excellent grounds for our belief. We should not, however, use this

---

4. For a recent formulation of this objection, see Keith Lehrer, *Knowledge,* pp. 123ff. The objection is sometimes traced to Frege's *Grundlagen.* In ''Frege's Epistemology,'' section II, I argue that Frege's own approach to knowledge is similar to that proposed here, and that the type of psychologism I am concerned to defend is not the type to which Frege objected.

observation to indict psychologistic epistemology. The psychologistic epistemologist claims that states of knowledge that p are distinguished from states of
mere true belief that p by the character of the processes which, in each case,
produce belief that p. This allows for an adequate treatment of the examples in
which we arrive at a belief defectively and later come to know it. In such
instances, the process which produces the original state of belief does not meet
the conditions required of processes which engender knowledge. Later states
of belief that p are produced by different processes which do meet those conditions. Putting the issue in terms of explanation, we may point out that, in the
examples which are allegedly troublesome, the explanation of the original state
of belief is different from the explanation of later states of belief, and this
difference enables us to recognize the later states as states of knowledge even
though, initially, the subject only had true belief. Properly understood, psychologistic epistemology avoids the genetic fallacy and allows for a distinction
between discovery and justification.

I conclude that an adequate theory of knowledge will be psychologistic. My
main goal in this chapter will be to show how the notion of a priori knowledge
should be understood in the terms of psychologistic epistemology. However,
before proceeding to the notion of apriority, I want to make some general
epistemological points which will be useful for my enterprise.

II

There is a simple normal form for a psychologistic account of knowledge. We
may introduce the term 'warrant' to refer to those processes which produce
belief "in the right way" proposing the equivalence

(1) X knows that p if and only if p and X believes that p and X's belief
that p was produced by a process which is a warrant for it.[5]

Obviously, (1) is only the first step towards an account of what knowledge is.
Psychologistic epistemology must proceed by specifying the conditions on warrants. In using (1) to frame a theory of mathematical knowledge I shall not
draw on any of the accounts of warrants which others have put forward, nor
shall I offer a general analysis of my own. I have several reasons for proceeding in this way. First, I reap the benefits of neutrality. The points I shall make
about mathematical knowledge rest only on the thesis that knowledge is to be
understood along the lines of (1). They do not presuppose any particular de-

5. Here I should emphasize that 'process' is to refer to a *token process*—a specific datable sequence of events—not to a process type. As we shall see below, of two processes which belong to
the same type, one may succeed in warranting belief and the other may not. We shall also see that
a token process which warrants belief in one situation may not warrant belief in a different counterfactual situation where background conditions are different.

velopment of (1). Second, I think that at this stage of epistemological inquiry our intuitive judgments about which processes count as warrants for the beliefs they produce are far more reliable than the verdicts of any analysis that I might offer. Third, currently available accounts of warrants are primarily motivated by a small number of examples of knowledge: perceptual knowledge has received far more attention than other kinds of knowledge. By considering mathematical knowledge from a psychologistic perspective, I hope to amass new data which a general account of warrants should accommodate. For these reasons I shall leave the notion of warrant unanalyzed.[6]

However, it will be useful to note some differences among warrants. One distinction I shall employ will be that between *basic* and *derivative* warrants. This distinction is easily motivated by reflecting on the ways in which we come to believe. Sometimes, when we make inferences from beliefs which we already hold to new conclusions, we go through a process which involves other states of belief. Those states of belief are causally efficacious in producing the belief which is the outcome of the process. At other times, it *appears* that matters are different. My present belief that the tree outside the window is swaying slightly can naturally be viewed as the product of a process which does not involve prior beliefs. On a simple account of perception, the process would be viewed as a sequence of events, beginning with the scattering of light from the surface of the tree, continuing with the impact of light waves on my retina, and culminating in the formation of my belief that the tree is swaying slightly; one might hypothesize that none of my prior beliefs play a causal role in this sequence of events. Comparing this example with the case of inference, we can tentatively advance a distinction. A process which warrants belief counts as a *basic* warrant if no prior beliefs are involved in it, that is, if no prior belief is causally efficacious in producing the resultant belief. *Derivative* warrants are those warrants for which prior beliefs are causally efficacious in producing the resultant belief.

I have used the examples of perception and inference as paradigms to motivate a distinction between basic and derived warrants, but, as my qualifying remarks about the perceptual case may already have indicated, I want to allow that matters may be more complex than they initially appear to be. Perhaps our naïve model of perception is wrong, and prior beliefs are included among the causal determinants of such perceptual beliefs as my belief about the swaying tree. This is an issue for psychology to decide. But, even if it should turn out that the perceptual beliefs which we naïvely take ourselves to acquire directly from perception—beliefs like my belief about the tree—have a more complicated causal history than appears at first sight, there will still be some state which is produced in us as a result of perceptual experience, independently of

6. The best available account of warrants seems to me to be that provided by Goldman. See his "Discrimination and Perceptual Knowledge" and "What Is Justified Belief?"

the action of prior belief, and we can regard basic warrants as processes which produce such states. Hence the distinction between basic and derivative warrants can still be drawn. Since my account is easier to formulate if we adopt the naïve model of perception which I used in the last paragraph, I shall continue to employ that model. The points I shall make will survive intact if the model is incorrect and if, as a result, the distinction between basic and derivative warrants needs to be redrawn.[7]

The utility of the distinction between basic and derivative warrants lies in the fact that it can be used to structure our inquiries into our knowledge of a given field. We can ask for a causal ordering of our beliefs which will show us how our knowledge could have been developed. However, it is important to understand that the distinction I have drawn and the causal ordering it induces are not prey to objections which have typically been directed against foundationalist theories of knowledge. Nothing in my account suggests that the beliefs which are produced by basic warrants are incorrigible or that the warrant itself discharges its warranting function independently of other beliefs.[8]

These last points require amplification and illustration. Let us begin by noting that there is a difference between the explanation of why a person has a particular belief and the explanation of why the belief is warranted, or, analogously, between what causes the belief and what causes the belief to be warranted. Given any case of knowledge, we explain the presence of the state of belief (which is the state of knowledge) by describing the process which produced it, a process which is in fact a warrant. However, to explain the state of knowledge *as a state of knowledge,* to show that the process is a warrant and that the subject knows, we shall typically have to do more. Our task will be to demonstrate that, in the particular situation in which the process produced belief, it was able to serve the function of warranting belief. The reason for this is quite straightforward. Processes which can warrant belief given favorable background conditions may be unable to warrant belief given unfavorable background conditions. Two kinds of background conditions are relevant here: features of the world external to the subject and features of the subject's beliefs may both affect the ability of a process to warrant belief.

Basic warrants are not exceptions to this general point. A process which does not involve prior beliefs as causal factors may produce a belief. Because it warrants belief, that process counts as a basic warrant. Yet the power of the

7. The account of basic warrants provided in the text will be adequate so long as perception is *direct.* So, for example, if the ecological realists are right, then there will be no need to redraw the distinction made here. See Michaels and Carello, *Direct Perception,* chapter 1.

8. The relation between psychologistic epistemology and the traditional dispute about foundationalism is illuminated by Hilary Kornblith's paper "Beyond Foundationalism and the Coherence Theory." My discussion in the next paragraphs is indebted to Kornblith's paper and to prior conversations with him.

process to warrant belief may depend on favorable background conditions. Change some features of the external world or of the subject's system of beliefs and, although the process may still be undergone by the subject (and may continue to operate without involving prior beliefs as causal factors), it may no longer warrant belief. Basic warrants, *qua* causal processes which engender belief, are independent of prior beliefs. Basic warrants, *qua* processes which engender warranted belief, are typically not independent of prior beliefs (or of features of the world external to the subject).

Examples of perceptual knowledge will help to make these matters clearer. Suppose that I am looking at some flowers on a table and that the circumstances of this inspection are perfectly normal. I come to believe that there are flowers before me, and (given my assumption about perception) I do so as the result of undergoing a process which is a basic warrant for the belief. I might undergo the same process under different circumstances: surrounded by a host of high quality fake flowers, the same genuine flowers could have reflected light into my eyes in the same way, and the belief could have been formed as in the normal case. But if I am unable to tell the real flowers from the fake ones it would be wrong to attribute to me the knowledge that there are flowers before me, for it is simply a fluke that I am right. Given these circumstances, my belief would not have been warranted. This example shows how circumstances external to the subject can affect the warranting power of a process. The parallel point about the role of background belief can be made by contrasting our everyday scenario with a different unusual case. Imagine that as in the original case all the flowers on the table are genuine, but that, prior to my inspection of the table, I have, for whatever reason, acquired the belief that my eyes are not functioning properly and that I am liable to mistake ordinary objects for quite different things. However, this belief does not influence me as I stand before the table. Behaving as I would if I did not have the belief, I look at the table and form the belief that there are flowers before me. The process which engenders my belief is the same as that which produced belief in the everyday situation. But it no longer warrants belief. Because I have perversely ignored my background belief about my perceptual powers, I do not know that there are flowers before me. Thus a process which, in standard circumstances, is a basic warrant for belief can be deprived of its power to warrant that belief by circumstances in which background beliefs are different.

Let me sum up the discussion of this section. In investigating our mathematical knowledge, it is helpful to introduce a name—"warrant"—for those processes which produce knowledge, and to distinguish between basic and derivative warrants. However, although basic warrants are processes which do not involve prior beliefs as causal factors, their power to warrant may be dependent on background circumstances, including the background beliefs of the subject. Because it can accommodate the latter point, my distinction between basic and derivative warrants can be used to structure the inquiry into mathematical knowledge and to formulate the apriorist program, without inheriting the traditional problems of foundationalist accounts of knowledge.

I shall now take up the topic which is the main concern of this chapter, the notion of a priori knowledge.

III

"A priori" is an epistemological predicate. What is *primarily* a priori is an item of knowledge.[9] Of course, we can introduce a derivative use of "a priori" as a predicate of propositions: a priori propositions are those which we could know a priori. In many contemporary discussions, it is common to define the notion of an a priori proposition outright, by taking the class of a priori propositions to consist of the truths of logic and mathematics (for example). But when philosophers allege that truths of logic and mathematics are a priori, they do not intend merely to recapitulate the definition of a priori propositions. Their aim is to advance a thesis about the epistemological status of logic and mathematics.

To understand the nature of such epistemological claims, we should return to Kant, who provided the most explicit characterization of a priori knowledge: "we shall understand by a priori knowledge, not knowledge which is independent of this or that experience, but knowledge absolutely independent of all experience."[10] Two questions naturally arise. What are we to understand by "experience"? And what is to be made of the idea of independence from experience? Apparently, there are easy answers. Count as a person's experience the stream of her sensory encounters with the world, where this includes both "outer experience," that is, sensory states caused by stimuli external to the body, and "inner experience," that is, those sensory states brought about by internal stimuli. Now we might propose that someone's knowledge is independent of her experience just in case she could have had that knowledge no matter what experience she had had. To this obvious suggestion there is an equally obvious objection. The apriorist is not ipso facto a believer in innate knowledge. So we cannot accept an analysis which implies that a priori knowledge could have been obtained given minimal experiences.

Many philosophers contend both that analytic truths can be known a priori and that some analytic truths involve concepts which could only be acquired if we were to have particular kinds of experience. If we are to defend their doctrines from immediate rejection, we must allow a minimal role to experience, even in a priori knowledge. Experience may be needed to provide some con-

---

9. The point has been forcefully made by Saul Kripke. See "Identity and Necessity," pp. 149–51, and *Naming and Necessity*, pp. 34–38.

10. *Critique of Pure Reason* B2–3.

cepts. So we might modify our proposal: knowledge is independent of experience if any experience which would enable us to acquire the concepts involved would enable us to have the knowledge.

It is worth noting explicitly that we are concerned here with the *total* experience of the knower. Suppose that you acquire some knowledge empirically. Later you deduce some consequences of this empirical knowledge. We should reject the suggestion that your knowledge of those consequences is independent of experience, because, at the time you perform the deduction, you are engaging in a process of reasoning which is independent of the sensations you are then having. Your knowledge, in cases like this, is dependent on your total experience: different total sequences of sensations would not have given you the premises for your deductions.

Let us put together the points which have been made so far. A person's experience at a particular time will be identified with his sensory state at the time. (Such states are best regarded physicalistically in terms of stimulation of sensory receptors, but we should recognize that there are both "outer" and "inner" receptors.) The total sequence of experiences X has had up to time t is *X's life at t*. A life will be said to be *sufficient for X for p* just in case X could have had that life and gained sufficient understanding to believe that p. (I postpone, for the moment, questions about the nature of the modality involved here.) Our discussion above suggests the use of these notions in the analysis of a priori knowledge: X knows a priori that p if and only if X knows that p and, given any life sufficient for X for p, X could have had that life and still have known that p. Making temporal references explicit: at time t, X knows a priori that p just in case, at time t, X knows that p and, given any life sufficient for X for p, X could have had that life at t and still have known, at t, that p. In subsequent discussions I shall usually leave the temporal references implicit.

Unfortunately, the proposed analysis will not do. A clearheaded apriorist should admit that people can have empirical knowledge of propositions which can be known a priori. However, on the account I have given, if somebody knows that p and if it is possible for her to know a priori that p, then, apparently, given any sufficiently rich life she could know that p, so that she would meet the conditions for a priori knowledge that p. (This presupposes that modalities "collapse," but I don't think the problem can be solved simply by denying the presupposition.) Hence it seems that my account will not allow for empirical knowledge of propositions that can be known a priori.

We need to amend the analysis. We must differentiate situations in which a person knows something empirically which could have been known a priori from situations of actual a priori knowledge. The remedy is obvious. What sets apart corresponding situations of the two types is a difference in the ways in which what is known is known. An analysis of a priori knowledge must probe the notion of knowledge more deeply than we have done so far.

IV

At this point, let us recall the equivalence (1) of Section II, which presents the general psychologistic approach to knowledge. My present aim is to distinguish a priori knowledge from a posteriori knowledge. We have discovered that the distinction requires us to consider the ways in which what is known is known. Hence I propose to reformulate the problem: let us say that X knows a priori that *p* just in case X has a true belief that *p* and that belief was produced by a process which is an *a priori warrant* for it. Now the crucial notion is that of an a priori warrant, and our task becomes that of specifying the conditions which distinguish a priori warrants from other warrants.

At this stage, some examples may help us to see how to draw the distinction. Perception is an obvious type of process which philosophers have supposed *not* to engender a priori knowledge. Putative a priori warrants are more controversial. I shall use Kant's notion of pure intuition as an example. This is not to endorse the claim that processes of pure intuition are a priori warrants, but only to see what features of such processes have prompted Kant (and others) to differentiate them from perceptual processes.

On Kant's theory, processes of pure intuition are supposed to yield a priori mathematical knowledge. Let us focus on a simple geometrical example. We are supposed to gain a priori knoweldge of the elementary properties of triangles by using our grasp on the concept of triangle to construct a mental picture of a triangle and by inspecting this picture with the mind's eye.[11] What are the characteristics of this kind of process which make Kant want to say that it produces knowledge which is independent of experience? I believe that Kant's account implies that three conditions should be met. The same type of process must be *available* independently of experience. It must produce *warranted* belief independently of experience. And it must produce *true* belief independently of experience. Let us consider these conditions in turn.

According to the Kantian story, if our life were to enable us to acquire the appropriate concepts (the concept of a triangle and the other geometrical concepts involved) then the appropriate kind of pure intuition would be available to us. We could represent a triangle to ourselves, inspect it, and so reach the same beliefs. But, if the process is to generate *knowledge* independently of experience, Kant must require more of it. Given any sufficiently rich life, if we were to undergo the same type of process and gain the same beliefs, then those beliefs would be warranted by the process. Let us dramatize the point by imagining that experience is unkind. Suppose that we are presented with experiences which are cunningly contrived so as to make it appear that some of our basic geometrical beliefs are false. Kant's theory of geometrical knowledge

11. More details about Kant's theory of pure intuition are given in my paper "Kant and the Foundations of Mathematics," and also in Chapter 3 below.

presupposes that if, in the circumstances envisaged, a process of pure intuition were to produce geometrical belief then it would produce warranted belief, despite the background of misleading experience.

So far I have considered how a Kantian process of pure intuition might produce warranted belief independently of experience. But to generate *knowledge* independently of experience, a priori warrants must produce warranted *true* belief in counterfactual situations where experiences are different. This point does not emerge clearly in the Kantian case because the propositions which are alleged to be known a priori are taken to be necessary, so that the question of whether it would be possible to have an a priori warrant for a false belief does not arise. Plainly, we could ensure that a priori warrants produce warranted *true* belief independently of experience by declaring that a priori warrants only warrant necessary truths. But this proposal is unnecessarily strong. Our goal is to construe a priori knowledge as knowledge which is independent of experience, and this can be achieved, without closing the case against the contingent a priori, by supposing that, in a counterfactual situation in which an a priori warrant produces belief that p, then p. On this account, a priori warrants are ultra-reliable; they never lead us astray.

Summarizing the conditions that have been uncovered, I propose the following analysis of a priori knowledge.

(2) X knows a priori that p if and only if X knows that p and X's belief that p was produced by a process which is an a priori warrant for it.

(3) α is an a priori warrant for X's belief that p if and only if α is a process such that, given any life e, sufficient for X for p,

(a) some process of the same type could produce in X a belief that p

(b) if a process of the same type were to produce in X a belief that p, then it would warrant X in believing that p

(c) if a process of the same type were to produce in X a belief that p, then p.

It should be clear that this analysis yields the desired result that, if a person knows a priori that p then she could know that p whatever (sufficiently rich) experience she had had. But it goes beyond the proposal of Section III in spelling out the idea that the knowledge is obtainable in the same way. Hence we can distinguish cases of empirical knowledge of propositions which could be known a priori from cases of actual a priori knowledge.

V

In this section, I want to be more explicit about the notion of "types of processes" which I have employed, and about the modal and conditional notions which figure in my analysis. To specify a process which produces a belief is to pick out some terminal segment of the causal ancestry of the belief. I think that, without loss of generality, we can restrict our attention to those segments which consist solely of states and events internal to the believer.[12] Tracing the causal ancestry of a belief beyond the believer would identify processes which would not be available independently of experience, so that they would violate our conditions on a priori warrants.

Given that we need only consider psychological processes, the next question which arises is how we divide processes into types. It may seem that the problem can be sidestepped: can't we simply propose that to defend the apriority of an item of knowledge is to claim that that knowledge was produced by a psychological process and that *that very process* would be available and would produce warranted true belief in counterfactual situations where experience is different? I think it is easy to see how to use this proposal to rewrite (3) in a way which avoids reference to "types of processes." I have not adopted this approach because I think that it short-cuts important questions about what makes a process the same in different counterfactual situations.

Our talk of processes which produce belief was originally introduced to articulate the idea that some items of knowledge are obtained in the same way while others are obtained in different ways. To return to our example, knowing a theorem on the basis of hearing a lecture and knowing the same theorem by following a proof count, intuitively, as different ways of knowing the theorem. Our intuitions about this example, and others, involve a number of different principles of classification, with different principles appearing in different cases. We seem to divide belief-forming processes into types by considering content of beliefs, inferential connections, causal connections, use of perceptual mechanisms, and so forth. I suggest that these principles of classification probably do not give rise to one definite taxonomy, but that, by using them singly, or in combination, we obtain a number of different taxonomies which we can and do employ. Moreover, within each taxonomy, we can specify types of processes more or less narrowly.[13] Faced with such variety, what characterization should we pick?

There is probably no privileged way of dividing processes into types. This is not to say that our standard principles of classification will allow *anything* to count as a type. Somebody who proposed that the process of listening to a lecture (or the terminal segment of it which consists of psychological states and

12. For different reasons, Goldman proposes that an analysis of the general notion of warrant (or, in his terms, justification) can focus on psychological processes. See section 2 of "What Is Justified Belief?"

13. Consider, for example, a Kantian process of pure intuition which begins with the construction of a triangle. Should we say that a process of the same type must begin with the construction of a triangle of the same size and shape, a triangle of the same shape, any triangle, or something even more general? Obviously there are many natural classifications here, and I think the best strategy is to suppose that an apriorist is entitled to pick any of them.

events) belongs to a type which consists of itself and instances of following a proof, would flout *all* our principles for dividing processes into types. Hence, while we may have many admissible notions of types of belief-forming processes, corresponding to different principles of classification, some collections of processes contravene all such principles, and these cannot be admitted as genuine types.[14]

My analysis can be read as issuing a challenge to the apriorist. If someone wishes to claim that a particular belief is an item of a priori knowledge then he must specify a segment of the causal ancestry of the belief, consisting of states and events internal to the believer, and type-identity conditions which conform to some principle (or set of principles) of classification which are standardly employed in our divisions of belief-forming processes (of which the principles I have indicated above furnish the most obvious examples). If he succeeds in doing this so that the requirements in (3) are met, his claim is sustained; if he cannot, then his claim is defeated.

The final issue which requires discussion in this section is that of explaining the modal and conditional notions I have used. There are all kinds of possibility, and claims about what is possible bear an implicit relativization to a set of facts which are held constant.[15] When we say, in (3), that, given any sufficiently rich life, X could have had a belief which was the product of a particular type of process, should we conceive of this as merely logical possibility or are there some features of the actual world which are tacitly regarded as fixed? I suggest that we are not just envisaging any logically possible world. We imagine a world in which X has similar mental powers to those he has in the actual world. By hypothesis, X's experience is different. Yet the capacities for thinking, reasoning, and acquiring knowledge which X possesses as a member of *Homo sapiens* are to remain unaffected: we want to say that X, *with the kinds of cognitive capacities distinctive of humans,* could have undergone processes of the appropriate type, even if his experiences had been different.[16]

Humans might have had more faculties for acquiring knowledge than they actually have. For example, we might have had some strange ability to ''see'' what happens on the other side of the Earth. When we consider the status of a particular type of process as an a priori warrant, the existence of worlds in which such extra faculties come into play is entirely irrelevant. Our investiga-

---

14. Strictly, the sets which do not constitute types are those which violate correct taxonomies. In making present decisions about types, we assume that our current principles of classification are correct. If it should turn out that those principles require revision then our judgments about types will have to be revised accordingly.

15. For a lucid and entertaining presentation of the point, see David Lewis, ''The Paradoxes of Time-Travel,'' pp. 149–51.

16. Of course, X might have been more intelligent, that is, he might have had better versions of the faculties he has. We allow for this type of change. But we are not interested in worlds where X has extra faculties.

tion focusses on the question of whether a particular type of process would be available to a person with the kinds of faculties people actually have, not on whether such processes would be available to creatures whose capacities for acquiring knowledge are augmented or diminished. Conditions (3b) and (3c) are to be read in similar fashion. Rewriting (3b) to make the form of the conditional explicit, we obtain: for any life e sufficient for X for p and for any world in which X has e, in which he believes that p, in which the belief is the product of a process of the appropriate kind, and *in which X has the cognitive capacities distinctive of humans*, X is warranted in believing that p. Similarly, (3c) becomes: for any life e sufficient for X for p and for any world in which X has e, in which he believes that p, in which his belief is the product of a process of the appropriate kind, *and in which X has the cognitive capacities distinctive of humans*, p. Finally, the notion of a life's being sufficient for X for p also bears an implicit reference to X's native powers. To say that a particular life enables X to form certain concepts is to maintain that, given the genetic programming with which X is endowed, that life allows for the formation of the concepts.

The account I have offered can be presented more graphically in the following way. Consider a human as a cognitive device, endowed initially with a particular kind of structure. Sensory experience is fed into the device and, as a result, the device forms certain concepts. For any proposition p, the class of experiences which are sufficiently rich for p consists of those experiences which would enable the device, with the kind of structure it actually has, to acquire the concepts to believe that p. To decide whether or not a particular item of knowledge that p is an item of a priori knowledge we consider whether the type of process which produced the belief that p is a process which would have been available to the device, with the kind of structure it actually has, if different sufficiently rich experiences had been fed into it, and whether, under such circumstances, processes of the type would warrant belief that p, and would produce true belief that p.

VI

At this point, I want to address worries that my analysis is too liberal, because it allows some of our knowledge of ourselves and our states to count as a priori. Given its psychologistic underpinnings, the theory appears to favor claims that some of our self-knowledge is a priori. However, two points should be kept in mind. First, the analysis I have proposed can only be applied to cases in which we know enough about the ways in which our beliefs are warranted to decide whether or not the conditions of (3) are met. In some cases, our lack of a detailed account of how our beliefs are generated may mean that no firm decision about the apriority of an item of knowledge can be reached. Second, there may be cases, including cases of self-knowledge, in which we have no clear pre-analytic intuitions about whether a piece of knowledge is a priori.

Nevertheless, there are some clear cases. Obviously, any theory which implied that I can know a priori that I am seeing red (when, in fact, I am) would be suspect. But, when we apply my analysis, the unwanted conclusion does not follow. For, if the process which leads me to believe that I am seeing red (when I am) can be triggered in the absence of red, then (3c) would be violated. If the process cannot be triggered in the absence of red, then, given some sufficiently rich experiences, the process will not be available, so that (3a) will be violated. In general, knowledge of any involuntary mental state—such as pains, itches, or hallucinations—will work in the same way. Either the process which leads from the occurrence of pain to the belief that I am in pain can be triggered in the absence of pain, or not: if it can, (3c) would be violated; if it cannot, then (3a) would be violated.

This line of argument can be sidestepped when we turn to cases in which we have the power, independently of experience, to put ourselves into the appropriate states. For, in such cases, one can propose that the processes which give us knowledge of the states cannot be triggered in the absence of the states themselves and that the processes are always available because we can always put ourselves into the states.[17] On this basis, we might try to conclude that we have a priori knowledge that we are imagining red (when we are) or thinking of Ann Arbor (when we are). However, the fact that such cases do not fall victim to the argument of the last paragraph does not mean that we are compelled to view them as cases of a priori knowledge. In the first place, the thesis that the processes through which we come to know our imaginative feats and our voluntary thoughts cannot be triggered in the absence of the states themselves requires evaluation—and lacking detailed knowledge of those processes, we cannot arrive at a firm judgment here. Second, the processes in question will be required to meet (3b) if they are to be certified as a priori warrants. This means that, whatever experience hurls at us, beliefs produced by such processes will be warranted. We can cast doubt on this idea by imagining that our experience consists of a lengthy, and apparently reliable, training in neurophysiology, concluding with a presentation to ourselves of our own neurophysiological organization which appears to show that our detection of our imaginative states (say) is slightly defective, that we always make mistakes about the contents of our imaginings. If this type of story can be developed, then (3b) will be violated, and the knowledge in question will not count as a priori. But, even if it cannot be coherently extended, and even if my analysis

---

17. In characterizing pain as an involuntary state one paragraph back I may seem to have underestimated our powers of self-torture. But even a masochist could be defeated by unkind experience: as he goes to pinch himself his skin is anesthetized.

does judge our knowledge of states of imagination (and other "voluntary" states) to be a priori, it is not clear to me that this consequence is counterintuitive.

In fact, I think that one can make a powerful case for supposing that *some* self-knowledge is a priori. At most, if not all, of our waking moments, each of us knows of herself that she exists.[18] Although traditional ideas to the effect that self-knowledge is produced by some "non-optical inner look" are clearly inadequate, I think it is plausible to maintain that there are processes which do warrant us in believing that we exist—processes of reflective thought, for example—and which belong to a general type whose members would be available to us independently of experience. Trivially, when any such process produces in a person a belief that she exists, that belief is true. All that remains, therefore, is to ask if the processes of the type in question inevitably warrant belief in our own existence, or whether they would fail to do so, given a suitably exotic background experience. It is difficult to settle this issue conclusively without a thorough survey of the ways in which reflective belief in one's existence can be challenged by experience, but perhaps there are Cartesian grounds for holding that, so long as the belief is the product of reflective thought, the believer is warranted, no matter how bizarre his experience may have been. If this is correct, then at least some of our self-knowledge will be a priori. However, in cases like this, attributions of apriority seem even less vulnerable to the criticism that they are obviously incorrect.

At this point we must consider a doctrinaire objection. If the conclusion of the last paragraph is upheld then we can know some contingent propositions a priori.[19] Frequently, however, it is maintained that only necessary truths can be known a priori. Behind this contention stands a popular argument.[20] Assume that a person knows a priori that p. His knowledge is independent of his experience. Hence he can know that p without any information about the kind of world he inhabits. So, necessarily, p.

This hazy line of reasoning rests on an intuition which is captured in the analysis given above. The intuition is that a priori warrants must be ultra-

18. I ignore the tricky issue of trying to say exactly what is known when we know this and kindred things. For interesting explorations of this area, see Hector-Neri Castañeda, "Indicators and Quasi-Indicators" and "On the Logic of Attributions of Self-Knowledge to Others"; John Perry, "Frege on Demonstratives" and "The Problem of the Essential Indexical"; and David Lewis, "Propositional Attitudes *De Dicto* and *De Se*." For further discussion of the issues stemming from this type of self-knowledge, see my "Apriority and Necessity."

19. In *Naming and Necessity*, Kripke tries to construct examples of contingent propositions which can be known a priori. For an evaluation of his examples see Keith Donnellan, "The Contingent A Priori and Rigid Designators," and my "Apriority and Necessity."

20. Kripke seems to take this to be the main argument against the contingent a priori. See *Naming and Necessity*, p. 38.

reliable: if a person is entitled to ignore empirical information about the type of world she inhabits then that must be because she has at her disposal a method of arriving at belief which guarantees *true* belief. (This intuition can be defended by pointing out that if a method which could produce false belief were allowed to override experience, then we might be blocked from obtaining knowledge which we might otherwise have gained.) In my analysis, the intuition appears as (3c).[21]

However, when we try to clarify the popular argument we see that it contains an invalid step. Presenting it as a *reductio,* we obtain the following line of reasoning. Assume that a person knows a priori that *p* but that it is not necessary that *p*. Because *p* is contingent there are worlds in which *p* is false. Suppose that the person had inhabited such a world and behaved as she does in the actual world. Then she would have had an a priori warrant for a false belief. This is debarred by (3c). So we must conclude that the initial supposition is erroneous: if someone really does know a priori that *p* then *p* is necessary.

Spelled out in this way, the argument fails. We are not entitled to conclude from the premise that there are worlds in which *p* is false the thesis that there are worlds in which *p* is false *and* in which the person behaves as she does in the actual world. There are a number of propositions which, although they could be false, could not both be false and also be believed by us. More generally, there are propositions which could not both be false and also believed by us in particular, definite ways. Obvious examples are propositions about ourselves and their logical consequences: such propositions as those expressed by tokens of the sentences "I exist," "I have some beliefs," "There are thoughts," and so forth. Hence the attempted *reductio* breaks down and allows for the possibility of a priori knowledge of some contingent propositions.

I conclude that my analysis is innocent of the charge of being too liberal in ascribing to us a priori knowledge of propositions about ourselves. Although it is plausible to hold that my account construes some of our self-knowledge as a priori, none of the self-knowledge it takes to be a priori is clearly empirical. Moreover, it shows how a popular argument against the contingent a priori is flawed, and how certain types of contingent propositions—most notably propositions about ourselves—escape that argument. Thus I suggest that the analysis illuminates an area of traditional dispute.

Finally, I want to consider a different objection to my analysis. This objection, like those just considered, charges that the analysis is too liberal. My

---

21. As the discussion of this paragraph suggests, there is an intimate relation between my requirements (3b) and (3c). Indeed, one might argue that (3b) would not be met unless (3c) were also satisfied—on the grounds that one cannot allow a process to override experience unless it guarantees truth. The subsequent discussion will show that this type of reasoning is more complicated than it appears. Hence, although I believe that the idea that a priori warrants function independently of experience does have implications for the reliability of these processes, I have chosen to add (3c) as a separate condition.

account apparently allows for the possibility that a priori knowledge could be gained through perception. We can imagine that some propositions are true in any world of which we can have experience, and that, given sufficient experience to entertain those propositions, we could always come to know them on the basis of perception. Promising examples are the proposition that there are objects, the proposition that some objects have shapes, and other similar propositions. In these cases, one can argue that we cannot experience worlds in which they are false and that any (sufficiently rich) experience would provide perceptual warrant for belief in the propositions, regardless of the specific content of our perceptions. If these points are correct (and I shall concede them both, for the sake of argument), then perceptual processes would qualify as a priori warrants. Given any sufficiently rich experience, some perceptual process would be available to us, would produce warranted belief and, *ex hypothesi,* would produce warranted *true* belief.

Let us call cases of the type envisaged cases of *universally empirical* knowledge. The objection to my account is that it incorrectly classifies universally empirical knowledge as a priori knowledge. My response is that the classical notion of apriority is too vague to decide such cases: rather, this type of knowledge only becomes apparent when the classical notion is articulated. One could defend the classification of universally empirical knowledge as a priori by pointing out that such knowledge required no particular type of experience (beyond that needed to obtain the concepts, of course). One could oppose that classification by pointing out that, even though the content of the experience is immaterial, the knowledge is still gained by perceiving, so that it should count as a posteriori.

If the second response should seem attractive, it can easily be accommodated by recognizing both a stronger and a weaker notion of apriority. The weaker notion is captured in (2) and (3). The stronger adds an extra requirement: no process which involves the operation of a perceptual mechanism is to count as an a priori warrant.

At this point, it is natural to protest that the new condition makes the prior analysis irrelevant. Why not define a priori knowledge outright as knowledge which is produced by processes which do not involve perceptual mechanisms? The answer is that the prior conditions are not redundant: knowledge which is produced by a process which does not involve perceptual mechanisms need not be independent of experience. For the process may fail to generate warranted belief against a backdrop of misleading experience. (Nor may it generate true belief in all relevant counterfactual situations.) So, for example, certain kinds of thought-experiments may generate items of knowledge given a particular type of experience, but may not be able to sustain that knowledge against misleading experiences. Hence, if we choose to exclude universally empirical knowledge from the realm of the a priori in the way suggested, we are building on the analysis given in (2) and (3), rather than replacing it.

In what follows, the distinction between the weaker and stronger notions of apriority will not greatly concern us. The reason for this is that the classical attempts to develop mathematical apriorism to be considered below endeavor to specify processes which do not involve perceptual mechanisms. Moreover, it will be relatively easy to see that the considerations which undermine these attempts could be applied to defeat someone who tried to dcfend a view of mathematical knowledge as universally empirical. The failures of mathematical apriorism cannot be eliminated by exploiting the distinction which I have just drawn.

VII

I have examined some general epistemological consequences of my analysis in order to forestall concerns that the framework within which I shall present and criticize mathematical apriorism is fundamentally misguided. Before I undertake my main project there is one further point which should be made explicit. In rejecting mathematical apriorism, I am not committed to any conclusions about the modal status of mathematical truths. It is quite consistent to claim that mathematical truths are necessary, even that we know that they are necessary, and to deny that our mathematical knowledge is a priori. Although I shall not argue for the necessity of mathematics, I believe that the thesis that mathematical truths are necessary is defensible and that my critique of mathematical apriorism is irrelevant to it.

From the perspective of much traditional thinking about "the a priori," these claims will sound absurd. The notions of apriority and necessity have been so closely yoked in philosophical discussion that 'a priori' and 'necessary' are sometimes used as if they were synonyms. They are not. We have already discovered that 'a priori' and 'necessary' are not even coextensive by finding examples of contingent truths which can be known a priori. What I now want to show is that necessity does not imply apriority.

The first point to note is that there are probably necessary truths which no human could know at all, let alone know a priori. One who believes that mathematical truths are necessary ought to accept this conclusion on the grounds that there are some mathematical truths which are too complicated for humans to apprehend. This point can be accommodated by abandoning the traditional thesis that all necessary truths can be known a priori in favor of the more cautious claim that all necessary truths which humans can know can be known a priori. Given our understanding of the notion of a priori knowledge, it would be surprising if this were true. Why should it be that, for any necessary truth which we can know, there is some process which could satisfy the constraints on a priori warrants and which could produce belief in that truth?

Saul Kripke has argued directly that there are necessary truths which are not knowable a priori. On Kripke's account of names the truth expressed by "If Hesperus exists then Hesperus = Phosphorus" is necessary. 'Hesperus' and 'Phosphorus' are alleged to be *rigid designators,* that is, in any world in which they designate, they designate what they designate in the actual world, namely Venus. As a result, if Venus exists in a particular possible world then 'Hesperus' and 'Phosphorus' both designate Venus in that world, so that "Hesperus = Phosphorus" is true in that world. On the other hand, if Venus does not exist in the world in question, then 'Hesperus' fails to designate in that world, so that "Hesperus exists" is false in that world. Either way, "If Hesperus exists, then Hesperus = Phosphorus" comes out true. So it is necessarily true that if Hesperus exists then Hesperus = Phosphorus. Apparently, however, we could not have known that if Hesperus exists, then Hesperus = Phosphorus, without the help of experience. So we have an example of a necessary truth which could not have been known a priori.[22]

Since some people may want to reject Kripke's view of names, or may want to try to argue that, contrary to appearances, the proposition expressed by the statement can be known a priori, I think it is worth noting that there is another way to question the thesis that necessary truths are knowable a priori. Consider any English sentence of form ⌜The F is G.⌝ We can *rigidify* this sentence by distributing tokens of 'actual' and 'actually' in it in appropriate ways, thereby obtaining a sentence which is necessary if the original sentence is true. So, for example, if ⌜The F is G⌝ is true then a rigidification of it, ⌜The actual F is actually G,⌝ is necessarily true. To see this, consider what happens when we evaluate ⌜The actual F is actually G,⌝ as used by us, in any possible world. In any world, ⌜the actual F,⌝ as we use it, refers to the referent of ⌜the F⌝ in the actual world, an object *d,* let us say. At any world, the phrase ⌜actually G,⌝ as we use it, has as its extension the extension of G in the actual world, a set α, let us say. ⌜The actual F is actually G⌝ is true in a world just in case the referent of ⌜the actual F⌝ at that world belongs to the extension of ⌜actually G⌝ at the world, that is, just in case $d \, \epsilon \, \alpha$. Clearly, $d \, \epsilon \, \alpha$ just in case ⌜The F is G⌝ is true (at our world, the actual world). So, if ⌜The F is G⌝ is true, then, as we use it, ⌜The actual F is actually G⌝ is necessarily true. We can generalize the point: appropriate insertions of 'actual' and 'actually' can yield a necessarily true sentence from any true sentence. (One trivial way to perform the trick is to prefix the entire sentence with 'actually.')

Suppose now that all necessary truths are knowable a priori. Let S be any true sentence. Any human must be able to know a priori the truths expressed by rigidifications of S. Now an extension of the argument used in Section VI to show that we know a priori that we exist will yield the thesis that each of us can know a priori that he is actual.[23] But if we know a priori that we are

---

22. *Naming and Necessity,* pp. 102–5, 108–10.
23. I give the extension in section IV of "Apriority and Necessity."

actual and we know a priori the truth expressed by some rigidification of S, then we can infer a priori the truth expressed by S. Consider, for concreteness, the example used in the last paragraph. If I know a priori that the actual F is actually G and I know a priori that I am actual, then, by putting my two pieces of knowledge together, I can know a priori that the F is G. Hence, since any sentence has rigidifications, the thesis that necessary truths are knowable a priori entails that anyone can know any truth a priori, and this surely achieves a *reductio* of that thesis, which hoped to draw an important division *within* human knowledge.

As I have already noted, it would be reasonable for us to claim only that necessary truths which we can know can be known a priori. But this restriction is of no avail with the present problem. Given that we can understand a sentence we can surely understand some rigidification of it. Hence, even on the restricted version, we would be forced to conclude that we can know a priori the truth expressed by some rigidification of S, whence the argument of the last paragraph would again produce disastrous consequences.

I conclude that there is no basis for supposing that the claim that p is necessary commits us to the thesis that we can know a priori that p (so that to deny the apriority of mathematics would be to abandon the necessity of mathematics). But there is another way to try to connect necessity and apriority which has sometimes been popular. Many writers have been tempted to suppose that our knowledge *that* a proposition is necessary must be a priori knowledge.[24] I want briefly to consider this suggestion.

There are complications which stem from the fact that we can obtain empirical knowledge that a proposition is necessary if we have empirical knowledge that the proposition is true and if we know that propositions of that kind are necessary. Other complications result from our ability to know things on the authority of others. These factors do not touch the central intuition. However our modal knowledge is extended and transmitted, there is a strong temptation to think that it must begin from items of a priori knowledge.

This conclusion results, I think, from one line of reasoning, which is rarely explicit but extremely pervasive. We imagine someone who has a piece of *primary modal knowledge,* knowledge that a proposition is necessary, which she has gained for herself and which does not result from the application of more general items of modal knowledge. What could warrant her belief? We are inclined to think that the warrant could not be any kind of perceptual process, and that perception could play no essential part in it. Perceptual processes appear to reflect only the features of the actual world; they seem to give us no access to other possible worlds. Hence, no perceptual process could warrant our belief that something is true of all possible worlds. Moreover, for the same

---

24. For example, Kant. Kant's terminology helps him to conflate knowledge of the truth of a necessary proposition with knowledge of its necessity. See, for example, *Critique of Pure Reason* B15.

reason, perceptual processes could not supply us with any essential information. Thus we conclude that primary modal knowledge must be a priori.

Whatever the merits of the picture of possible worlds and our access to them
which this line of reasoning presents,[25] I think that there is a problem in the
last step of the argument, a problem which my analysis of apriority makes clear.
Let us assume that warrants for items of primary modal knowledge do not
involve the processing of perceptual information. For concreteness, let us suppose, as I think some champions of the argument would like us to believe, that
primary modal knowledge is obtained by some clearly non-perceptual process
such as abstract reflection or experimentation in imagination. It does not follow
that primary modal knowledge is a priori. Condition (3b) of my analysis of 'a
priori warrant' brings out the important idea that a priori warrants have to be
able to discharge their warranting function, no matter what background of disruptive experience we may have. But the fact that a process is non-perceptual
does not rule out the possibility that the ability of that process to warrant belief
might be undermined by radically disruptive experiences. I can imagine experiences which would convince me that my own efforts at experimentation in
imagination (for example) were an extremely unreliable guide to anything at
all. Hence, the last step in the popular argument illegitimately conflates non-
perceptual sources of knowledge with sources of a priori knowledge.

Since I think that this argument provides the *only* basis for the thesis that
primary modal knowledge must be a priori, I conclude that another traditional
effort to salvage a connection between necessity and apriority has failed. There
is no apparent reason to deny that there are many truths known to be necessary
but little a priori knowledge.

I have now assembled the general epistemological notions which I shall need
to show what is wrong with mathematical apriorism and to develop my own
account of mathematical knowledge. Let us now forsake the abstract epistemological perspective of this chapter and focus on the specific case of mathematics.

25. The argument under consideration presupposes a particular view of possible worlds and our
access to them: the possible worlds are imagined as laid out like stars in a galaxy; perception gives
us access to one of them, and it is thus supposed to be impossible for us to use perception to arrive
at features which hold for all of them; reason, however, is able to transcend the interstellar spaces.
Champions of possible worlds semantics for modal logic will insist that this picture is not forced
on them. See Kripke, *Naming and Necessity*, pp. 43ff, and Robert Stalnaker, "Possible Worlds,"
pp. 65–70.

2
The Apriorist Program

I

The most common way of presenting mathematical apriorism is to contend that mathematical knowledge is based on proof. In twentieth-century discussions of mathematics, a particular apsychologistic conception of proofs has been dominant, and this conception has served to conceal the fundamental epistemological issues. I shall try to expose the mistakes in this conception, to offer an account of proof which brings into the open the central presuppositions of apriorism, and thus to prepare the way for specific criticisms.

A proof is a linguistic type, whose tokens may be printed in books, inscribed on blackboards, or uttered by mathematicians.[1] The conception I want to attack proposes to characterize the types which count as proofs in *structural* terms.[2] Formal logic offers us the precise notion of proof within a system. A proof in a system is a sequence of sentences in the language of the system such that each member of the sequence is either an axiom of the system or a sentence which results from previous members of the sequence in accordance with some rule of the system. We can use this relativized notion of proof to define an absolute notion. Call those mathematical theories which mathematicians currently accept *standard theories*. Call formalizations of those theories which logicians would currently accept *standard formal theories*. Then a proof is a proof in some standard formal theory.

The trouble with this is that we ought to believe that there are proofs which are not proofs in standard formal theories. It would be presumptuous to think that the discovery of new mathematical axioms has come to an end. Nor can

---

1. Of course, some proofs may have no tokens at all.

2. For presentation of a conception of this kind, see Mark Steiner, *Mathematical Knowledge*, chapter 3. It is slightly ironic that Steiner should advance this approach to proof, since his book is notable for its clear and correct emphasis on the need to take epistemological issues seriously and its recognition that much twentieth-century philosophy of mathematics has neglected epistemology.

we simply extend the notion of a standard formal theory to embrace those theories which will *in fact* be adopted by our successors. For their endeavors may always remain incomplete, and, more fundamentally, we should worry whether social acceptance is sufficient to make a proof within a system a proof *simpliciter*. Perhaps our descendants may decide that mathematics is too hard, and take to adopting inconsistent systems or, less dramatically, to elevating to the status of axioms results which mathematicians currently struggle to prove. At best, *correct* or *reasonable* social practice can determine which sequences are proofs. Yet now we must ask what makes the adoption of a theory or system correct or reasonable. Turning the question on ourselves, what are the characteristics of proofs in standard formal theories which make us select them as proofs?

The structuralist approach offers us a definition by enumeration. It is as though we had asked the question "What is an acid?" and been greeted by a list of the chemical formulas of the acids. Even if we had been satisfied that the enumeration was complete, it would still not have told us what we wanted to know. Just as we would like some insight into what makes an acid an acid, so too we would like insight into what makes a proof a proof. Apsychologistic epistemologies provide pat answers to these questions. They suggest that the axioms of genuine proofs are "basic a priori principles" and that the rules to be employed in proofs are "elementary a priori rules of inference." Our examination of general epistemological issues in Chapter 1 denies the adequacy of this kind of response or of the usual apsychologistic ways of articulating it. To herald some particular set of principles as "basic a priori principles" from which genuine proofs can begin is to advance a thesis about *how* those principles can be known, a thesis which will require elaboration and defense. Similarly to identify some rules of inference as especially privileged for use in proofs is to make claims about how our a priori knowledge can be extended. Once we adopt a psychologistic approach to knowledge, we can no longer rest content with the account of proof which I have given so far, an account which figures in many twentieth-century discussions on the topic.

We should abandon the structuralist approach to proofs in favor of a *functional* characterization. To put the point simply, proofs are sequences of sentences which do a particular job and, if we have a predilection for formal proofs, it is because we think that formal proofs do the job better than informal ones. What job do proofs do? The apriorist emphasis on the importance of proofs in mathematics reflects a traditional answer: proofs codify psychological processes which can produce a priori knowledge of the theorem proved. If we are to embed the popular thesis that mathematical knowledge is a priori because it is based on proof in an adequate epistemology, then I submit that this is the answer which we should adopt.

The central idea of this answer is to distinguish proofs by characterizing the notion of following a proof. To follow a proof is to engage in a particular kind of psychological process. The proof (which is, we shall continue to suppose, a linguistic type) serves as a pattern for the process. Combining some of our previous terminology, let us say that a statement is a *basic a priori statement* if it can be known a priori by following a process which is a basic warrant for belief in it. The statements from which proofs begin must be basic a priori statements. From these starting points, proofs proceed by means of *apriority-preserving* rules of inference. A rule of inference tells us that it is permissible to infer a statement of a particular form (the conclusion form) from statements of particular forms (the premise forms). A rule is apriority-preserving just in case there is a type of psychological process, consisting in transition from beliefs in instances of the premise forms to the corresponding instance of the conclusion form, unmediated by other beliefs, such that, if the instances of the premise forms are known a priori, then the transition generates a priori knowledge of the instance of the conclusion form. As their name suggests, apriority-preserving rules of inference reflect psychological transitions which can extend a priori warrants while preserving apriority. We can now define a proof as a sequence of statements such that every member of the sequence is either a basic a priori statement or a statement which follows from previous members of the sequence in accordance with some apriority-preserving rule of inference. To follow the proof is to undergo a process in which, sequentially, one comes to know a priori the statements which occur in the proof, by undergoing basic a priori warrants in the case of basic a priori statements and, in the case of those statements which are inferred, by undergoing a psychological transition of the type which corresponds to the rule of inference in question. I maintain that this is the conception of proofs which results when the thesis that proofs provide a priori mathematical knowledge is detached from its usual apsychologistic setting and placed in the more adequate epistemological context presented in Chapter 1.[3]

Let us briefly consider how this conception of proofs can account for various features of mathematical practice. The thesis that proofs are proofs in standard formal systems will now be regarded as embodying a substantive epistemological claim, the contention that the axioms and rules of these systems have the

---

3. From Frege on, philosophy of mathematics has been dominated by writers who do not give detailed defenses of their claims about basic a priori statements and apriority-preserving rules of inference. In the case of Frege, this is the result of belief that fundamental epistemological issues have been settled. (See my ''Frege's Epistemology'' for discussion of this point.) After Frege, the popularity of apsychologistic epistemology has buried even more deeply questions about how proofs give us knowledge (a priori knowledge). Thus, in the works of Russell, Poincaré, Weyl, Brouwer, Heyting, Carnap, Hilbert, Bernays, Kreisel, and many recent philosophers and logicians, epistemological theses are simply assumed. These epistemological assumptions then color the discussions of technical issues. I am not claiming that the writers I have cited present their views in the way described in this section. They do not. However, I do claim that when they are interpreted from the perspective of an adequate epistemology, their apriorist doctrines must fit the form outlined here.

right epistemic attributes. Obviously, we can envisage the discovery of more axioms or rules with these attributes, and so we can entertain the idea that our successors may extend the class of accepted proofs. Similarly, we can view our predecessors as sharing with us certain epistemic goals, and as setting forth sequences, which, by our lights, do not count as proofs, in attempts to achieve those goals. We may even suppose that, in some cases, earlier mathematicians followed the processes which are depicted in our proofs, even though they failed to present the detailed character of their reasoning. In a similar vein, we may suggest that, for the contemporary expert as for the great figures of the past, it is usually unnecessary to write out a complete (formal) proof. Presented with major parts of the pattern, the *cognoscenti* can fill in the details: they undergo a more intricate sequence of psychological transitions than those explicitly represented on paper, and, in effect, the proofs they follow are the formal proofs which underlie the abbreviated arguments of mathematical books and articles.

Our analysis of the apriorist notion of proof enables us to provide a canonical form for apriorist theories of mathematical knowledge. The apriorist typically claims that all known statements of mathematics can be known a priori by following proofs. To defend this claim, one must defend some thesis of the following form:

(4) There is a class of statements A and a class of rules of inference R such that
   (a) each member of A is a basic a priori statement
   (b) each member of R is an apriority-preserving rule
   (c) each statement of standard mathematics occurs as the last member of a sequence, all of whose members either belong to A or come from previous members in accordance with some rule in R.

Many of the fiercest battles in the foundations of mathematics have been fought on the question of whether particular versions of (4c) are true. In some cases, the disputes have been conducted against the background of assumptions of forms (4a) and (4b), which have been accepted without any careful examination of the underlying epistemological issues. This approach obviously runs the risk of chasing wild geese. If one sets out to provide a foundation for mathematics, conceiving of the enterprise as one of revealing the apriority of mathematical knowledge, then the theses of forms (4a) and (4b) are as crucial to the significance of the program as (4c).

I shall offer two different kinds of criticism of mathematical apriorism. One of these, the more important criticism, will be directed against claims of form (4a): I shall consider the various ways in which apriorists have committed themselves to the existence of basic a priori statements and argue that processes traditionally regarded as basic a priori warrants are not a priori warrants at all. The other criticism will develop a worry that philosophers have sometimes had about foundationalist programs. Is the fact that some mathematical theorems only admit of extremely long proofs compatible with the assumption that those theorems can be known a priori? I shall take up this latter criticism immediately, focussing initially on whether establishing some version of (4) would suffice for mathematical apriorism.

II

Let us imagine that we have conceded that some version of (4) is correct. Does this suffice to show that every standard mathematical statement can be known a priori? Seemingly so. For let S be any true standard mathematical statement. By (4c) there is a sequence of sentences all of whose members belong to A or come from previous members by one of the rules in R. We can show by induction, using (4a) and (4b), that every statement in the sequence is knowable a priori. *A fortiori,* S is knowable a priori. Hence every truth of standard mathematics is knowable a priori.

We can oppose to this argument the worry about long proofs. There are true standard mathematical statements S such that the shortest proof of S would require even the most talented human mathematicians to spend months in concentrated effort to follow it. Can we really suppose that S is knowable a priori? After all, anyone who had followed a proof of S would reasonably believe that he might have made a mistake. (Many of the best contemporary mathematicians are concerned that some important theorems, published in the last two decades, may have proofs containing hitherto unnoticed errors. These concerns are by no means neurotic.) So we might conclude that our knowledge of S is inevitably uncertain, and therefore not a priori.

Faced with this pair of contrary suggestions, we seem to have three options: (i) we can accept the inductive argument and the point about long proofs, concluding that no version of (4) can be correct; (ii) we can accept the inductive argument and reject the point about long proofs, thereby concluding that (4) suffices to establish mathematical apriorism; (iii) we can reject the inductive argument, concluding that (4) does not suffice to establish apriorism. Let us first consider (iii).

It is a familiar fact that mathematical induction can lead us into trouble. To cite a standard example, 1 is a small number and the operation of adding 1 is smallness-preserving (that is, adding 1 to a small number yields a small number). Any natural number can be obtained as the last member of a sequence whose first member is 1 and which has the property that each subsequent member is obtained by adding 1 to its predecessor. Using mathematical induction (in an exactly parallel fashion to that suggested in the first paragraph of this section) we can conclude that $10^{10^{10}}$ is a small number (or, for that matter, that any natural number is small).

What goes wrong here is easily traced to the vagueness of 'small.' At this point it is reasonable to ask whether the predicates of propositions 'known a priori' and 'knowable a priori' are similarly vague. Now someone who is sensitive to the worry about long proofs may want to answer this question affirmatively by adopting what I shall call the "Decay Picture" of what goes on when we follow a long proof. According to this picture, we should think of basic a priori warrants as processes which, against the background of any possible experience, provide a very high degree of support for the statements they warrant, so high in fact that it is correct to ascribe knowledge of the statements to those who undergo the processes, no matter what lives they may have had. The Decay Picture also supposes that there are special kinds of inferences, such that, if we perform one of these inferences on premises which enjoy a particular degree of support (as the result of our having arrived at belief in them in a way that gives them that degree of support), then the conclusion has a degree of support only slightly less than that possessed by the worst supported premise. In particular, if we use one of these processes to infer a statement C from a set of premises P, then, no matter what our background experience may have been, if the degree of support for each member of P was clearly above the threshold for the ascription of knowledge, then the degree of support for C will be above the threshold. The special modes of inference preserve apriority in an exactly analogous fashion to that in which the operation of adding 1 preserves smallness among numbers. Hence, if the Decay Picture is adopted, (iii) can be defended: one can accept both (4) and the point about long proofs, blaming the apparent tension between them on a misapplication of mathematical induction.

From the point of view of the apriorist, this is not an optimal position, for it only allows the apriority of a (vaguely defined) subset of mathematical truths. Apriorists will attempt to challenge the contention that long proofs do not yield a priori knowledge of their conclusions. Now it is hard to resist the observation from which the contention derives. Eminent mathematicians, as well as beginning students, become more convinced of the correctness of their proofs when others endorse their reasoning. Hume puts the point eloquently:

> There is no Algebraist nor Mathematician so expert in his science, as to place entire confidence in any truth immediately on his discovery of it, or regard it as anything but a mere probability. Every time he runs over his proofs, his confidence encreases; but still more by the approbation of his friends; and is rais'd to its utmost perfection by the universal assent and applauses of the learned world.[4]

Hume's point is psychological. What the apriorist will want to deny is not the accuracy of its delineation of our feelings of confidence, but the drawing from this psychological premise of epistemological conclusions.

There are two ways to defend the view that Hume's point is epistemologi-

4. *Treatise of Human Nature*, p. 180.

cally insignificant. The first is to suggest that the uncertainty is an accidental feature of the situation, stemming from the incomplete character of the proofs we normally follow. The second is to propose that it is possible for us to know a proposition a priori without knowing it for certain. I shall try to show that both these approaches lead nowhere.

The fundamental point is that the failures of nerve to which Hume alludes are not only widespread but also perfectly reasonable. We know ourselves to be fallible. We know that our attention may lapse and that we sometimes mis-state what we previously proved. Hence we are *reasonably* concerned, as we arrive at the end of a long proof, that an error may have crept in. There are many occasions on which the possibility of error does not matter: although we are aware of our fallibility, we still assert the conclusion, and are rightly credited with knowledge on the basis of following the proof. I suggest that the psycho-logical observation that we *feel* uncertain is explained by the epistemological observation that it is *reasonable* for us to feel uncertain, and, despite the fact that the reasonable uncertainty does not normally affect our knowledge, I shall argue that it does indicate that our knowledge is not a priori.

Reasonable uncertainty about the conclusion is not just a feature of the kinds of proofs which mathematicians commonly construct and follow. If proofs were presented in complete formal style, the increase in length would exacerbate the rational worry that, at some point, one's attention may have lapsed or one may have misremembered some result established earlier. Several philosophers have wondered how compelling a full formal proof of an arithmetical identity would be.[5] Arithmetical identities are by no means the worst cases. Some theorems of analysis, widely used in mathematical physics, never receive general proofs which are rigorous even by the standards of *informal* rigor which mathemati-cians accept. Stokes's theorem is a good example.[6] Even if our minds do not boggle at the thought of a mathematician following a formal proof of one of these theorems, we should readily admit that the activity would *probably* be error-ridden, and that it would be highly unreasonable for the mathematician to dismiss the possibility of a mistake.

Why should this matter? I have already admitted that the fact that it is rea-sonable to wonder whether one has made a mistake does not normally under-mine knowledge based on following a proof. But someone who wants to main-

5. Notably Wittgenstein (*Remarks on the Foundations of Mathematics*). For a clear discussion of Wittgenstein's concerns, see Steiner, *op. cit.*, chapter 1.

6. Stokes's theorem states that, for a vector field A and any boundary B, bounding any surface S, $\int_B A \cdot ds = \int_S (\nabla \times A) \cdot dS$. The usual proofs of this theorem (which typically skip steps galore) take it for granted that B is a well-behaved closed curve (at least C^1 everywhere, and, standardly, C^∞ everywhere). But engineers and physicists standardly apply the theorem to boundaries which are not so well behaved (boundaries with corners, for example). Nobody doubts that the theorem is *true* in these cases—but, so far as I know, no one has taken the trouble to prove it, even according to the standards of informal proof characteristic of vector analysis.

tain that *a priori knowledge* is compatible with rational uncertainty must defend a stronger claim. One can easily go astray here, by conflating a priori knowledge with knowledge obtained by following a non-empirical process. Consider the following example, advanced by Saul Kripke:

> Something can be known, or at least rationally believed, *a priori*, without being quite certain. You've read a proof in the math book; and, though you think it's correct, maybe you've made a mistake. You often do make mistakes of this kind. You've made a computation, perhaps with an error.[7]

In this case, a nonempirical process engenders belief. However, the statement known (believed) is not known (rationally believed) a priori. The process producing it does not meet condition (3b) on a priori warrants. Experiences which cast doubt on the accuracy of the book (by appearing to expose errors in many "theorems," let us say), and in which eminent mathematicians denied the conclusion, would interfere with the ability of the process to warrant the belief. If I check through a proof in a book, thinking I see how the inferences go, and if the proof is very complex, then, under circumstances in which there is weighty evidence against both book and theorem, it would be unreasonably arrogant and stubborn of me to form the belief.

I think that the same point holds no matter how we interpret the process of following a proof. The reasonable doubt which arises when we follow complicated proofs can be exploited by circumstances in which we receive *criticism* rather than applause from the learned world. Reasonable uncertainty is typically compatible with knowledge because of the kindly nature of background experience. Transform the quality of our lives, and that knowledge could no longer coexist with the uncertainty.

I have thus argued against choosing option (ii) of the three possible responses we considered above. The intuition that long proofs do not produce a priori knowledge cannot be countered with the reply that the proofs we usually follow are not completely formal. For formalization would only exacerbate our rational uncertainty. Nor can one block the inference from uncertainty to the absence of apriority. Rational uncertainty does not preclude knowledge but it does rule out a priori knowledge. So the optimal apriorist response does not seem to be available.

Ironically, a classical apriorist attempt to overcome the difficulty presented by long proofs generates an interesting picture of what goes on in following a lengthy chain of reasoning which rivals the Decay Picture presented above. In several passages in the *Regulae*, Descartes laments the fact that extended deductions are uncertain because they exceed the scope of what we can simultaneously represent to ourselves.[8] He proposes that we can remedy this predica-

---

7. *Naming and Necessity*, p. 39.
8. See especially Rules VII and XI.

ment by continually rehearsing the reasoning to ourselves, so that, eventually, we are able to grasp the entire chain of inferences in a single act of apprehension. While I shall reject his cure as based on an overly optimistic view of our capacities for self-improvement, I think that Descartes offers an acute description of the conscious psychological process of following a proof.

I shall call the picture Descartes presents the "Storage Picture." Descartes imagines us as gaining certain knowledge of the first principles and proceeding with our deductions. As the reasoning develops, we are no longer able to keep in mind the evidence for the first principles; instead, we have to "store" these principles, believing them now on the basis of the *recollection* that they were once established. Similarly, some of our subconclusions have to be stored in the same way, simply because we cannot continue to attend to the reasoning which led up to them while articulating a further chain of reasoning from them. Thus, as we pursue the proof our grounds for beliefs we arrived at earlier shift. Instead of believing them on the basis of the processes which originally warranted them, we believe them because we recall that we underwent an appropriate type of process.

A diagram may help to make clear what Descartes envisages. Let us suppose, for the sake of simplicity, that the only rules of inference with which we are concerned are one-premise rules. Then, before taking a Cartesian course in consciousness raising, the idea is that mathematicians instantiate the following flow chart.

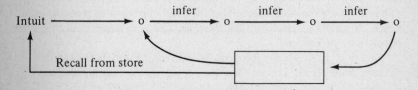

Here I assume (with obvious arbitrariness) that, in the ordinary state of practice, mathematicians lack the power to hold in mind anything more than one intuitive and three inferential steps. Descartes's thesis is that we can attain certainty until we have to take a step of recalling from store. Thus, given the assumptions underlying the diagram, uncertainty would enter in proofs which are more than four lines long.

Let us now translate the thesis into the terms I have been using. We suppose, with the apriorist, that when we follow a proof we begin by undergoing a process which is an a priori warrant for belief in an axiom. This process serves as a warrant for the belief so long as it is present to mind. As we proceed with the proof, there comes a stage when we can no longer keep the process and the subsequent reasoning present to mind: we cannot attend to everything at once. In continuing beyond this stage, we no longer believe the axiom on the basis of the original warrant, but rather because we recall having apprehended its truth in the appropriate way. However, this new process of recollection, although it normally warrants belief in the axiom, does not provide an a priori warrant for the belief. So, when we follow long proofs we lose our a priori warrants for their beginnings.

Descartes thinks that we can alleviate our predicament by continually running over the proof until we become able to apprehend all the steps at once. He believes that the repetition will *at least* enable us to expand the number of steps of the particular proof we rehearse which we can contemplate at once—and he may also think that it will enlarge our capacity for simultaneous attention. This proposal, like the Storage Picture itself, is consonant with facts of mathematical experience. Yet it is reasonable to wonder if the possibilities of self-improvement are unlimited. Can we really suppose that Descartes's training program would enable us simultaneously to attend to all the steps in a proof of Stokes's Theorem or the theorem that there is no general method for solving the quintic in radicals? The answer is clearly "No," and the fault lies not in Descartes's proposal but in the nature of our cognitive capacities. The representational ability of any system which could perform so staggering a feat would be enormous—and I suspect that cognitive science will demonstrate some day that we don't measure up.

If this is correct, then the right response to the puzzle about long proofs will be (i). There are systematic reasons for thinking that there is no true instance of (4), the normal form for mathematical apriorism. The heart of the trouble is (4b). If we adopt the Storage Picture then there will be no apriority-preserving rules. We are lured into thinking that there are such rules—*modus ponens,* for example—because we focus on the issue in a particular way. We imagine that we have an a priori warrant for premises of the appropriate forms and then conceive of the transition to the conclusion as preserving apriority. What is overlooked in our imaginative representation is the point on which the Storage Picture insists, to wit the impossibility of preserving our a priori warrant for the premises through a sequence of these transitions, while, simultaneously, attending to each of the inferential steps.

I shall not try to decide between the Decay Picture and the Storage Picture. My goal in this section has been to aggravate a common anxiety about long proofs, and to show how to provide pictures of what occurs when we follow proofs which will account for the phenomenological data. It is quite possible that a more refined psychology will expose the limitations of the sketches I have offered, but it would be highly surprising if psychology were to bring aid and comfort to apriorism. For we have found reasons to reject the apriorist's preferred option, (ii), and, whether we defend (i) or (iii), apriorism is in trouble. Accepting (iii) and the Decay Picture, the most that the apriorist could hope for would be a defense of the apriority of part of mathematics. The same result follows if (i) and the Storage Picture are correct.

The preceding discussion casts some light on the recent controversy surrounding computer-assisted proofs. With the advent of results which are based on information generated by computers and which, given the cognitive capacities of *Homo sapiens,* seem to be inaccessible to the kinds of proofs mathematicians have traditionally offered, it has been suggested that there has been an important change in the character of mathematical knowledge.[9] I do not wish to deny that there are some epistemologically important differences between computer-assisted proofs and ordinary proofs. However, we cannot capture the differences by alleging that, while all our mathematical knowledge used to be a priori, there are now parts of mathematics which are not a priori. For there are many theorems of traditional mathematics whose proofs are so long that they cannot lead us to a priori knowledge. Computer-assisted proofs are merely a new variation on an old theme. From this perspective, the new worries about flaws in computer-assisted proofs are continuous with previous anxieties of an everyday kind: mathematicians commonly complain that, as they look at each step of a long proof, they are certain of its correctness, but that they still suspect that an error lurks somewhere. My goal has been to show that there is nothing pathological about the complaint and that, when it is properly understood, it exposes trouble for the apriorist.

III

The difficulty discussed in the last section is not the most fundamental problem for mathematical apriorism in that it allows for some a priori mathematical knowledge. In the next two chapters I shall try to expose a more serious weakness. Mathematical apriorism is committed to the thesis that we have (or can have) basic a priori knowledge of mathematical truths, and I shall charge that there is no prospect of an explanation of how this is possible.

Apriorist proposals about the character of the basic a priori warrants for mathematical knowledge divide naturally into three main types: *conceptualist, constructivist,* and *realist.* This taxonomy is based on division according to theories of mathematical *truth.* The most fundamental cleavage is between thinkers who take mathematical statements to be true in virtue of our concepts and those who claim that such statements owe their truth to the mathematical facts. Apriorists in the former class will account for the apriority of basic items of mathematical knowledge by suggesting that the knowledge stems from our understanding of the statements. Apriorists in the latter class will attempt to

9. See T. Tymoczko, "The Four-Color Problem and Its Philosophical Significance," and K. Appel and W. Haken, "The Solution of the Four Color Map Problem." Tymoczko's conclusions have been cogently criticized by Paul Teller ("Computer Proof") and by Michael Detlefsen and Mark Luker ("The Four-Color Theorem and Mathematical Proof"). I believe that my discussion provides a clear way of summing up some points made by Tymoczko's critics.

describe some a priori mode of access to mathematical reality. Their endeavors fall into one of two subclasses: either mathematical reality is identified as a construct of the human mind, or it is alleged to exist independently of the mental activities of mathematicians.

Obviously, one can take a stand on the character of mathematical truth, occupying a position within my taxonomy, without any commitment to apriorism. My point is that it is helpful to divide varieties of apriorism according to their proposals about mathematical truth. The reason for this is that the theory of mathematical truth dictates the form of the problem of characterizing basic a priori warrants. For the conceptualist apriorist the question is "How do we have a priori access to conceptual relations?" Constructivists must answer the question "How do we have a priori access to our mental constructions?" Realists face the demand "How do we have a priori access to mind-independent mathematical reality?" In brief, my goal in the next sections is to examine the ways in which apriorists of different stripes have approached these questions and to argue that their proposals fail.

Many of the major contributors to the philosophy of mathematics are not easily placed with respect to my scheme of classification. The reasons are various, and do not reflect any inadequacy in the taxonomy. The influence of apsychologistic epistemology, the vagueness of traditional usages of 'a priori,' and the precision to which questions about instances of (4c) naturally lend themselves, frequently divert attention from the fundamental epistemological issues. Thus some writers assume that the epistemological questions have been answered, concentrating their efforts on producing derivations from axioms whose epistemic attributes they never question.[10]

It is worth pointing out that different views about the character of basic a priori warrants frequently accompany divergent specifications of which truths are basic to mathematics (and even of which truths belong to mathematics). Apriorists have variously claimed that truths of logic, definitions, geometrical axioms, arithmetical axioms, or the axioms of set theory are the basic truths on which all mathematics is founded. I shall not be interested in deciding whether any of these supplies an adequate basis. Conceding the choice of basis, I shall focus on the issue of whether there is a type of process which could provide basic a priori knowledge of the favored principles.

Finally, let me confront an objection to my proposed procedure. By characterizing varieties of apriorism in terms of their approaches to mathematical truth, have I stacked the deck by excluding some (perhaps the most promising) versions? Apparently, there are apriorists, or at least thinkers with similar views,

10. In particular, many of the writers cited in note 3 above seem to fall into this category. Those who comment on their proposals typically ignore the underlying epistemological questions, so that, in twentieth-century philosophy of mathematics, discussions of the merits of logicism, formalism, intuitionism, and so forth are often conducted without any examination of the epistemological presuppositions of these enterprises.

who deny that some (or any) mathematical statements have truth values. Formalists adopt this doctrine, and, though Hilbert is usually hailed as a formalist, his writings are full of yearnings to show the certainty of mathematics and to "cleanse [its] fair name."[11] There is a simple reply to the objection. Insofar as we can talk about mathematical *knowledge* we must be able to talk about mathematical *truth*. Since mathematical apriorism is a thesis about the nature of mathematical knowledge, nothing will be excluded by using the taxonomy introduced above. While this response is correct as far as it goes, it misses a deeper point. Formalists may suggest that their account allows for analogs of mathematical knowledge and of apriorist theses about that "knowledge." When mathematicians do their job properly they inscribe theorems of standard formal systems, and they do so as the result of inscribing proofs in those systems: under these circumstances we may say that they "know" the theorems. The "knowledge" may be said to be a priori if no sufficient life would make it reasonable for the mathematician to abstain from the inscriptional practice in question. However, given this development of formalism, it appears that my approach to apriorism is not defective. The formalist analog of apriorism will be defensible only if we can support the apriority of certain kinds of metamathematical knowledge. For if we are to credit the mathematician with something akin to knowledge, then it must be reasonable for her to believe that the system with which she is working is consistent. (There may be other necessary rational beliefs, but I shall only consider the minimal condition of consistency.) To sustain the apriority of mathematical "knowledge," one will thus have to show that there is an a priori warrant for belief in the consistency of standard formal systems. Hence the formalist analog of apriorism will be committed to apriorism about metamathematics, and the latter apriorism will take one of the forms distinguished in our taxonomy. Hilbert's case is typical. At the heart of Hilbert's program is the goal of showing how we can know a priori that certain systems are consistent. This goal compels him to adopt a standard apriorist theory of basic (meta)mathematical knowledge.[12]

11. For references, and a presentation of Hilbert's views, see my paper "Hilbert's Epistemology."
12. In fact, the situation is more complicated than I have portrayed it here: Hilbert's metamathematics corresponds to part of formal mathematics, so that, in a sense, he also defends the idea that we have a priori mathematical knowledge. (See "Hilbert's Epistemology" for details.)

    I should note that formalists are not committed to apriorism or any analog of apriorism. H. B. Curry is an example of a formalist who believes that one shows empirically that the constraints on adequate formal systems apply to systems in which mathematicians are interested. (See his *Outline of a Formalist Philosophy of Mathematics*.)

3
Mathematical Intuition

'Intuition' is one of the most overworked terms in the philosophy of mathematics. Frege's caustic remark frequently goes unheeded: "We are all too ready to invoke inner intuition, whenever we cannot produce any other ground of knowledge."[1] In this chapter, I shall examine whether we can elaborate mathematical apriorism in either a realist or a constructivist way by supposing that there is some special process which can yield basic a priori mathematical knowledge. Our investigation will hardly break with tradition if we label these putative processes as "intuitions."

I shall begin my examination of apriorist positions by considering the constructivist version of apriorism. The advantage of this starting point is the existence of a relatively clear approach to the epistemological issues, namely Kant's theory of pure intuition. By considering this theory, we shall be able not only to discern the difficulties which face constructivist apriorism, but also to arrive at some general points which can be deployed against the murkier proposals of realists.[2]

We have already encountered a simplified version of Kant's theory of mathematical knowledge, since the notion of pure intuition was used to introduce the conception of a priori warrant in Chapter 1. Let us recapitulate. Kant proposes that we construct figures in thought, inspect them with the mind's eye, and thus arrive at a priori knowledge of the axioms from which our proofs begin. The theory is clearest in accounting for our geometrical knowledge, and it is hardly surprising that when he is pressed for an example Kant turns to geometry. However, even if we waive questions about how Kant would provide an explanation of basic a priori mathematical knowledge in areas other

1. *The Foundations of Arithmetic*, p. 19.

2. The following discussion condenses and extends material presented in my paper "Kant and the Foundations of Mathematics."

than geometry, there are some features of his proposal which are *prima facie* puzzling. It is hard to understand how a process of looking at mental cartoons could give us knowledge, unless it were knowledge of a rather unexciting sort, concerned only with the particular figures before us. Hilbert and the intuitionists, who follow Kant in claiming that the fundamental mode of mathematical knowledge consists in apprehension of the properties of mentally presented entities, fail to explain how mathematics is anything more than a collection of trivial truths, concerned only with the properties of those mental entities which mathematicians chance to have discerned or those mental constructions which they happen to have effected.[3] We appear to face a dilemma. If mathematical statements are not merely reports of the features of transient and private mental entities, it is unclear how pure intuition (the process in which we inspect the entities and read off their properties) generates mathematical knowledge. If, on the other hand, mathematical statements merely describe transient and private mental entities, then such statements express different propositions for different people and they express different propositions at different times. Mathematical statements are no longer viewed as having a permanent, intersubjective content. Moreover, it is quite unaccountable why these statements should prove so important in our transactions with the physical world. It appears that we shall either fail to explain mathematical knowledge or else be driven to conclude that such knowledge is trivial.

Kant develops an ingenious response to this dilemma. He denies that the function of pure intuition is merely to lead us to knowledge of the properties of particular figures. By constructing figures in pure intuition, we are supposed to become aware of principles which necessarily characterize all our experience. Our minds are regarded as imposing a structure on experience. Construction of mathematical entities highlights this structure, so that, by inspecting our mental constructions, we discover the features which characterize possible experience.

Although Kant's own proposal is tied to a sensuous notion of pure intuition—we draw mental pictures and look at them—it is relatively easy to generalize it. The heart of the theory consists of two claims: mathematical truths are true in virtue of the structure which our psychological constitution imposes on all experience; by apprehending the features of mentally presented mathematical entities, we can disclose to ourselves the structural properties of the mind in virtue of which mathematics is true, and, by doing so, we can arrive at a priori mathematical knowledge. However we choose to articulate these fundamental claims, we are potentially vulnerable to three types of problems. The first is the *irrelevance problem:* how do we distinguish between those properties of the presented entities which reflect the structure of the mind and

---

3. Heyting comes very close to acknowledging this. See his *Intuitionism: An Introduction*, p. 3. For the problem as it arises for Hilbert, see my "Hilbert's Epistemology."

those which are accidental? The second is the *practical impossibility problem:* how do we determine that sequences of presentations which we cannot *in practice* achieve are *in principle* possible for us? Finally, there is the *exactness problem:* how can we resist the challenge that the presented entities do not have *exactly* the properties we take them to have?

To illustrate these problems I shall briefly consider how they arise for Kant's original proposal. Kant believes that we can gain a priori knowledge about the general properties of triangles by drawing and inspecting a particular triangle. But how do we come to generalize over the right properties and avoid generalizing over the wrong ones? How, for example, do I have the right to conclude, on inspecting a scalene triangle, that the sum of the lengths of two sides of a triangle is greater than the length of the third side but not that all triangles are scalene? Were Kant to suggest that we should only generalize over those properties which are determined by the *concept* of triangle, then the process of constructing mental diagrams would simply be a vehicle for disclosing conceptual relations and Kant's position would become a conceptualist version of apriorism.[4] Thus Kant must conclude that the presented triangle has three types of property: those properties determined by the concept of triangle; those properties which reflect the structure we necessarily impose on experience; and those properties which result from accidental decisions made in the construction. For his account to succeed we need a method of discriminating properties of the two latter types, so that we can legitimately generalize over the former and avoid generalizing over the latter. But to be able to do this is to have precisely that knowledge of the structure of experience for which Kant is attempting to account!

The practical impossibility problem arises in a similar way.[5] Kant claims that pure intuition can yield the knowledge that line segments are infinitely divisible. Now it is evident that we cannot attain this knowledge by observing a line segment infinitely divided. So what Kant must intend is that we give ourselves a sequence of presentations, showing a continued process of subdivision. Since there are practical limits on our ability to do this, we shall face an awkward question: are these limits reflections of a structural property of experience? To resolve this issue we need, again, that same insight into the structure of experience which pure intuition was supposed to provide.[6]

The exactness problem is even more straightforward. Just as our powers of ordinary perception are limited and fallible, so too are our powers of mental perception. Because of this we cannot assume that mental perception will give

4. Intuitionists sometimes veer in this direction. See Heyting, *Intuitionism: An Introduction*, pp. 13–14, and A. Troelstra, *Principles of Intuitionism*, p. 12.

5. This example stems from Charles Parsons's illuminating paper "Infinity and Kant's Conception of the 'Possibility of Experience.'"

6. For the intuitionists and for Hilbert, this problem arises with respect to the principle of mathematical induction.

us exact knowledge even of the features of the particular figures we construct. We should concede that we might be unable to distinguish a straight line from one that is very slightly curved. The concession is dangerous. For imagine that we follow Kant's procedure to arrive at belief in a geometrical truth. The warranting power of the procedure can be undermined by experiences involving deceptive measurement which seem to show that the statement is only a close approximation to the truth. Given such experiences it would be rational for us to suppose that our (mental) visual acuity had failed us, and thus to inhibit formation of the belief. Therefore the process Kant describes fails to meet condition (3b) on a priori warrants.

One might think that these difficulties arise for Kant's theory simply because his construal of the process of intuition is too crude. Perhaps mental picturing is just not an appropriate way of gaining mathematical knowledge. We should note, however, that Kant's identification of the process of intuition has the advantage of picking out a process which plays a recognizable role in our mental life. Furthermore, we know enough about such processes to assess their credentials as a priori warrants. If someone proposes that intuition be divorced from the sensuous then we have a right to ask for a description of the process of intuition which will enable us to iden/ify it and to determine whether it can serve as an a priori warrant for mathematical beliefs. Lacking a description of this kind, it is unclear that the proposal amounts to a theory of mathematical knowledge at all, much less to a theory of a priori mathematical knowledge.

In any case, to becloud the notion of intuition would not necessarily solve the problems which undermine Kant's theory. Consider first the irrelevance problem. This arises because Kant's method for avoiding the trivialization of mathematics rests on the thesis that the figures constructed in pure intuition have three types of properties: those which are determined by the concepts used in the construction; those which reflect the structure which the mind imposes on experience; and those which are accidental. Our difficulty is to distinguish the second type of property from the third, for it appears that we shall only be able to make this distinction if we have some independent access to the features which the mind imposes. Denying that intuition is sensuous does not evade the difficulty. So long as the intuited objects have the three types of properties the problem will continue to vex us. We might try to respond by denying that these objects have any accidental properties, but this does not work. In the first place, the objects we intuit will (at least) have the accidental properties of being intuited by us at some times and not at others. Second, what is relevant is our ability to *recognize* that a particular property is not accidental. Even if intuited objects lacked accidental properties, we would need to know this, and it is hard to see how intuition could help us to this knowledge or could override the contrary suggestions of an uncooperative experience.

Similarly, we cannot solve the practical impossibility problem by merely denying that intuition is a sensuous process. The problem retains its sting so long as it is maintained that our knowledge of some mathematical principles depends on our recognition of the possibility of a sequence of intuitions which we cannot in practice give ourselves. Whether intuition is sensuous or not, the brevity of human existence appears to place a finite upper bound on the number of intuitions which we can give ourselves. To defend the process of intuition as a source of all a priori mathematical knowledge one will be forced to argue that there are no mathematical axioms which we could only know a priori by giving ourselves an indefinitely long sequence of intuitions. It appears that constructivist apriorism can only escape this difficulty by embracing finitism.

We began this section by posing a dilemma for the constructivist: either mathematics is trivial or there is an apparent problem in explaining how apprehension of mental constructions yields mathematical knowledge. Kant's clever attempt to solve this dilemma involves the idea of a gap between our constructions and the underlying mathematical reality which they represent, and it is the presence of this gap which generates the two difficulties we have been considering, the *irrelevance problem* and the *practical impossibility problem*. Even if we liberate the notion of intuition from Kant's interpretation of it as a sensuous process, these problems will persist, so that any attempt to construct a theory of a priori mathematical knowledge along constructivist lines will either face the same fate as Kant's or else will have to embrace the other horn of the dilemma. Before we consider whether there is any solace for the apriorist in this latter approach, I want to examine the third problem which was raised against Kant. This problem will lead us into issues which will be important for later discussions.

II

At first sight the exactness problem seems to be a difficulty which we could avoid by abandoning Kant's construal of intuition. The root of the problem, however, is the fact that our mental perceptual powers are not superior to our ordinary perceptual powers: given an appropriate recalcitrant experience, mental perception of a mental construction would no more give us the right to override the suggestions of experience than perception of a figure on a blackboard. The process of pure intuition does not measure up to the standards required of a priori warrants not because it is *sensuous* but because it is *fallible*. Once this is recognized, we can see how to present a more general version of our criticism.

Talk of the fallibility of the process of pure intuition is relatively imprecise. What is crucial to the objection against Kant is that it is reasonable for us to believe that the process of pure intuition might lead us to false beliefs. Let us introduce some terminology. A type of process which generates belief will be said to be *dubitable* if there is a life given which it would be reasonable to believe that some processes of the type engender false beliefs. Suppose now that we are investigating the status of a process α as an a priori warrant for belief that p. I shall assume that α belongs to a type of process, the *availability type of* α, which we have identified, such that a process of that type would be available given any sufficient experience. Then I take the exactness problem to stem ultimately from the thesis

> (5) If the availability type of α is dubitable and, if there are lives which would suggest the falsity of p, then there are sufficient lives given which the available processes of the same type as α would not warrant belief that p.

The basic idea is very simple. If I can have grounds for worrying whether a type of process yields false belief then, under some circumstances in which experience suggests that the belief I have formed by undergoing a process of the type is false, I would not be warranted in the belief. We need simply imagine that my life consists in part of experiences which cast doubt on the reliability of the process type and in part of experiences which call into question the conclusion I have formed.[7]

If, for a given process α, we can establish the appropriate instances of the antecedents of (5) then we can conclude that α is not an a priori warrant for belief that p. For it will follow that there are (sufficient) lives given which the available processes of the type of α will not warrant belief that p, and, hence, α will not satisy condition (3b) of our analysis of 'a priori warrant.' Precisely this strategy was used in developing the exactness problem: I used the similarity of mental perception and ordinary perception to support the dubitability of pure intuition, and I took it for granted that there are possible experiences which could suggest the falsity of geometrical axioms. Were we to have such experiences and also to have experiences which called into doubt the reliability of pure intuition, it would be unwarranted for us to form beliefs in the questionable axioms on the basis of pure intuition. Thesis (5) encapsulates this idea and presents its general form, thereby pointing toward a wider application of our criticism.

At this point, I anticipate two kinds of criticism. The first concentrates on a specific feature of my deployment of (5) against Kant. I have just acknowledged that I took it for granted that there are possible experiences which could suggest the falsity of geometrical axioms. Do I have any right to this assumption? Surely one of the strengths of Kant's theory is that it subverts the thesis that there are possible experiences which would suggest the falsity of mathe-

---

7. Let me make an assumption explicit. I suppose that the two kinds of experiences can be joined in a single life. Thus I imagine someone being presented both with evidence against the belief and with evidence against the universal reliability of a type of process. Under these circumstances, it would be unreasonable for the person to use a process of the type to override the countervailing evidence. For our purposes here, I think the assumption that the two recalcitrant bits of experience can be joined together is harmless.

matical truths. For Kant not only claims that mathematical truths *are* necessarily true but also that they must necessarily *appear* to be true. The very point of the thesis that mathematical truths describe the structure which the mind imposes on experience is to deny that experience could mislead us about mathematics. Indeed, we might credit Kant with appreciating the much vaunted unimaginability of the falsity of true mathematical statements, and providing a primitive psychological explanation for the phenomenon.

The objection fails. We are not entitled to suppose that an experience suggesting the falsity of a mathematical truth must consist in some relatively direct presentation of a situation in which it is cunningly made to appear that the statement is incorrect. There are indirect ways in which experience may suggest to us that the revision of our beliefs is in order. To illustrate the point, it will help to revert, temporarily, to Kant's account of geometry. Let us consider the statement that the sum of the angles of any triangle is 180°, a statement Kant takes to be a priori. Let me draw a (rough) distinction between three kinds of misleading experiences which could challenge our belief in the statement.[8] A *direct challenge* will consist in a perceptual experience of a figure which, judged by our very best criteria, appears to contradict the statement. A *theoretical challenge* will consist in a sequence of experiences which suggest that a physics-cum-geometry which does not include this statement will provide a simpler total description of the phenomena than a physics-cum-geometry which does. A *social challenge* will be a sequence of experiences in which apparently reliable experts deny the statement, offer hypotheses about errors we have made in coming to believe it, and so forth. (This division of challenging experiences is useful but obviously rough: I do not intend to suggest that it is exhaustive or that the boundaries of the categories are precise.) Now Kant's thesis that our psychological constitution dictates the geometrical structure of experience may rule out the possibility of direct challenges, but it is quite compatible with the existence of theoretical or social challenges. So there is no reason to reject the assumption on which the exactness problem rests.

The same strategy can be generalized against those who believe the more general Kantian thesis that the mind imposes mathematical structure on possible experience. We can make analogous divisions to those just introduced. Thus if someone attempts to block application of (5) against the credentials of an alleged a priori warrant, on the basis of the claim that experience could not even make it *appear* to us that particular mathematical statements are false, we can respond by insisting that not only direct challenges, but theoretical and social challenges as well, must be considered. In the geometrical case, the history of the investigation of space makes it easy for us to describe a theoretical challenge; in other cases, it may be harder to envisage how a theoretical challenge

8. I am conceding the necessity of the truth, and so I am not supposing that there could be *veridical* challenges. Obviously, if this concession were not made, Kant's case would be even weaker.

might occur. Yet, if worse comes to worst, we can always fall back on the possibility of social challenges. We need not think of a social challenge as an experience in which apparent authorities simply deny what we assert. We can imagine that the experts demonstrate their expertise by producing verifiable solutions to problems which baffle us, that they produce plausible arguments against our contentions (arguments whose flaws are too well hidden for us to detect), and that they offer convincing psychological explanations of our "mistake." In such cases, experience would suggest the falsehood of the mathematical statements in question.

The second objection I want to consider is based on a worry that (5) is too strong. In a sense, I think that the worry is justified: we shall see that (5) can be deployed as a very effective weapon against apriorism. I would claim, however, that (5) brings out an underlying assumption of apriorism, an assumption of which apriorists may not have been aware. A priori warrants have to be able to warrant belief against the background of any experience, and this means (as (5), in effect, claims) that they must belong to indubitable types. To see why this is so, it may be helpful to recall our discussion of the ways in which background experience may defeat the ability of processes which would normally warrant belief. If you have reason to believe that your senses sometimes play tricks on you, then if you also have reason to think that the perceptual belief which you are inclined to form is false, your perceptual process (which may, in fact, be perfectly normal) does not warrant the belief. (5) generalizes the point. Once we have grounds for believing that a type of process can lead us astray, then we should agree that recalcitrant experience could tip the scales against it. To put the point another way, if we override the contrary suggestions of experience, forming our beliefs on the basis of a nonempirical process, when we have reason to suspect the reliability of processes of that type, then we are falling prey to irrationality and dogmatism.

I have attempted to show that the heart of the exactness problem is a point with potential for wide-ranging criticisms of apriorism. I shall now try to show how it applies to constructivist versions of apriorism which take the opposite horn of our dilemma to that chosen by Kant.

Claiming that mathematics describes the properties of transient and private mental entities is not an attractive position for the apriorist. I shall simply note in passing that there are apparently severe problems in our having a priori knowledge with a common content and in our having the same a priori knowledge at two different times. The heart of constructivist apriorism is that we can have a priori knowledge of the features of the construction which is currently present to mind. If this should fail then the position is bankrupt. Now in the Kantian case we discovered that this thesis was vulnerable: Kant would be wrong to insist that pure intuition yields a priori knowledge of the exact features of the diagrams we draw. Our discussion led us to a result which constrains constructivist apriorism: the constructivist apriorist must maintain that we have a means of apprehending the properties of our constructions which can escape any suspicion that it can lead us astray. Do we have such a means?

It might be thought that we do. After all our constructions are *ours,* and we are easily lured into believing that they must be transparent to us. Yet this is to underrate the potential recalcitrance of experience. Constructivists do not tell us much about the ways in which we apprehend our constructions—although they would probably dismiss Kant's straightforward attempt to be informative as far too crude. But, for any type of method we can envisage, there is a kind of experience which would threaten our confidence in it. Any of us could be confronted by experiments which demonstrated (or convincingly appeared to demonstrate) a correlation between the performance of particular types of construction (introspectively reported by subjects) and particular kinds of brain states. We could then be shown tests which revealed that our judgments were sometimes at variance with those predicted on the basis of the correlations. If we were also offered a diagnostic explanation, apparently identifying a flaw in the neural mechanism which is taken to instantiate our detection of our constructions, our experience would indeed threaten the reliability of our means of apprehending the features of constructions. Thus constructivists cannot simply *assume* that we have a priori knowledge of our present constructions. They must show that the process of apprehending those constructions can resist the threat of such experiences.

At this point, I shall leave constructivist apriorism with the conclusion that even its most minimal central thesis is not uncontroversially true. In the next section, I shall strengthen my attack on this central thesis by developing further an analogous argument against realist apriorism.

III

Like constructivists, many mathematical realists are fond of making reference to mathematical intuition. My examination of whether intuition, interpreted along realist lines, can yield basic a priori mathematical knowledge will be briefer than the corresponding investigation of constructivist appeals to intuition. There is a straightforward reason for the difference. Realist epistemology of mathematics rarely provides more than fragmentary metaphors,[9] and the absence of detail makes it hard to assess the promise of the account of knowledge. Nevertheless, I shall try to show that the realist's notion of intuition cannot meet the standards demanded of basic a priori warrants.

The central tenet of mathematical realism is the thesis that mathematical statements are true or false in virtue of the features of mind-independent math-

---

9. An exception is Penelope Maddy's "Perception and Mathematical Intuition." For discussion, see note 16 below.

ematical reality. The thesis is standardly articulated by supposing that there is a realm of mind-independent mathematical objects (sets, numbers) whose properties mathematicians attempt to describe. When realism is articulated in this way, I shall call the resulting position *Platonism. Prima facie,* realism and Platonism are distinct positions. Indeed, I would contend that there are defensible non-Platonist forms of realism.[10] However, because Platonism has been the most popular version of realism, I shall focus on attempts to work out a Platonist theory of a priori mathematical knowledge. I think it will be evident that the criticisms I level against this theory could be applied equally to other realist apriorist approaches.

With good reason, Platonists suppose that mathematical objects are not ordinary physical objects. Apart from the fact that there are probably not enough physical objects to serve the Platonist's turn, it would be implausible to suppose that the truth of mathematical statements should depend on the fate of particular physical objects. So the mathematical objects are taken to be *abstract:* they do not have spatio-temporal locations and, on some views at least, they do not enter into causal relations with other entities. The picture which emerges is of a universe of mathematical objects which is explored by the mathematician in a parallel way to that in which the natural scientist investigates the physical realm.

How does the mathematical explorer chart the features of this abstract realm? It is customary to pursue the analogy with natural science. Just as the natural scientist has, in sense perception, a basic mode of access to the objects she wishes to describe, a mode of access which produces knowledge of those objects, so too it is suggested that the mathematician has a basic mode of access to the objects which interest her, and, by exercising it, she comes to basic mathematical knowledge. Since mathematical objects are abstract rather than physical, it is usually held that the mathematician's source of knowledge is not sensory perception. However, the source is supposed to be like sense perception in being the kind of process which generates beliefs as output without taking beliefs as input; the source is supposed to provide noninferential knowledge. Let us call it "mathematical intuition," bearing in mind that its workings are quite different from those of the processes hypothesized by Kant and other constructivists.

A forthright statement of the position at which we have arrived can be found in a celebrated passage by Kurt Gödel.

> But, despite their remoteness from sense experience, we do have something like a perception also of the objects of set theory, as is seen from the fact that the

---

10. Realist approaches to mathematical truth distinct from standard Platonism have been offered by Michael Resnik ("Mathematical Knowledge and Pattern Recognition" and "Mathematics as a Science of Patterns: Ontology") and by Michael Jubien ("Ontology and Mathematical Truth"). The account offered in Chapter 6 may also be viewed as a type of realism.

axioms force themselves upon us as being true. I don't see any reason why we should have less confidence in this kind of perception, i.e. in mathematical intuition, than in sense perception, which induces us to build up physical theories and to expect that future sense perceptions will agree with them and, moreover, to believe that a question not decidable now has meaning and may be decided in the future.[11]

In this passage, Gödel does not claim that mathematical intuition yields a priori knowledge. Yet there is an obvious way in which his remarks encourage supporters of apriorism. Mathematical intuition is a nonempirical process. Hence anyone who confuses nonempirical processes which actually warrant belief with a priori warrants will read Gödel as upholding apriorism. Even if one does not make this conflation, the mathematical intuitions which Gödel hypothesizes will be processes which are available given any sufficient experience, so that condition (3a) will be met.

Gödel's proposal encounters an important theoretical challenge, which has dominated recent discussion of Platonism. In a lucid essay,[12] Paul Benacerraf points out that there is an apparent tension between the Platonist's view of mathematical truth and three other plausible theses: (a) knowledge of mathematical objects requires a causal relation between those objects and the knowing subject; (b) on the Platonist's account there can be no causal relations between mathematical objects and other entities; and (c) we know some mathematics. (Here, (a) is taken to be a consequence of the causal theory of knowledge; (b) results from the Platonist's characterization of mathematical objects as abstract.) Benacerraf's point casts doubt on the ability of Gödel's hypothetical process to generate knowledge. Platonists have struggled to avoid the problem.[13] Instead of reviewing the tangle of issues which has resulted, I shall press criticisms which are orthogonal to that raised by Benacerraf. Even if we concede the general possibility of a knowledge-generating process like that envisaged by Gödel, there are still two important questions to be asked: (i) do we have the capacity for undergoing such processes? and (ii) could this type of process yield *a priori knowledge?*

How do we tell if we have a faculty of Gödelian intuition? Platonists tell us very little about the character of intuitions. Gödel's remarks are typical: intuition is introduced by analogy with sense perception, and that is the end of the matter. The situation appears to contrast with Kant's appeal to intuition, for Kantians can tell us how to give ourselves geometrical intuitions, directing us to draw figures and inspect them with the mind's eye. Platonists can retort that the contrast is superficial. We can tell a person how to perform a process by

11. "What is Cantor's Continuum Problem?" p. 271.

12. "Mathematical Truth." See also Jonathan Lear, "Sets and Semantics"; Jubien, "Ontology and Mathematical Truth."

13. Steiner, *Mathematical Knowledge,* chapter 4; Maddy, "Perception and Mathematical Intuition."

reducing that process to a sequence of more primitive processes which our pupil already knows how to perform. If the process we are trying to describe is not reducible in this way we may be able to offer no helpful instructions. Platonists may contend that this is precisely their predicament when they are asked what intuitions are like, adding for good measure that Kantian descriptions of intuition are little better, in that Kantians would be pressed to explain the notion of "inspecting with the mind's eye."

Although the Platonist has a point, his confessed inability to describe intuition gives ground to scepticism. How can Platonists respond to those who wonder whether they have performed intuitions, or whether *anyone* has performed intuitions? There seem to be two possible strategies. One can appeal to the testimony of mathematicians, or one can argue that there must be some process of the type envisaged. The former route is more direct. If the sceptic can be convinced that mathematicians recognize in themselves processes of intuition which acquaint them with basic facts about mathematical objects, then she will reasonably conclude either that she lacks the ability to intuit (thus regarding herself as similar to someone who is blind or deaf) or that she has failed to identify the exercise of the ability in herself. However, this method of countering scepticism with mathematical authority has been less popular than the strategy of arguing that there has to be a process which provides basic knowledge of mathematical objects.

The case for Platonist epistemology rests heavily on the argument for Platonist ontology. Platonists standardly argue that we are compelled to regard mathematics as reporting the facts about mind-independent abstract objects if we are to account for mathematical truth. The notion of mathematical intuition is then introduced by following the route we traced at the beginning of this section. If this is the whole story about mathematical intuition then we can draw some important conclusions. First, processes of intuition are theoretical entities, in the sense that they are entities in whose existence we believe because we hold a particular theory, in this case a theory about the nature of mathematics and our mathematical knowledge. The theory can be challenged, either on the grounds that it faces severe difficulties in accounting for the phenomena with which it is supposed to deal, or because a better theory is available.[14] Second, what we know about processes of intuition is what the theory tells us about them, and that is not very much. By adopting the indirect strategy of arguing that a notion of intuition is needed to complement an ontology which is forced upon us, the Platonist abandons the idea that mathematicians can recognize the processes of intuition which they perform. In effect, the Platonist replies to the sceptic by sympathizing: "Like you, I can't identify processes of intuition in myself, but I've given compelling reasons for thinking that they exist; and I

---

14. Benacerraf's essay "Mathematical Truth" can be interpreted as advancing the former criticism. My aim, in Chapter 6, is to give substance to the latter.

suppose that they go on in you, just as they go on in me." As we shall see below, this is a damaging concession if the Platonist wants to be an apriorist.

Can the concession be avoided? The most promising idea for the Platonist is to allow that the sceptic is not atypical, that she is not defective or unaware of what passes in her mental life, but that intuition is the prerogative of talented (maybe exceptional) mathematicians. Perhaps we can use the testimony of mathematicians to show that, for *some* people, possibly a tiny minority, intuition serves as a mode of basic knowledge. This approach exploits the ambiguity of 'intuition.' When mathematicians talk about intuition, they usually do so in the context of discussing problem-solving. Great creative mathematicians—such as Euler, Riemann, and Ramanujan—are frequently praised for their powers of intuition. To admire the intuitions of a Riemann (or, at a humbler level, those of a promising student) is to recognize an ability to obtain an unusual and fruitful gestalt on a problem. Intuition of this type is frequently a *prelude* to mathematical knowledge. By itself it does not warrant belief, although it may play an important heuristic role and also serve as *part* of a warranting process. Moreover, this kind of intuition is normally exercised in the solution of research problems, not in the knowledge of axioms. The talented mathematician looks at a recalcitrant puzzle from a new point of view, "intuiting" that a particular manoeuvre will help with the summation of a series or the evaluation of an integral, that a problem in number theory reduces to a result in the theory of functions. The secret of his success is not taken to be some special ability to discern features of mathematical reality. We do not think of the mathematician as gazing on the mathematical objects and coming up with the fruitful idea. The intuitions of which mathematicians often speak are not those which Platonism requires.

Nevertheless, there is one kind of common remark which does appear relevant to the Platonist's program. Recall Gödel's claim that "the axioms force themselves upon us as being true." We might suppose (as Gödel does indeed seem to suppose) that the presence of a feeling of familiarity with basic principles, a sense of their obvious correctness, signals the fact that our belief in them has been generated by an intuition of mathematical reality. This state of "at-homeness" might thus be used to identify the occurrence in us of intuitions. However, the fact that the axioms of set theory (let us say) seem obvious to us does not guarantee that our belief in those axioms is generated by a process in which we directly apprehend mathematical objects. Even if we were to accept the Platonist's view about the nature of mathematical reality, we might adopt a different explanation of the phenomenon. Perhaps the feeling of evidence results from the exercise of those conceptual abilities which we have acquired in learning to talk about sets. Or perhaps it stems from indoctrination we received in our mathematical youth. I shall explore both of these rival hypotheses in more detail below. For the present, I want simply to note that remarks about the "intuitive evidence of mathematical first principles" are open to different explanations and that Platonist epistemology advances one particular hypothesis. Intuitions are entities which are introduced by an epistemological theory, and what the theory tells us exhausts our knowledge of them. There is no reason to believe that anyone (including Gödel and other great mathematicians) has some special knowledge of what it is like to have them.

IV

I shall now argue that this inaccessibility of intuitions, the very ignorance of their nature which makes Platonist epistemology so nebulous, militates against the thesis that they are a priori warrants. In the passage quoted above, Gödel contends that we should not think of intuition as being less reliable than ordinary sense perception. Our discussion in Section II made it clear that more than this would be required if intuitions were to count as a priori warrants. My arguments against the Platonist apriorist will turn on a principle related to that which I used against the Kantian. Let us say that a type of process is *suspect* just in case there are possible lives given which a person could carry out some process of the type but given which that person would be aware of his inability to discriminate the type of process performed from other processes known to generate false beliefs. Suppose that we are investigating the status of a process α as an a priori warrant for belief that p, and that we have identified the availability type of α. Then, parallel to (5), I maintain

> (6) If the availability type of α is suspect and, if there are sufficient lives which would suggest the falsity of p, then there are sufficient lives given which the available processes of the same type as α would not warrant belief that p.

The rationale for (6) is that if α belongs to a suspect type and if there are sufficient lives suggesting the falsity of p, a life, giving one reason to wonder whether the available process of the type was of a kind capable of engendering false beliefs and also suggesting the falsity of p, would deprive that process of its power to warrant belief that p. Reliance on that process, in the face of adverse experience, would be undercut by the legitimate question of whether one was not committing the mistake familiar from other cases. Recognizing that one could not distinguish the process at hand from another process, known with the advantage of hindsight to have yielded false belief, one would be irrational to form belief that p on the basis of the process.[15]

---

15. The rationale for (6) is obviously akin to that offered for (5). In both cases we envisage worlds in which the subject has evidence against p and attempts to override the evidence by using an available process. In the case of (5), the subject has grounds for believing that some processes of that type engender false beliefs. In the case of (6), the subject recognizes her inability to discriminate processes of the type from processes known to engender false beliefs. Given either type of situation—either "epistemic limitation"—it is irrational for the subject to override the contrary evidence.

To apply (6) against Platonist apriorism, I need to show both that the availability types of Platonist intuitions are suspect and that there are sufficient lives suggesting the falsity of mathematical statements, given the Platonist's interpretation of those statements. The former point can easily be made by recalling some episodes from the history of mathematics. On several occasions in the past, mathematicians have hailed some principles as intuitively evident, giving them the same status that we give to the axioms of set theory. It has then turned out that those principles are false. The most familiar example is that of Frege, Dedekind, and Cantor, each of whom advanced a universal comprehension principle, taking any property to determine a set. This is by no means the only case of its kind. Many mathematicians of the eighteenth century believed in the self-evidence of a "law of continuity," which states that what holds up to the limit holds at the limit. The existence of such cases is disconcerting. For, granting that mathematicians can and do undergo Platonist intuitions, we must ask ourselves whether or not the processes which contemporary mathematicians undergo are substantially different from those undergone by our misguided predecessors. Posing this question makes it evident that, for us, with our background of experience, it is reasonable to believe that we cannot discriminate intuitions from processes known (after the fact) to yield false beliefs. It follows that the availability types of the intuitions which the Platonist claims that we perform are suspect.

In discussing mathematical intuition, Gödel himself raises the question of the import of the paradoxes of naïve set theory. Immediately following the passage I have quoted, he writes: "The set-theoretical paradoxes are hardly any more troublesome for mathematics than deceptions of the senses are for physics." The similarity between the paradoxes and sensory illusions seems to me to be correct. If we suppose that there are indeed mathematical intuitions, then the ability of such processes to yield knowledge is not impugned by our incapacity to discriminate them from processes which generate false beliefs, any more than the possibility of perceptual knowledge is precluded by the difficulty of discriminating veridical from nonveridical sensory processes. Where our discriminatory shortcomings *do* matter is in cases where experience suggests that the belief we have formed is mistaken, and this applies both to perception and to mathematical intuition. The existence of deceptions of the senses is not an obstacle to our knowledge of physics; it is a stumbling block for the thesis that the sensory processes which actually warrant our beliefs could continue to do so, whatever experience we were to have. Similarly, the set-theoretic paradoxes do not challenge the possibility of mathematical knowledge, but they do threaten apriorism.[16]

16. For two different reasons, I have not offered any explicit criticism of Maddy's recent attempt to defend Platonist apriorism. First, Maddy's position is a variant of the view I call "conceptualism" rather than a doctrine like Gödel's. Maddy believes that we are able to develop certain neurophysiological mechanisms which enable us to perceive (impure) sets. Once these mechanisms are in place, they are able to generate beliefs in set-theoretic axioms. Insofar as Maddy provides

The only issue that remains to be resolved is whether or not there can be experiences which suggest the falsehood of mathematical statements, given the Platonist's interpretation of those statements. Platonists have two options. They may either propose that we can find out the truth values of mathematical statements by straightforwardly empirical means (observation, experimentation, and so forth) or they may deny that such empirical means can ever lead us to ascertain whether mathematical statements are true or false. Adopting the former proposal appears to favor the idea that experience could mislead us by suggesting that some (true) mathematical statements are false: if observation and experimentation can reveal the truth values of mathematical statements, then we could apparently arrange for deceptive experiences to offer false mathematical suggestions. The alternative approach, which denies that experience can expose mathematical reality, avoids this apparent consequence, but at the cost of forsaking a primary merit of Platonist ontology, namely its ability to account for the application of mathematics in the sciences.[17] However, on *either* proposal, there is at least one type of experience which can suggest to us that the beliefs we have formed on the basis of intuition are incorrect. Mathematical beliefs are vulnerable to social challenges. Such challenges pose a sufficient threat to make it unreasonable for us to form beliefs on the basis of intuition, when we recognize that we cannot discriminate our intuitions from processes which misled our predecessors. In fact, I think that Platonists are unable to resist the conclusion that there are direct and theoretical challenges to mathematical statements. However, showing this would be more complicated, and the existence of social challenges is enough to make my point.

The argument of this chapter can be presented as follows. Intuition has been a favorite process of mathematical apriorists. The apriorist can either take a bold line, identifying intuition with some process which we recognize as occurring in our mental life, or he can leave intuition as a process characterized only by its role in giving us mathematical knowledge. The former approach takes the risk that, when the nature of the process is exposed, it will be seen not to meet the standards required of a priori warrants. (Kant's proposal was vulnerable in just this way.) Yet it will not do to retreat into vagueness, for our acknowledged ignorance of the character of the process is itself a handicap to its functioning as an a priori warrant. I conclude that intuition, whether constructivist or Platonist, whether clearly specified or ill-defined, will not do the job which the apriorist demands of it.

an account of the explicit knowledge of set theorists (rather than an account of some "tacit knowledge" which all those who have acquired "set-detectors" possess), that account is close to the view considered in the next chapter. Second, and more crucial, Maddy explicitly allows that the beliefs generated by exercising her alleged neurophysiological mechanisms can be false. Given this admission and the analysis of a priori knowledge I have offered, it is easy to see that the knowledge generated is not a priori.

17. In Chapter 6, I shall suggest that Platonism may be less successful in providing this type of account than it is usually taken to be.

4
Conceptualism

I

I shall now turn my attention to the last of the three versions of mathematical apriorism which I distinguished in Chapter 2. Conceptualists claim that we have basic a priori knowledge of mathematical axioms in virtue of our possession of mathematical concepts. In the recent history of the philosophy of mathematics, conceptualists have typically embedded their proposal within an apsychologistic epistemology.[1] This move is mistaken, not only because of the inadequacies of apsychologistic epistemology, but because it makes conceptualism unnecessarily vulnerable to criticism. Although I shall eventually conclude that conceptualism cannot save apriorism, I shall attempt to show that, given an adequate treatment of fundamental epistemological issues, we can do justice to some of the ideas which motivate conceptualism.

Let us begin by recognizing the force of those ideas. Consider the statement that all groups contain a unit element. It is natural to regard the truth of this statement as determined, in some sense, by the concept of a group. If someone were to disagree with us about the statement, then we should wonder whether his usage of 'group' diverged from ours, finding it hard to envisage that anyone should understand 'group' as we do and yet dissent from the statement. Similarly, were we to be asked to explain why we believe the statement to be true, we would respond by citing our understanding of 'group.' The point of departure for conceptualism is a desire to take these intuitive responses at face value and to sustain them.

However, one of the most famous episodes in recent philosophy is the onslaught on doctrines of this kind, an onslaught which has been led by Quine.

---

1. This applies to the positivists and many of their successors. Earlier conceptualists, such as Locke and Frege, espoused a psychologistic version of conceptualism. For a defense of this interpretation of Frege, see my "Frege's Epistemology."

Repudiating the notions of conceptual truth, meaning, and analytic truth, Quine and his followers have launched a barrage of criticisms against philosophical deployments of these notions. The central thrust of the criticisms is that the notions belong to a bad theory, a theory which purports to offer explanations, but in fact explains nothing. My aim in this chapter is twofold: I shall try to disentangle genuine problems of conceptualism from complaints which only apply to apsychologistic (positivist) versions of the doctrine; and I shall attempt to reveal the source of the feeling (widespread among crypto-conceptualists) that Quine's approach fails to do justice to the intuitive ideas from which conceptualism begins.

To say that a statement such as "All groups contain a unit element" is true in virtue of the concept of group (or in virtue of the meaning of 'group') invites criticism. Can we make sense of the notion of conceptual truth or truth in virtue of meaning? Here is a natural proposal.[2] Consider any language L which is used by a community of speakers. An *adequate linguistic description* for L is a set of statements in some metalanguage (which may include L itself) which provides a complete description of all the syntactic and semantic facts about L. We envisage adequate linguistic descriptions as exposing the linguistic capacities of users of L, as making clear in what an understanding of L consists. A sentence S of L is *true in virtue of meaning in L* just in case the metalinguistic statement "S is true in L" is a consequence of an adequate linguistic description of L.

Quine's writings contain, or have inspired, three main criticisms of this proposal, two of which I take to be misguided. I shall consider the best of the objections first. The explication of truth in virtue of meaning just presented can be articulated in two different ways. To say that S is true in virtue of meaning in L might be to claim that "S is true in L" follows from *some* adequate linguistic description for L, or that this statement follows from *any* adequate linguistic description for L. The claims are only equivalent if adequate linguistic descriptions agree on their consequences concerning ascriptions of truth to object-language sentences. Quine has reasons for believing that, in the interesting cases of languages used by ordinary speakers, agreement will not be forthcoming. Nor can we shrug off the problem by suggesting that there will only be relatively minor divergences between the deliverances of different adequate linguistic descriptions. Quine contends that a speaker's understanding consists in a set of dispositions to verbal behavior, and that the set of dispositions to verbal behavior can be adequately captured by widely divergent linguistic descriptions. To put the point in its starkest form, the constraints on adequate linguistic descriptions are so weak that, for any true sentence S of a natural language L, there will be adequate linguistic descriptions which yield the me-

---

2. The proposal follows ideas of Carnap. See his monograph *On the Foundations of Logic and Mathematics* and *Meaning and Necessity*.

talinguistic consequence "S is true in L" and adequate linguistic descriptions which do not yield this metalinguistic consequence. If it is correct, this point undermines the proposal for presenting a *useful* notion of truth in virtue of meaning. For, if we claim that a sentence S is true in virtue of meaning in L if *some* adequate linguistic description for L implies that S is true in L, then *every* true sentence will be true in virtue of meaning. On the other hand, if we require that *any* adequate linguistic description for L must imply that S is true in L, then *no* sentence will be true in virtue of meaning.

This criticism is a serious one, and, to rebut it, conceptualists must show how to constrain the choice of adequate linguistic descriptions. There are two ways in which one might set about this task. The first would be to argue that Quine's equation of a speaker's understanding with dispositions to verbal behavior is incorrect and that there are semantic facts which are not reflected in such dispositions. Unfortunately, adoption of this approach invites the charge of mystery-mongering. Quine will quite reasonably insist that hypotheses about the semantic features of a language are justified only by their ability to explain aspects of the behavior of speakers of the language. The second course of action is to argue that there are kinds of linguistic behavior which can only be accounted for by supposing that the class of semantic facts is richer than Quine would allow. These facts would then filter out unwanted linguistic descriptions, and one would thus avoid the threatened collapse of the notion of truth in virtue of meaning. I shall suggest shortly that this is a more promising strategy for the conceptualist, and that a crucial point in its development is the adoption of a psychologistic epistemology.

Before I pursue this response, I want to set aside the two misguided objections to which I alluded above. In several places, Quine contends that a proper understanding of the phrase 'true in virtue of' will enable one to appreciate the poverty of the thesis that some statements—logical laws, for example,—are true in virtue of meaning. For example, in *Philosophy of Logic*, after apparently formulating the notion of truth in virtue of meaning as I have done above, Quine continues by offering an analysis of 'true in virtue of.'

> How, given certain circumstances and a certain true sentence, might we hope to show that the sentence was true by virtue of those circumstances? If we could show that the sentence was logically implied by sentences describing those circumstances, could more be asked? But any sentence logically implies the logical truths. Trivially, then, the logical truths are true by virtue of any circumstance you care to name—language, the world, anything.[3]

The rhetorical second question attributes to Quine's opponent the enterprise of explaining 'true in virtue of' and offers an explanation which is taken to be unassailable. But the intended goal was not to give so general an explanation but to distinguish sentences true in virtue of meaning from other true sentences.

3. *Philosophy of Logic*, p. 96.

Moreover, the criterion suggested by Quine is at odds with that which the conceptualist (both on my account and on Quine's earlier reconstruction) would favor. To say that S is true in virtue of meaning in L is not to say that S itself is a consequence of an adequate linguistic description for L, but to claim that "S is true" is a consequence of such a description. Let us note, in passing, that S may not even belong to the language in which the linguistic description is formulated. But suppose that the metalanguage used to describe L does indeed contain L. Can we argue that Quine's reformulation of the conceptualist's criterion is an insignificant change, so that the consequences Quine draws from it tell against the explication proposed above? We cannot. Assume that S is a logical truth. Then Quine is correct to point out that S is a consequence of any sentence of the metalanguage we care to choose. But "S is true," the metalinguistic statement whose status interests us, is *not* a consequence of any metalinguistic sentence we choose. (If the linguistic description contains " 'S is true' if and only if S" then "S is true" will be a consequence of the linguistic description, so that S will be true in virtue of meaning.)[4] Once we apply the conceptualist's criterion in its proper form, Quine's attempt at trivialization is blocked.

A different objection, which also fastens on the 'true in virtue of' locution, occurs in the writings of a number of philosophers influenced by Quine. The objection alleges that the idea of truth in virtue of meaning should be rejected because the phrase 'truth in virtue of meaning' fails to pick out any type of *truth*. The premise is that there is only one notion of truth, that of "correspondence to fact," and that every sentence is true in virtue of some fact, that feature of the world which makes it true. W. D. Hart puts the point concisely, charging that philosophers who hope to avoid commitment to abstract entities by claiming that mathematical statements are analytic must show how "analyticity [is] or provide [s] a species of truth not requiring reference."[5]

This argument derives its plausibility from two factors: the notion of truth in virtue of meaning was used by some early positivists in attempts to avoid ontological commitments, and there is a pervasive temptation to oppose the notion to the idea of truth in virtue of fact. Conceptualists can agree that there is a sound core to the thesis that every true sentence is true in virtue of "correspondence to fact." One may concede that, for any true sentence, we can explain its truth by showing how its constituent expressions refer and how the referents meet the conditions elaborated in the theory of truth for the language. The

4. What this means is that if an adequate linguistic description meets Donald Davidson's well-known constraint on theories of meaning (advanced in "Truth and Meaning" and subsequent essays), namely that the Tarski biconditionals be generated, then the truths of logic will count as true in virtue of meaning. Of course, this is a result which the conceptualist will welcome.

5. "On an Argument for Formalism," pp. 44–45. See also Benacerraf, "Mathematical Truth," pp. 676–79. 'Analyticity' is, of course, a standard contemporary term for truth in virtue of meaning. I have generally avoided using this term because, for some readers, it may carry apsychologistic connotations which I want to avoid.

concession is at odds with the endeavors of some philosophers to avoid difficult ontological questions by bypassing the notion of reference. But it is open to the enlightened conceptualist.[6] To suggest that the truth of a statement can be explained by deriving it from an adequate linguistic description is not to assert that this is the *only* way of accounting for the truth of that statement. An example may help here. The conceptualist who believes that "All groups contain a unit element" is true in virtue of meaning can accept a *referential* explanation of the truth of that sentence: 'group' has as its extension a set of mathematical structures which is a subset of the extension of the predicate '. . . contains a unit element.' He will deny, however, that the referential explanation tells the whole story. From his perspective, the relation between the extensions is not simply a brute fact but is itself the consequence of semantic features of the language. The *concept* of a group determines groups as structures containing a unit element. We do not abandon the referential explanation of the truth of the sentence. We deepen it by showing why the referential relations obtain.

II

We can therefore ignore the charge that the notion of truth in virtue of meaning somehow violates an important feature of the concept of truth. Let us now turn to the central question of whether there are phenomena of language use which narrow the class of adequate linguistic descriptions and thus save the conceptualist from the criticism we uncovered above. In addressing this issue, it is easy to beg the question against Quine. Quine's scepticism about the existence of a class of semantic facts which is sufficiently rich to salvage the concept of truth in virtue of meaning is not to be turned back by simply noting that people make certain kinds of semantic judgments. For, insofar as these judgments are classified in neutral language, Quine will admit that they are made and will propose his own account of them; if these judgments are supposed to involve the conceptualist's preferred notions, however, Quine will complain that the judgments of ordinary speakers are innocent of such philosophical theorizing. So, for example, Quine has been emphatic on the point that ordinary remarks about synonymy do not embody the technical notion of synonymy beloved of conceptualists.[7]

I think that the best way to reply to Quine is to invoke one of the central ideas which motivate conceptualism, the idea that our knowledge of some state-

---

6. Thus, for example, Frege contends that the truths of mathematics are analytic while allowing (indeed insisting) that mathematical expressions refer. A similarly enlightened version of conceptualism can be found in Carnap's later writings (and in those of his philosophical successors), and, I would suggest, in Maddy's "Perception and Mathematical Intuition."

7. This kind of mistake seems to underlie many of Jerrold Katz's attempts to defend the analytic-synthetic distinction. See, for example, "Recent Criticisms of Intensionalism."

6

ments results from our understanding of the language.[8] Historically, the notions of conceptual truth (analyticity, truth in virtue of meaning, and so forth) arose from particular epistemological problems: Locke, Hume, Kant, and others contended that some parts of our knowledge could only be adequately explained by tracing them to our grasp of concepts. In their hands, the explanation took forms which twentieth century philosophers find objectionable. Locke's account, for example, trades on identifying concepts with private mental images. The remedy is not to fashion an apsychologistic epistemology (as the positivists tried to do), but to avoid the faulty psychological assumptions of the old theory.

Conceptualist doctrine can be accommodated within psychologistic epistemology by adopting the following picture. When we learn our language a complex set of dispositions is set up in us. In virtue of the presence of these dispositions, which comprise our linguistic ability, we become able to entertain certain beliefs. Let us now suggest that exercise of our linguistic ability generates in us particular beliefs and that it warrants those beliefs. So, for example, we might propose that, in learning the language which enables us to formulate to ourselves statements of group theory, we acquire a complex set of dispositions, and that exercise of these dispositions can generate and warrant belief that all groups have unit elements. If we like, we can think of our linguistic training as setting up neurophysiological states in us, and, in virtue of the presence of these states, as providing a capacity whose exercise produces warranted belief. We can use this idea to defend the notion of an *elementary* conceptual truth. Elementary conceptual truths can be identified as those truths which can be known through exercise of linguistic ability, and we can go on to identify the class of conceptual truths as the closure of the class of elementary conceptual truths under logical consequence. Logical consequence itself might also be specified by reference to our linguistic abilities. Thus we would achieve the distinction which conceptualists have wanted to draw. (I shall take no stand on the issue of whether the states and capacities which are hypothesized here as constitutive of linguistic ability are to be thought of in terms of epistemological notions—whether, for example, we should credit ourselves with some kind of "tacit knowledge" of semantic representations. That issue should be settled empirically, by determining whether epistemological theorizing of the type envisaged by Noam Chomsky, Jerrold Katz, and Jerry Fodor will help us to chart our linguistic abilities in illuminating ways.)[9]

8. There are some hints of this idea in one of the best early responses to Quine (H. P. Grice and P. F. Strawson, "In Defense of a Dogma"). For a more extended version of the idea, see Michael Dummett, *Frege: Philosophy of Language*, pp. 614–21. I believe that these presentations are handicapped by failure to adopt a psychologistic epistemology.

9. See my paper "The Nativist's Dilemma" for discussion of claims about tacit knowledge of linguistic rules. I now believe that Section II of that paper makes too strong a claim. We can indeed account for parts of our explicit knowledge by hypothesizing that we have a complex system

As thus presented, conceptualism bears no commitment to any particular psychological view about our linguistic ability. Its burden is simply that linguistic training induces psychological changes, and these changes make available processes which can generate warranted belief. There is no suggestion that our linguistic training sets up in us a private museum of ideas, a suggestion which Quine rightly scorns. The proposal is simply to account for certain kinds of knowledge in ways which are compatible with our natural explanations of them. Asked how we know that all groups contain a unit element, we might well respond by appealing to our understanding of the expressions 'group,' 'unit element,' and so forth. Our version of conceptualism takes such responses at face value and interprets them as behavioral phenomena which can be used to curtail the class of adequate linguistic descriptions.

There are two kinds of objections which I anticipate. The first is that there is something amiss with the suggested account even as a *potential* explanation of parts of our knowledge. The other is that better accounts are available. I think that the first of these can be met relatively easily. There are other areas of our knowledge which seem to be correctly explained in ways which are parallel to the conceptualist account. Consider, for example, our knowledge of the syntactic features of sentences of the languages we speak. Faced with a sentence we have never seen before and a linguist's query, we can arrive at a correct assessment of its grammaticality, and we credit ourselves with *knowledge* of its grammaticality. How is this knowledge obtained? The following answer suggests itself. In learning the language we acquired a complex of abilities. The exercise of these abilities regularly and reliably generates true beliefs about syntactic properties of expressions of the language. On the present occasion, a process in which the abilities are exercised generates a true belief and, because the process is of a type which regularly and reliably produces true belief, it warrants the belief generated. So our exercise of abilities, set up in us in our youth, produces syntactic knowledge.[10] The conceptualist proposes to adapt this story to semantics.

To defend the legitimacy of this style of explanation is not to show that it is forced upon us. Conceptualists will have to face Quinean criticisms that we do not need to hypothesize some semantic ability to explain our knowledge that all groups contain a unit element (for example). (Note that the worry is no longer that conceptualism is nonexplanatory, but that there is a simpler, rival explanation.) What kind of explanation for this knowledge can a Quinean offer? Let us eliminate, at the outset, the idea that the knowledge is to be explained by appeal to past observation (intuition?) of groups and inductive gen-

---

of linguistic abilities, but I was wrong to dismiss the possibility that there might be empirical reasons for explaining those abilities in terms of "tacit knowledge." I have not yet seen any such compelling empirical reasons, but the possibility should not be precluded.

10. For a fuller version of the story, see "The Nativist's Dilemma." Chomsky's preferred story would serve my purposes equally well.

eralization. We know enough about the way in which we achieve our knowledge to recognize that this is wrong. Quine's preferred account would not be so crudely empiricist. He would deny the possibility of separating from the long series of events in which we absorbed the lore of our ancestors a specific program of linguistic training; or, more exactly, he would suggest that the knowledge for which the conceptualist seeks to account is of a piece with other items of knowledge which are founded in the testimony of our elders.[11]

Conceptualists can respond by citing two features of our linguistic behavior, for which the envisaged Quinean explanation so far fails to account. In the first place, we do distinguish those items of knowledge which are warranted by remembering the testimony of others and those items of knowledge which are warranted without appeal to others. We do not simply declare that we were told that all groups have unit elements. Instead of deferring to another authority, we cite our present understanding of 'group.' If there is no genuine difference here, Quine at least owes us an account of the *illusion* of a difference. The second point presents a deeper challenge. Our practice is to attribute to ourselves *knowledge* of those statements which the conceptualist hails as true in virtue of meaning. If this knowledge is to be obtained by reliance on the testimony of others, then our teachers must have known what they passed on. But now we face the problem of providing a Quinean explanation for *their* knowledge. Appeal to the testimony of ever more remote ancestors must stop at some point, and, at this point, conceptualists will challenge Quine to find an alternative to the apparently unsatisfactory proposal that the knowledge is obtained by inductive generalization from experience.

We are looking for the origin of knowledge of those statements conceptualists classify as conceptual truths, and a natural suggestion is that such knowledge is coeval with the introduction of the language used to express it. We imagine one of our predecessors using the term 'group' (or some cognate) for the first time, and *stipulating* that groups shall be structures containing unit elements. Here we seem to find a clear example of the type of knowledge which the conceptualist envisages: the original user knows, on the basis of understanding the newly introduced term 'group,' that groups contain unit elements. Perhaps we can even extend the idea to our own case by counting as knowledge based on exercise of linguistic ability our knowledge of those statements which we would be prepared to use in parallel stipulative fashion. It seems that Quine's response to these ideas is to allow a place for stipulation—he explicitly concedes that "legislative postulation institutes truth by convention."[12] However, the thrust of one of his most famous arguments is that

11. See, for example, the closing sentences of "Carnap and Logical Truth" (*The Ways of Paradox*, p. 132).

12. *The Ways of Paradox*, p. 118. I shall explain below how Quine is able to make this apparent concession.

explicit stipulations could not account for everything the conceptualist wants to identify as knowledge based on understanding.[13] Furthermore, as I shall argue below, Quine presents subtle reasons for thinking that legislative postulation cannot serve as a source of a priori knowledge.

Let me summarize the course of our discussion so far. I believe that the most promising way to develop conceptualism is to suppose that linguistic training sets up in us abilities whose exercise can lead us to knowledge of some truths (the elementary conceptual truths). This approach need not commit itself to a particular psychological story about the operation of the abilities in question, and its potential for explanation can be defended by appealing to the parallel account of our syntactic knowledge. Quine would respond by insisting that we cannot separate any specifically linguistic training which would differentiate our knowledge of the alleged conceptual truths from knowledge of other items of ancestral lore. Conceptualists can counter by pointing to the apparent difference between reliance on the authority of others and appeal to one's own understanding, and by challenging Quine to provide an alternative explanation of the roots of our knowledge of conceptual truths. In some cases, they can plausibly regard such knowledge as having its origin in an act of explicit stipulation, and thus contend that there are some uncontroversial cases in which knowledge results from the exercise of linguistic ability. Finally, they may propose that cases of reliance on authority may be separated from examples in which the knower appeals to her own understanding by the presence in the knower, in the latter cases, of a disposition to engage in such explicit stipulation.

I do not want to pretend that this provides a conclusive rejoinder to Quine's critique of truth in virtue of meaning, but I do wish to claim that it brings out into the open the motivating ideas of conceptualism, exposing the source of the suspicion that Quinean objections do not touch those ideas. At this point I shall concede to the conceptualist, without further argument, the thesis that there are linguistic abilities whose exercise can produce knowledge of conceptual truths. My aim will be to determine whether the exercise of these abilities can generate *a priori knowledge*. To proceed in this way does not diverge from Quine's most fundamental position. For, as I interpret him, Quine aims to show that the notion of conceptual truth cannot do the work demanded of it, specifically that so-called conceptual truths are not a priori, and with this conclusion I shall agree. Thus I hope that, by making a concession which seems prima facie un-Quinean, it will be possible to bring some of Quine's most important ideas into clearer focus.

13. The argument is that which Quine derives from Lewis Carroll at the end of "Truth by Convention" (*The Ways of Paradox,* pp. 103–6). For an attempt to respond to this argument, see David Lewis, *Convention.*

III

Before beginning my investigation of whether our linguistic abilities can provide us with a priori knowledge, it will be worth looking briefly at two concrete cases in which the version of conceptualism I have presented can avoid specific Quinean criticisms. Consideration of these cases will show clearly that psychologistic epistemology increases the resources of conceptualism, and it may also forestall the complaint that my version of conceptualism thrives on ignoring Quine's central criticisms.

In his dispute with Carnap, Quine supposes that the thesis that truths of logic are true in virtue of meaning (the "linguistic doctrine of logical truth") is intended to explain our reactions to utterances in which people appear to deny the laws of logic. We try to translate the seemingly deviant logician in ways which will avoid attributing difference in doctrine. Quine insists that the linguistic doctrine "leaves explanation unbegun":[14] why should our translation of others as assenting to sentences which are false in virtue of meaning convince us that we have mistranslated? He concludes that we may just as well explain our practice by noting that the laws of logic are obvious, and by taking the enterprise of translation to be governed by the maxim "save the obvious."

The criticism succeeds because of Carnap's avoidance of psychologistic epistemology. We need an explanation of why people cannot (normally) be so badly wrong as to assent to sentences which are false in virtue of meaning. My version of conceptualism answers the need. To translate the deviants at face value presupposes ascribing to them the same set of linguistic abilities present in us. In our case, exercise of the abilities generates belief in the laws of logic. So we have a choice: we can either suppose that the deviants have acquired a different set of linguistic dispositions (and try to translate them nonhomophonically) or that something prevents the normal exercise of the dispositions in their case. Except in highly exotic circumstances, we would have no reason for adopting the latter alternative, and so it is no surprise that our normal practice is to seek other translations when we confront apparent logical deviance.

This account improves on Quine's bald assertion that the laws of logic are obvious. To assert that something is obvious is to give no reason for it, and, usually, to admit that no further reasons can be given. "Obvious" truths are a mixed bag, including perceptual reports as well as laws of logic. Asserting that logic is obvious provides only a partial account of our translational practice. We should try to fathom what makes the obvious obvious. Enlightened conceptualists should claim that, just as some perceptual beliefs are obvious in being generated through the exercise of our perceptual powers, so too the laws of logic are obvious because our beliefs in them are generated by exercising our linguistic abilities.

14. *The Ways of Paradox*, p. 113.

At one point, Quine comes close to appreciating the point. After admitting that 'obvious' has no explanatory value, but insisting that "the linguistic doctrine of elementary logical truth likewise leaves explanation unbegun," he continues as follows:

> I do not suggest that the linguistic doctrine is false and some doctrine of ultimate and inexplicable insight into the obvious traits of reality is true, but only that there is no real difference between these two pseudo-doctrines.[15]

This passage is curious because it contrasts an *epistemological* suggestion (the suggestion that we know the laws of logic through insight into the traits of reality) with a thesis about what makes the laws of logic *true*. To explain our translational practice we need a theory which explains why we find it hard to see how others could fail to believe the laws of logic. Because the linguistic doctrine has no bearing on this issue, it had to be inadequate to the task. What Quine does is to satirize a theory of the wrong type by comparing it with a bad theory of the right type. Conceptualism can answer his criticism by providing an epistemological extension of the doctrine he rejects.

The second specific objection I wish to consider is the attack on analyticity which Quine presents in the course of his celebrated argument for the indeterminacy of translation. Quine concedes that the statements traditionally classified as analytic have a "typical feel." He proposes a "behavioristic ersatz" for analyticity by taking a sentence to be *stimulus-analytic* for a subject if the subject would assent to it (or nothing) given any stimulus.[16] Quine views the legitimate concept of stimulus-analyticity as an inadequate reconstruction of the traditional concept of analyticity, even if we focus on sentences which are stimulus-analytic for an entire community. The trouble is that stimulus-analyticity covers equally sentences like "There have been black dogs" and "All groups contain a unit element."[17] However, we can differentiate sentences of these types by attending to epistemological features. A speaker will explain why she believes that there have been black dogs by pointing out that she has seen some. She will give a quite different response when asked why she believes that all groups contain unit elements. Consistent with the demand for linking concepts of theoretical semantics to overt linguistic behavior, we can refine the concept of stimulus-analyticity.

The mistake which Quine has made is to employ a crude version of the thesis that matters of meaning must turn on dispositions to verbal behavior. Quine's semantics takes as fundamental the behaviorally respectable idea of patterns of assent and dissent to single sentences, and he seems to allow as legitimate only those semantic properties of a sentence which are specifiable in terms of patterns of assent and dissent to it. So he arrives at stimulus-analyticity as the

15. *Ibid.*, p. 113.
16. *Word and Object*, p. 55.
17. *Ibid.*, p. 66.

best approximation to the classical notion of analyticity. This approach over-looks the possibility that there might be genuine semantic properties of a sen-tence which are only specifiable in terms of patterns of assent and dissent to it *and to other sentences*. Our dispositions to verbal behavior are not revealed solely in our affirmations and denials but in our explanations and justifications as well. Or, to put the point in Quine's preferred terms, patterns of assent and dissent to sentences of the form "I believe that . . . because ———" are indicators of semantic properties of the sentence in the first place, properties which are not specifiable simply in terms of patterns of assent and dissent to the embedded sentence.

I conclude that central Quinean objections to the notion of truth in virtue of meaning can be resisted by embedding that notion in a psychologistic episte-mology. Let us now see if the traditional doctrine that we have a priori knowl-edge of conceptual truths can survive the reformulation.

IV

There is an apparently straightforward argument which suggests that knowledge which is based on the exercise of linguistic abilities is a priori. This argument underlies the traditional doctrine that analytic truths are a priori. Suppose that a person's knowledge that *p* is generated by a process of exercising her lin-guistic ability, specifically by exercising dispositions she acquired in learning those parts of her language which, in combination, enable her to express the thought that *p*. Then, given any experience which enables her to entertain the thought that *p*, it appears that that experience will have to set up in her similar dispositions and so make available to her the same type of process. Moreover, it seems that any such process will produce true belief (sentences stating that *p* will be true in virtue of meaning), and that there is no reason why it should not produce warranted belief. For, if we concede in ordinary circumstances that people are warranted in using their linguistic understanding, it is hard to see how there could be grounds for denying that this procedure can warrant belief in counterfactual situations where, although experience is different, the linguis-tic understanding remains.

In its traditional guise, the argument is simpler, and perhaps more compel-ling. Suppose that a person knows that *p* by recognizing relations among the constituent concepts. Then a sufficient experience for *p* is one which enables him to possess those concepts. Given a sufficient experience, it would be pos-sible for him to discern the relations among his concepts in the way he actually does. Were he to proceed in this way he would obtain true belief. Finally, he would arrive at warranted belief, for no experience could be relevant to the warranting power of a process which consisted in recognizing conceptual rela-tions.

Let me call this argument, in both its traditional presentation and in my own reformulation, the *naïve conceptualist argument*. Because of its initial air of plausibility, one might suspect that the concession made to conceptualism in the last sections was too rash. Once we allow the legitimacy of the notion of linguistic abilities and their exercise, we seem to be swept into the thesis that conceptual truths are knowable a priori. I shall show that this is incorrect. There are acute problems with the naïve conceptualist argument.

The first point I shall address is an identification which underlies both versions of the argument. Can we assume that any life which enables one to entertain the thought that p must also give one the full range of linguistic abilities associated with expressions which could be combined to state that p (all the constituent concepts)? Initially, the identification appears trivial. When we think of a life which enables someone to entertain a thought that all A's are B's (to focus, for the moment, on thoughts of a particular form), we tend to imagine lives which provide the person with full criteria for identifying A's and B's, so that, if "All A's are B's" is true in virtue of meaning, we suppose that it is possible for the person to deploy the criteria to recognize that all A's are B's. Whether our imagination focusses on a restricted range of cases depends on the interpretation we give to the phrase "entertaining the thought that p." Recent work in philosophy of language (specifically in the theory of reference) has undermined the claim that a speaker needs (nontrivial) criteria of identification if she is to use a name to pick out its referent or use a predicate to refer to its extension.[18] Applying this work, we might argue that it is perfectly possible—indeed common—for us to be able to form the belief that all A's are B's, or, equivalently, to entertain the thought that all A's are B's, without our being able to specify descriptions which pick out the A's and the B's, and so, in particular, without our knowing those criterial specifications in virtue of which, on the picture provided by the naïve conceptualist argument, the conceptual connection is to be made. If the conceptualist should dig in her heels and contend that the kind of situation just envisaged is, strictly speaking, *not* one in which we can form the belief in question, then she will be vulnerable to the charge that the ordinary idea of being able to form the belief is being abandoned in favor of a technical notion, whose incorporation within the analysis of apriority is entirely ad hoc, and designed only to favor the conceptualist's pet thesis.

It is worth looking at this issue more concretely. Consider an example which Hilary Putnam has frequently used.[19] Most people use 'elm' without being able to provide any description which would distinguish elms from other kinds of trees. (Many of those who use 'elm' would be likely to confuse elms with

18. See Saul Kripke, *Naming and Necessity,* Keith Donnellan, "Proper Names and Identifying Descriptions," Hilary Putnam, "The Meaning of 'Meaning.' "
19. See *ibid.*

other trees, even when viewing them under optimal conditions.) Despite this, the people in question can make statements in which they refer to elms. Given our ordinary notions, it would be misleading to deny that such people lack the ability to form beliefs about elms: we naturally attribute to them a capacity for wondering whether anything can be done to stop the death of elms in North America, and so forth. Experiences which would suffice for the entertaining of the thoughts expressed by sentences of the form "All elms are . . ." do not need to acquaint subjects with descriptions which could be used to individuate the elms. Hence the naïve conceptualist argument can be challenged on the grounds that there are sufficient lives which fail to induce a set of linguistic abilities rich enough for the unfolding of the process which the conceptualist envisages.

There is an intricate response which the conceptualist can try. The starting point is to maintain that the objection of the last paragraph subtly mixes the idea of being able to use an expression without having an *identifying* description, with the idea of being able to use an expression while lacking *any* associated description. In ordinary cases, we might well agree that the differences between those who cannot distinguish elms from beeches and the experts who can give botanical descriptions are too slender to deny to the former a title to beliefs which we readily attribute to the latter. The conceptualist asks us to consider more extreme cases. Would we be content to count people as referring to elms if they did not associate with 'elm' the description "a kind of tree"? Is it feasible that someone would be able to form beliefs that elms are P (for various properties P) without having the linguistic ability to generate the belief that elms are trees?[20] Putnam has argued for an affirmative answer to the first question. Several of his examples are directed at showing that our beliefs about the referents of our terms—even our most central, "stereotypical" belief—may be badly mistaken. The history of the sciences is filled with cases in which, on Putnam's account, thinkers referred to a particular kind while having wildly incorrect views of the criteria for membership in the kind. A simple extension of the view yields an affirmative answer to the second question as well. If we think that someone can use 'elm' to refer to elms, while having an erroneous idea (or even no idea at all) about the kind of thing an elm is, then on what basis are we to deny that he lacks the ability to form beliefs that elms are P? We have the ability, and one obvious explanation of our ability is to see it as derivative from our ability to refer to elms. Granted that the other person has the latter ability, by what right do we deny his capacity to form the beliefs?

To develop the strategy considered in the last paragraph, the conceptualist has to find some way to limit the new ideas about reference. Perhaps she can argue that Putnam misdescribes the cases from the history of science which are used to buttress the claim that we can refer to things about which we are woe-

20. Dummett raises this type of question (*Frege: Philosophy of Language,* p. 99).

fully ignorant. Or perhaps she can contend that, for someone to be able to form beliefs that elms are P, more is required than a simple ability to refer to elms. The view would be that, although the ignoramus can refer to elms, his saying to himself "Elms are P" does not constitute the entertaining of the same thought as a similar private utterance on the part of an expert. Criteria for identifying belief-content are more, stringent than criteria for coreference. Now there can be no doubt that this view has, historically, been extremely popular. The issue is whether it can be sustained in the face of new insights about reference. Instead of pursuing the issue, I shall leave it an open question. For, as we shall quickly see, conceptualists have a way of defending themselves which does not presuppose a particular answer.

The trick is to turn the claims about reference against themselves. If it is possible to refer to elms whether or not one associates any particular description with 'elm,' then it is possible to preserve the same reference even though one has added explicit stipulations which, in part, determine the referent of 'elm.' In effect, the conceptualist may argue as follows.

> The challenge to my position is that it is possible for someone to entertain the thought that all elms are trees without associating with 'elm' the description "a kind of tree." Let us concede this possibility. Now envisage a world in which someone comes to refer to elms in the suggested way. In this world, the person may decide to perform an act of explicit stipulation, declaring that she will use 'elm' to refer to whatever it is she has been referring to, subject only to the proviso that elms are to be trees. On the basis of her act of stipulation, she may come to know that elms are trees. You may protest that this act of stipulation does not enable her to know the proposition she used to express by using the sentence "All elms are trees." But this would be shortsighted of you. For if you respond in this way then you will have to abandon the claim that coreference is enough for belief identification, a thesis on which your earlier criticism of my program depended. The reason that you will have to give up this claim is that, on your account, the uses of 'elm' before and after the stipulation are coreferential. So there are two possibilities: either you take a liberal attitude towards the individuation of beliefs, allowing that people can share the same beliefs provided only that they can refer to the same things, or you can be more restrictive. If you adopt the former approach, then I will agree that there are lives sufficient for p which do not set up the full range of linguistic abilities with respect to expressions used in stating that p, but I will claim that the loss can be made up by acts of stipulation. On the other hand, if you suppose that not every life which enables one to refer to the entities to which we refer in stating that p will enable one to form the belief that p, then your original objection evaporates.

I conclude that the conceptualist can respond to an initial challenge that the processes which are supposed to generate a priori knowledge would not be available given certain sufficiently rich lives. Nevertheless, in elaborating a response to this challenge, I have exposed some features of conceptualism which will prove troublesome when we examine whether the favored processes can warrant belief independently of experience. I shall use the preceding considerations as a background for my central objection, a criticism which derives ultimately from Quine.

V

Defenders of analyticity have often construed the main thrust of Quine's most famous attack, "Two Dogmas of Empiricism," as arguing that the concept of analyticity is undefinable in notions Quine takes to be unproblematic. Seen in this way, the attack allows a number of plausible countermoves: one might respond by denying the need for a definition or by rejecting Quine's delimitation of "unproblematic" concepts. I locate Quine's central point elsewhere. The importance of the article stems from its final section, a section which challenges not the existence of analytic truths but the claim that analytic truths are knowable a priori.[21] The argument is encapsulated in the following passage:

> . . . it becomes folly to seek a boundary between synthetic statements, which hold contingently on experience, and analytic statements, which hold come what may. Any statement can be held true come what may, if we make drastic enough adjustments elsewhere in the system. . . . Conversely, by the same token, no statement is immune to revision. Revision even of the logical law of the excluded middle has been proposed as a means of simplifying quantum mechanics; and what difference is there in principle between such a shift and the shift whereby Kepler superseded Ptolemy, or Einstein Newton, or Darwin Aristotle?[22]

I shall use my reformulation of conceptualism and my analysis of apriority to elaborate the argument presented here.

Quine connects analyticity to apriority *via* the notion of unrevisability. If we can know a priori that p then no experience could deprive us of our warrant to believe that p. Hence statements which express items of a priori knowledge are unrevisable, in the sense that it would never be rational to give them up. But "no statement is immune from revision." It follows that analytic statements, hailed by Quine's empiricist predecessors and contemporaries as a priori, cannot be a priori; or, if analyticity is thought to entail apriority, there are no analytic statements.

21. Putnam has taken a similar view of Quine's article. See his papers "Two Dogmas Revisited" and "There Is at Least One A Priori Truth."

22. *From a Logical Point of View*, p. 43.

The obvious way for the conceptualist to respond is to deny Quine's claim that no statement is immune from revision. Here it is pertinent to ask what Quine means by 'statement.' If we interpret 'statement' as 'sentence,' then Quine is asserting what nobody has ever denied. We can and do jettison linguistic expressions, coining new words to say what we used to say, and the conceptualist will agree that sentences currently used to express conceptual truths could be given up. There is a more interesting reading. To say that no statement is immune from revision is to assert that, for any sentence S which we currently use to express something we believe to be true, we can envisage a rational development of our corpus of beliefs, culminating in a set of accepted sentences meeting one of the following conditions: (a) some sentence in the set is properly translated as the negation of S, as S is currently used; (b) no sentence in the set is properly translated as S, as S is currently used.[23] Developments which culminate in sets satisfying (a) may appropriately be called *strong* revisions of S; those which lead to the satisfaction of (b) will be *weak* revisions of S. Now when the conceptualist is confronted with this interpretation, she will deny that there can be strong revisions of conceptual truths. If S, as currently used, is true in virtue of meaning then to translate a rational being as assenting to the negation of S would, *ipso facto,* be mistranslation. (This is a generalization of the point about the translation of "deviant logicians," which we considered briefly in Section III.)[24] Hence the only possible revisions of conceptual truths will be weak revisions. However, the conceptualist will claim that weak revisions are epistemologically irrelevant. Any life which is sufficiently rich will allow for the development of language to state the truth expressed by S. Thus a weak revision can only occur if our experience is not sufficiently rich, and the existence of such experiences poses no threats to our a priori knowledge of the truth expressed by S.

However, Quine's point cannot be met so simply. Quine is concerned to recognize the existence of a special kind of weak revision, one in which beliefs are given up because we rationally abandon a particular way of talking and

23. This distinction between two ways in which statements can be abandoned is made by Grice and Strawson, albeit in terms which Quine would disavow. In a note to "There Is at Least One A Priori Truth," (pp. 166–67), Putnam also draws this distinction. However, in a *further* note (pp. 167–70), he seems to undercut the significance of the distinction, claiming that if some statements are only revisable by dropping particular concepts then we can obtain from those statements *conditional* statements which are (absolutely) unrevisable. The idea is that, if P can only be revised by dropping certain concepts (the concepts of X, Y, Z, say), then the statement ⌜If the concepts of X, Y, Z are retained then P⌝ is unrevisable. However, this does not work, for the latter statement might be rejected by dropping concepts which occur in it but not in P, the concept of a concept, for example. Hence, I do not think that Putnam succeeds in showing that all kinds of revisability reduce to strong revisability of statements.

24. As my account shows, Quine's position in "Two Dogmas" is compatible with the dicta of *Philosophy of Logic* concerning deviant logics (pp. 80–83). These remarks have often puzzled Quine's critics.

thinking. To put the issue in the terms the conceptualist prefers, there are experiences which would lead us to discard particular concepts by showing us that those concepts were useless for the normal purposes of explanation and description. Beliefs may suffer revision by undergoing demotion. Concepts which were formerly employed in scientific theorizing are dropped, or linger on solely for use in story-telling and intellectual history. There can be weak revisions of S which suffice to enable one to form the belief which used to be expressed by S, but which make the formation of that belief unreasonable. Experience can undermine our favored modes of thought and expression. To translate Quine's point into my terms, the warranting power of processes in which linguistic dispositions are exercized can be subverted by lives which deprive one of the warrant to employ the language in question.

I shall illustrate the point with an example, an example which was originally used by Mill with similar aims.[25] Chemists of the early nineteenth century were inclined to introduce the notion of an acid by stipulating, among other things, that acids contain oxygen. When they asserted ''Acids contain oxygen,'' they could defend their assertions by appealing to their understanding of the terms. Conceptualists will take the defense at face value, supposing that learning the language of nineteenth-century chemistry involves acquisition of a set of linguistic abilities whose exercise warrants belief that acids contain oxygen. Encountering a substance which appeared to behave in the same ways as other known acids, chemists named it ''muriatic acid'' and expected to be able to liberate oxygen from it. When successive attempts to obtain oxygen from this substance (hydrochloric acid) failed, the chemical community recognized that the continued classification of the substance as an acid would lead to a simpler chemical theory than adherence to the old definition of 'acid.' Accordingly, they abandoned the old definition, and the statement ''Acids contain oxygen'' was revised. The revision is weak, for no sentence in the resultant corpus of beliefs is properly translated as the negation of this sentence in its former usage.

During this episode the ability to employ the old concept of acid was not lost but the use of that concept became unreasonable. If some traditionalist had continued to insist that acids contain oxygen, and had appealed to his understanding of 'acid' to support his claim, it would not be correct to say that he knew that acids contain oxygen. This is not simply because, given our continued employment of the *word* 'acid' with a different sense, the attribution of knowledge in that form of words would be misleading. Rather, in the light of the experimental evidence available to the traditionalist, his continued use of the old language is unjustified and his belief no longer warranted. His assertion commits him to linguistic and conceptual practices which he should not rationally adopt. The moral is this: while appeal to linguistic understanding can serve as a *local* justification for belief, empirical discoveries are relevant to the

25. See *A System of Logic*, p. 91. I discuss Mill's aims and his deployment of the example in sections I–II of ''Arithmetic for the Millian.''

continued success of the appeal. Exercising our linguistic ability is not an a priori warrant if experience can undermine the use of the language.

Before I develop the point further, I want to connect it with the issues addressed in the previous section. When considering an episode from the history of science like the one just described, there are alternatives to the reconstruction which I have given. The interpretation adopted above is the most favorable to the conceptualist. The point is that even on the most favorable interpretation, the process which the conceptualist regards as generating a priori knowledge fails to do so. Someone sympathetic to the ideas about reference adduced by Putnam will object to conceptualism at an earlier stage. Instead of supposing that 'acid' shifts its referent during the period discussed, one might argue that 'acid' always did refer to the class of things we take to be acids. As a result, it would be held that, even in the mouths of traditionalists, "All acids contain oxygen" is false, and that the common defense of the statement by appeal to linguistic understanding reflects the practice of drawing on those widespread beliefs (stereotypes) [26] with which the community inculcates standard patterns of usage in its young. Plainly, to adopt an interpretation of this type is to revoke a concession made to the conceptualist in Section II. We agreed to take seriously the idea of knowledge produced by the exercise of linguistic ability and to distinguish such knowledge from the general body of lore transmitted from generation to generation. If we retract our agreement then conceptualism cannot get off the ground. If we stick by the agreement then we arrive at the reconstruction of the episode from the history of chemistry which I originally gave. Either way the conceptualist loses.

The most promising response for conceptualists to make is to deny that, in the circumstances envisaged, the traditionalist is unwarranted in the belief he continues to express with the sentence "All acids contain oxygen." The reply may be articulated as follows.

> To suppose that the traditionalist is unwarranted is to introduce inappropriate pragmatic considerations. The belief expressed in the traditionalist's sentence may be pointless or uninteresting, but it is, nevertheless, true. If he stipulates that he will continue to use 'acid' in the old way, then the belief he expresses by 'All acids contain oxygen' is true in virtue of his stipulation, and the fact that it may be confusing or futile for him to continue to declaim this sentence is irrelevant to the issue of whether he is warranted in doing so.

The conceptualist's insistence on a distinction between pragmatic and genuinely epistemological considerations is at odds with a fundamental Quinean insight. To attribute knowledge to another is to recognize her as endorsing, on rational grounds, some part of the results of inquiry, a contribution to "total science."

26. See Putnam, "The Meaning of 'Meaning.'"

Those who cling to outworn distinctions are failing to recognize the goals and standards of the ongoing cognitive enterprise just as much as those who continue to espouse theories which have been rationally rejected. Even conceding to the conceptualist that there is a difference between knowledge based on the exercise of linguistic ability and knowledge based on the testimony of the elders, we must still recognize the need to fashion our language in accordance with the aims of inquiry. To fail to do so is *ipso facto* to fall short of knowledge.

Let us continue to grant to conceptualism its preferred idiom. Then the point I have been making can be put simply as the thesis that there are principles governing our use of concepts. Our aims are normally to communicate with others—and thus to use words coreferentially with others—and to talk about what there is in a revealing way. These aims can easily conflict with private stipulation. Usually, we let our references adjust to those of fellow speakers, allowing that the descriptions we have initially used to characterize an intended referent may be either misleading or even incorrect. Yet there are occasions when conformity is not in order. A scientist may introduce a new expression— or give new meaning to an old expression—by engaging in an act of explicit stipulation (Quine's "legislative postulation"). These acts are not constrained by the linguistic practice of our fellows, but they are constrained nonetheless. To stipulate that, whatever they are, acids are to contain oxygen is to presuppose the view that introducing the concept of acid in this way will help us in describing and understanding reality. If experience shows or suggests that our method of introduction will not achieve these goals then the stipulative act is unreasonable, and our performance of it cannot lead us to knowledge. Stipulation is always possible. But it is not always rational.

We can see that stipulation provides no genuine epistemological magic by recognizing a disastrous consequence of the conceptualist's position. Suppose that we were to allow that the traditionalist is warranted in his belief that all acids contain oxygen, on the basis of his stipulation that he will continue to use 'acid' in the old way. Then it would become easy for us to increase our store of a priori knowledge. We currently justify scientific laws by citing the results of numerous experiments. There is no need for us to burden ourselves with these details. We could tailor our scientific concepts, stipulating that the sentences we currently accept are to be true by virtue of meaning. In the wake of our stipulation, we could then defend our assertions by citing our understanding of the language, and, if the conceptualist is right, we would then be able to know a priori everything we want to assert. Of course, conceptualists will deny that we obtain a priori knowledge of the truths which previously constituted our empirical science. On their view, what would have occurred is that we would have replaced empirical science with a different corpus of truths known a priori. But to suppose that is to concede that we do not need *empirical* science at all.

Conceptualism makes a priori knowledge come too cheap.[27] Were we to amend our scientific concepts in the way suggested, we would allegedly achieve a priori knowledge, but at an obvious cost. The risk that what we know will prove useless would be greatly increased. Moreover, evidence which we used to cite in support of scienfitic laws would now be relevant to showing the applicability of our concepts. Similarly, what would previously have been viewed as a falsification of a law would be construed as a demonstration of the inapplicability of some concept(s). It should be clear that the epistemological gain is negligible.

We have arrived at a response to the naïve conceptualist argument. The concession that there are processes which exercise our linguistic abilities and which warrant our belief in "conceptual truths" is compatible with the denial that these processes are a priori warrants. Background experiences can deprive us of our right to use the language we do, and thus undermine the warranting ability of these processes. As with the processes of intuition whose merits we examined above, we find that the conceptualist's favored processes do not meet condition (3b) of our analysis of 'a priori warrant.'

VI

So far, I have been discussing conceptualism completely generally, without attending to it as a thesis about mathematical knowledge. It is now time to apply what we have found to the special case which interests us. Exercise of linguistic ability to warrant belief in a mathematical statement could be subverted by an experience which called into question the rationality of using the concepts involved. But one might wonder how this could happen. We understand (roughly) how experience can subvert the employment of scientific concepts. Someone stipulates that a predicate is to refer to the set of things meeting a particular condition, and we then find that nothing meets that condition (or even approximately meets the condition), or that the set is heterogeneous, in the sense that we can frame simple laws governing the members of classes which intersect the set but no simple laws governing exactly the members of the set. The latter kind of discovery prompted revision of the concept of acid; the former type of discovery precipitated the repudiation of the concept of phlogiston as "that which is emitted in combustion." Could similar discoveries occur in mathematics, and, if so, how?

27. Interestingly, this was Quine's original worry. Most of "Truth by Convention" is devoted to elaborating the point. (See *The Ways of Paradox*, pp. 77–102, especially the summary at the top of p. 102.) Similar points were made by earlier thinkers. See, for example, Kant's reply to Eberhard (quoted in L. W. Beck, "Can Kant's Synthetic Judgments Be Made Analytic?" pp. 13–14); Locke, *Essay Concerning Human Understanding*, vol. 2, pp. 226–29; Mill, *A System of Logic*, pp. 148–50.

To detail the ways in which experience can give us reason to reform our mathematical language would be to embark on the project of presenting an empiricist theory of mathematical knowledge, and that is not my present purpose. So I shall merely indicate how analogous pressures to those found in scientific cases can bear on mathematical concepts. Let us begin by recalling an important point from Section I. The claim that a statement is true in virtue of meaning should not be interpreted as a dismissal of the view that the expressions occurring in that statement refer (that is, that they pick out something actually existing). Thus someone who believes that basic truths of mathematics are true in virtue of meaning is not absolved from the task of saying what the referents of mathematical terms are, or, to put it differently, what mathematical reality is like. Once this point is appreciated, then it is easy to see that there will be parallels with the scientific cases. We might find that our chosen concepts failed to pick out any aspect of mathematical reality or that they did not allow for the formulation of simple descriptions of it.

Let me illustrate this possibility with two examples. Consider, first, an example of the latter kind. Someone might stipulate that a group is to be a set closed under an associative operation (multiplication) such that division is unique wherever it is possible.[28] The stipulation would pick out the structures we call groups but it would also select infinite structures which are not groups. Algebraic investigation of the finite structures selected might easily lead one to the discovery of some important common properties: the existence of unit elements, inverses, and so forth. Recognizing the existence of these properties and the simple laws which flow from them, one would then have reason to discard from the extension of 'group' those infinite structures which lack the properties. The situation here is exactly parallel to that in the case of 'acid.' Initial ideas about which kinds of properties are usefully employed as criteria are subject to revision as one struggles to frame simple laws.

My second example involves an extreme case, the case of inconsistent stipulation. We can easily imagine someone stipulating that his conception of set, or of the universe of sets, is to be characterized, in part, by the existence, for any predicate, of a set whose members are exactly the things satisfying the predicate. A letter from some latter-day Russell would undermine this stipulation. Given the experience of receiving such a letter, it would be as irrational for our imagined set theorist to declare that he knows (a priori) that every predicate defines a set as it would be for a defender of the phlogiston theory to insist that he knows (a priori) that whatever is emitted in combustion is phlogiston.

Now of course we do not believe that the concepts of current mathematics are inconsistent or that they are askew in the way exhibited by my deviant

28. Obviously, far more bizarre kinds of stipulation are possible. My point is to show that even a relatively reasonable kind of stipulation can be undermined.

concept of group. The point of the examples is to show how, in principle, we could be led to believe that our linguistic practices require reform. To defeat the claims of the conceptualist, all we need to show is that there could be experiences which suggested a need for overhauling our concepts. Consider again the case of 'acid.' We ·can imagine that the experimental evidence was contrived, that the traditional criterion would in fact serve the scientific aims of explanation and description. Nonetheless, if it is reasonable for the traditionalist to believe that those purposes cannot be served, it is irrational for him to continue his practice of stipulation. An elaborate deception is just as effective as a revelation of reality. The point carries over to the mathematical case. If experience could mislead us into rational belief that our concepts are inadequate in one of the ways I have described, then our appeal to our understanding would no longer warrant our mathematical beliefs. The work of previous sections should have made it clear that such misleading experiences are possible. Perhaps we can envisage how there could be theoretical challenges to the thesis that a particular mathematical concept is adequate. Certainly, we can imagine what social challenges would be like.

As with theories of mathematical intuition, we find that conceptualism fails to yield a defensible version of mathematical apriorism. I conclude that mathematical apriorism is false, and that its falsity stems from the fact that the processes which apriorists have variously considered as the generators of mathematical knowledge would fail to warrant belief in the presence of suitable background experiences. The next chapter will use some of the points made in criticism of apriorism to begin the development of a different approach to mathematical knowledge.

5
Toward a Defensible Empiricism

Mathematical apriorism has been defended by some of the most acute figures in the history of Western thought. In rejecting the doctrine, we would do well to understand what has made it so attractive, and to see if we can preserve particular insights of the theories which we have found wanting. Hence I shall begin my development of a rival position by trying to isolate the ways in which mathematical apriorism breaks down, hoping thereby to see if any of its motivating ideas can be preserved.

I anticipate complaints that the central theses of apriorism—the doctrines which the apriorist really wanted to defend—survive my criticism unscathed. It may be thought that I began by pinning an overly strong thesis on apriorism, that nobody ever intended to claim that mathematical knowledge is a priori in the sense given by my analysis. Moreover, on several occasions I have failed to provide clear examples of experiences which could undermine our mathematical beliefs in a way which is independent of the testimony of others, and this, some may suggest, makes the refutation of apriorism trivial. I shall try to show that apriorism cannot be patched up so easily, and, by pointing to its flaws, to prepare the way for an alternative view.

The charge that my argument against apriorism presupposes too strong a notion of apriority is relatively easy to rebut. Previous chapters have shown, systematically, that the processes which apriorists take to generate our mathematical beliefs would be unable to warrant those beliefs against the background of a suitably recalcitrant experience. If apriorists are to escape this criticism on the grounds that the anaysis of apriority is too strong, then they must allow that it is not necessary for an a priori warrant to belong to a type of process members of which could warrant the belief in question given any sufficient experience. To make this concession is to abandon the fundamental idea that a priori knowledge is knowledge which is independent of experience. The apriorist would be saying that one can know a priori that p in a particular way, even though, given appropriate experiences, one would not be able to know that p in the same way. But if alternative experiences could undermine one's knowledge then there are features of one's current experience which are relevant to the knowledge, namely those features whose *absence* would change the current experience into the subversive experience. The idea of the support lent by kindly experience is the obverse of the idea of the defeat brought by uncooperative experience. To reject condition (3b), the condition of my analysis on which the central arguments above have turned, would be to strip apriorism of its distinctive claim.[1]

It is more difficult to close off an alternative line of escape for the apriorist, the denial that what I have called *social challenges* pose a serious threat to apriorism. The apriorist may try to contend that all that prevents particular processes from achieving the status of a priori warrants is the reasonableness of modesty in the face of criticism. Apriorism, or something like it, could then be salvaged by drawing a ring around certain kinds of experiences which, although sufficient for particular beliefs, are not to be considered in assessing the apriority of the beliefs. A priori knowledge, or *approximate* a priori knowledge, would be knowledge obtainable in the same way given any sufficient experience except those of a particular kind. The problem for the apriorist is to specify the kind of experience to be excluded in a way which will salvage the thesis that truths of mathematics can be known a priori (or approximately a priori) while still giving point to the thesis. It would obviously be futile if the principle of exclusion ruled out so many experiences that vast portions of our knowledge were hailed as (approximately) a priori—or even if analogs of the principle would produce this effect. However, you might think that the apriorist can succeed. After all, the aim is simply to exclude those experiences which defeat us through appeal to the contrary authority of others.

Several factors cast doubt on this strategy. Consider first the general way in which my argument against apriorism proceeds. In each case which I have

---

1. I would contend that the analysis of a priori knowledge given in Chapter 1 provides the only clear account of the epistemological notion of apriority which is currently available. Hence if someone wishes to protest that my analysis stacks the deck against the apriorist, it is incumbent upon him to provide an alternative. Given the arguments for psychologistic epistemology, rehearsed in Chapter 1, it seems that any such account will have to take the form of specifying conditions on a privileged class of processes which could serve as a priori warrants for belief. If these conditions do not include the constraint that the processes in question be able to sustain knowledge independently of experience, then I think the distinctive idea of epistemological apriority will have been abandoned. If they do include that constraint, then apriorism will be vulnerable in just the way I have taken it to be.

I suspect that the truth of the matter is that apriorists have not recognized the precise theses to which they are committed. 'A priori' is a term which has been used quite casually in twentieth-century philosophy. When the term is analyzed then, I claim, apriorist doctrines no longer look attractive.

discussed I argue that the processes favored by the apriorist will not escape reasonable doubt: so *if* there could be experiences suggesting the falsity of a proposition (or experiences which undermined its constituent concepts) then the process could not be used rationally to override such experiences. I appeal to social challenges as a general way of justifying the antecedents of the relevant conditionals. Thus the strategy proposed would leave the central part of my argument untouched. This means that in any case in which the antecedent of the relevant conditional can be established *without* appealing to social challenges my critique of apriorism will retain its force. I shall indicate below why I think it plausible that the antecedents of the relevant conditionals can be established without using the blanket method of citing the possibility of social challenges.

A different source of difficulty is the fact that, as I pointed out above, cases in which someone establishes her expertise and then contradicts one of the statements we accept only make up one extreme type of social challenge. It is easy to see that the attack on our beliefs may involve the production of intricate arguments whose flaws are too well hidden for us to detect. Or we can imagine that the scenario does not involve encounters with others at all: it is conceivable that we could become reasonably convinced by our experience that the ingestion of certain substances had enabled us to solve baffling theoretical puzzles and that, during one of these episodes, we had discovered a counterexample to a mathematical axiom; the notes we composed at the time could even remain to remind us of it. Thus there is a whole range of experiences, continuous with the simplest sort of social challenge, which the apriorist must debar. I suggest that so much will have to be excluded that the notion of (approximate) apriority will be trivialized.

For the sake of argument, let us suppose that the apriorist has managed to find a way to characterize social challenges and explicitly to rule them out. Will there still be appropriate experiences suggesting the falsity of mathematical statements (or the illegitimacy of mathematical concepts) so that the dubitability of the apriorists' favored processes can be exploited? I believe that there will. Apriorists hope that the difficulty of imaginative experiment reflects the impossibility of such experiences. In the foregoing, I have tried to dash that hope by invoking the generally applicable notion of a ''social challenge.'' Even if I forego this tactic, it would remain open to me to describe what I have called ''theoretical challenges.'' To implement this strategy would require greater exercise of imagination and it is, admittedly, difficult to entertain some of the possible ways in which the course of inquiry could go. In choosing to emphasize social challenges, I do not deny the possibility of theoretical challenges, but simply acknowledge that these are more tricky to describe. Furthermore, the idea that experience could not suggest the falsity of mathematical axioms thrives on the absence of an alternative to apriorist theories of mathematical knowledge. If we had a clearer view of how mathematical knowledge could be grounded in empirical procedures (ordinary observation, for example) we should be able to give more substance to the notion that experience might confuse us. Hence, as I outline a rival theory, the case against apriorism will be strengthened. We shall see the ways in which mathematical principles and concepts have been revised in the past, and we shall investigate the grounds for such revisions. A little imagination will then enable us to see how similar modifications might happen to contemporary claims.

Finally, if the apriorist should insist that social challenges to mathematical statements are somehow unimportant, then it is appropriate to reply by pointing out the social character of most of our knowledge. There is very little that we know without reliance on the testimony and support of others. Even in the case of empirical science, most of the knowledge of each individual is based, not on direct experience, but on the communications of others. Few of us have performed the delicate experiments, and not many more have studied the experimental results. We read that "experiment has shown that . . ." and we are, reasonably, satisfied. Indeed, the happy few who actually adjusted the apparatus and watched the instruments are dependent on their colleagues, albeit in different ways. Their knowledge is sustained, in part, by community approval of their techniques and background assumptions. To point to our interactions with others as both a potential source of knowledge and a potential means of defeating our own beliefs is to emphasize a pervasive feature of our epistemic situation, one which should not be forgotten in the case of mathematics. I shall elaborate on this idea below.

For all that has been said so far, it would be possible to maintain that one of the sources attributed with the power of engendering a priori mathematical knowledge could actually produce our mathematical knowledge—although, in the light of my criticisms, the claims about apriority would have to be retracted. I think that this approach is implausible. Why would anybody want to adopt the theories about the basic warrants for mathematical beliefs which have been reviewed above? Two answers occur to me: because of a desire to defend apriorism and out of sheer desperation. With the collapse of apriorism the first motive disappears. The source of the second answer is the apparent difficulty of giving *any* account of mathematical knowledge. We invoke some mysterious intuition of abstract objects, or try to squeeze mathematical knowledge out of our ability to detect the properties of our constructions or our understanding of our language, because we think of these as the only alternatives to a perceptual theory of our mathematical knowledge and because we reject that as a nonstarter. I shall try to show that we need not be so desperate.

II

Let us start with three obvious points. First, we originally acquire much of our mathematical knowledge from teachers, on whose authority we accept not only basic principles but also conceptions of the nature of mathematical reasoning.

Second, some of this knowledge is acquired with the help of perceptions. Our early training is aided by the use of rods and beads; later, we appeal to diagrams. Third, mathematics has a long history. The origins of mathematical knowledge lie in the practical activities of Egyptians and Babylonians (or, perhaps, people historically more remote). Later developments in mathematics are no longer tied to these practical activities, and it is only at rare moments that mathematicians again seem to show a concern for the physical properties of ordinary things. Yet the mathematics of the present is continually influenced by the mathematics of the past. Mathematics changes by response to the problems which have already been posed and the solutions which have already been achieved: Pappus and Diophantus are sources of inspiration for Fermat, who, in turn, inspires Gauss, Kummer, and Kronecker; Euclid and Descartes set the stage for Newton and Leibniz, and the methods developed by Newton and Leibniz are extended and modified by Euler, Lagrange, Cauchy, and Weierstrass.

These observations suggest a theory of mathematical knowledge which will avoid the pitfalls of apriorism and of crude empiricism, the theory which I sketched in the introduction and which I shall elaborate in the rest of this book. I propose that a very limited amount of our mathematical knowledge can be obtained by observations and manipulations of ordinary things. Upon this small basis we erect the powerful general theories of modern mathematics. Responding to the practical problems and methods of the Babylonians, the Greeks developed theories which would systematize the solutions already obtained. Their knowledge was based on the prior empirical knowledge of their predecessors, and, in its turn, it served as the basis for the knowledge of their successors. At each stage in the ensuing story, the knowledge of individuals is generated from the knowledge of teachers, who pass on what the mathematical community has so far learned. The knowledge of the community is itself the product of a long series of episodes, extending back to the simple observations with which mathematical knowledge began. For obvious reasons, I shall call a theory of mathematical knowledge constructed along these lines an *evolutionary theory* of mathematical knowledge.

There are many questions which need to be answered if an evolutionary theory is to be defended. We need to know how mathematical knowledge begins and how it is extended. Plainly, evolutionary theories tend to invert the usual view of the epistemological order: the statements we hail as axioms (which apriorists take to be epistemologically basic) will be taken to be warranted by inference from previous beliefs. Nevertheless, some of the motives for apriorism can be accommodated by an evolutionary theory.

No apriorist is likely to deny the correctness of the three observations I took to be obvious, but it is probable that apriorists will question my view of their significance. There is a time-honored strategy for dealing with the points adduced: one denies that they indicate anything except extraneous psychological features of the genesis of mathematical belief. A typical reaction to the role of the mathematics teacher would be to insist that, although the teacher's testimony may be the *original* source of knowledge, this source is quickly replaced by another. Unless the student is dull, she will recognize for herself (using one of the modes of basic knowledge which apriorists favor) the truth of the statements made by the teacher. Similarly, the rods and beads of childhood mathematics are simply props which the student can later abandon. The historical development of mathematics shows the same story writ large. Egyptian surveyors and Babylonian bureaucrats gained mathematical knowledge on the basis of experience. The Greeks showed how their knowledge could be independently grounded, thereby transforming mathematics—or, perhaps, producing a new science, *pure* mathematics.

Apriorists are on to something. Although authorities are the primary source of an individual's knowledge, the community supplements the primary source with *local justifications,* providing the student with ways of looking at mathematical principles which make them seem obvious. So it comes to *appear* that the mathematician, seated in his study, has an independent, individual means of knowing the basic truths he accepts. I propose that this is only appearance. What occurs in the mathematician's study is just an extension of the training process.

Most of our beliefs, both about mathematical and non-mathematical topics, are causally overdetermined. Although it is useful to oversimplify in dealing with some epistemological issues, we must recognize that it is an oversimplification to ask for *the* process which produced a particular state of belief. A mathematician's belief in some axiom of elementary arithmetic is probably produced by a number of different causal processes: recollections of reading texts and hearing lectures, perceptual recognition that the axiom holds for an initial segment of a sequence of stroke symbols, and, perhaps, some further processes. Apriorists would not deny the point. Their claim would be that the mathematician's recollections are irrelevant. Even if the processes of recollection had not occurred, they would contend that some process of the kind they favor could have occurred and, had it occurred, would have warranted belief. It is exactly here that the mistake is made. Apriorists discard the wrong processes as irrelevant props, and place too heavy an epistemological burden on processes which can, in fact, serve a useful function as local justifications. We found above that the ability of the favored processes to warrant belief depends on the presence of appropriate background beliefs. Because of this, processes of recollection turn out to be indispensable in a way that processes of mental visualization (or other alleged a priori warrants) do not. We can obtain warranted mathematical belief through the use of mental visualization if our memory and perception sustain suitable beliefs (if, for example, they sustain the belief that others do not disagree). We can also obtain warranted mathematical belief through the use of memory and perception alone: this occurs whenever our knowledge of mathematical truths is based on experiences of reading books or hearing lectures (or on recollections of such experiences). There is an asymmetry here and the asymmetry is important.

As I understand the notion, a local justification is a type of process which is a dispensable but useful aid to knowledge. More formally:

(7) A *local justification* for a statement S is a type of psychological process, α, such that
  (i) provided that suitable background beliefs are warranted, processes of type α can warrant belief in S;
  (ii) processes of type α cannot themselves warrant the suitable background beliefs;
  (iii) processes of the types which warrant the suitable background beliefs can warrant belief in S.

Our examination of apriorist proposals, coupled with the evolutionary theory sketched above, suggests that the alleged a priori warrants may be local justifications in the sense of (7). Once this point is appreciated, the significance of my critique of apriorism becomes clearer. *Gedankenexperimente* furnish obvious examples of local justifications. When Galileo or Einstein asks us to imagine what would happen in idealized situations, exploiting our ideas about, for example, symmetry, he may lead us to engage in processes which warrant us in particular beliefs. The warranting ability of the processes depends on our possession of empirically warranted background beliefs, and, of course, the principles which Galileo and Einstein hope to inculcate could themselves be warranted in the same way. Thought-experiments serve important pedagogical goals. They fix belief far more firmly than simple authoritative assertion could do, they aid the memory, and they connect beliefs. But they do not engender a priori knowledge. I suggest that some of the processes favored by apriorists play exactly the same role.

Two of the apriorist approaches considered above begin by recognizing the existence of a process which does function in our mathematical lives. Some mathematical statements are defended by appeal to definition. Others are upheld by the use of pictorial representation. Only in the case of intuition of abstract objects are we invited to consider a process which has no independent claims to function in mathematics: Platonic intuition is dragged in to serve the epistemological needs of a plausible theory of mathematical truth. Both conceptualists and (Kantian) constructivists do better.

Large parts of the language of mathematics are framed in such a way that "first principles" in some areas can be justified by appeal to definition. Individual mathematicians presuppose an appropriate grounding of the relevant concepts, and offer local justifications of their assertions. The appeal to linguistic understanding is not an a priori warrant, but, in the context of an experience which supports the propriety of the linguistic practice, it does provide knowledge. Likewise, there are mathematical disciplines in which principles can be defended (or, equivalently, in which the applicability of concepts can be demonstrated) by appeal to pictorial representation. Under pressure to elaborate a theory of a priori mathematical knowledge, Kantians have typically supposed that the pictorial representation takes place in the mind's eye. Once we have forsaken the search for this type of theory, we can recognize that visualization in imagination is a poor vehicle for the depiction of some principles, and that we do better to resort to external diagrams which are both more readily surveyable in detail and more durable than mental images. Perhaps Kant and his successors were right to think that pictorial representation can sometimes be carried out without employing external aids, but they were wrong both to deny the epistemic kinship of imaginative visualization with sense perception and to overrate the extent to which imaginative visualization is possible.

I have tried to show how an evolutionary theory of mathematical knowledge might provide a perspective on the shortcomings of apriorism and on the features of mathematical practice which make apriorism appear plausible. Let us now take a brief look at the prospects and problems of the evolutionary approach.

III

On the approach I have recommended, the knowledge of individul mathematicians is to be explained by the knowledge passed on to them by authorities. In many cases, the authority of teachers will entirely account for an individual's mathematical knowledge: some people, probably the majority, only assert mathematical statements which they have not been explicitly taught when those statements have been obtained from statements which were explicitly taught by applying rules which have been explicitly taught. However, I want to disavow any relativistic view of mathematical "knowledge." Not every set of widely held beliefs counts as authoritative knowledge. For teachers and textbooks to serve as vehicles of *knowledge*, the teachers and the textbook authors must know what they transmit. To refer an individual mathematician's knowledge to the authority of another is only to begin the explanatory story. Completion of the story awaits an account of the knowledge of the authority, and this, in turn, will usually require an account of the authority's authorities, and so on. I claim that the story can be completed in a way which will recognize *knowledge* as being acquired, transmitted, and extended.

In practice, of course, what we need is not a detailed version of the story, but a general idea about how the story goes. To explain the knowledge of a particular contemporary mathematician, we do not need to specify the ancestral chain of her authorities. We no more need to know who those authorities were and the idiosyncrasies of their states of mathematical knowledge than we need to know the individual peculiarities of ancestral organisms when we provide an evolutionary account of the presence of some trait in a member of a current species. What we require in the latter case is an idea about how the trait (perhaps in a rudimentary form) might originally have arisen and how natural selection might have led to its fixation. We need to know how the general laws which govern evolution apply to the particular example which interests us. Similarly, in the case of mathematical knowledge, we need a specification of how the principles which govern the development of mathematical knowledge apply to enable the mathematician to have a warrant for the statements she accepts.

If I am right, then *some* crucial issues in the epistemology of mathematics are issues which philosophers have largely ignored. The central question is "How does mathematical knowledge evolve?" As with evolutionary questions in other areas, this question breaks into two parts. An adequate answer requires both an explanation of the *origins* of mathematical knowledge and an account of the *growth* of mathematical knowledge. In the rest of this chapter, I shall take a brief look at the problems which arise in each of these areas, and outline my strategy for overcoming them.

My solution to the problem of accounting for the origins of mathematical knowledge is to regard our elementary mathematical knowledge as warrranted by ordinary sense perception. In this way our remote predecessors acquired the first items of mathematical knowledge. We emulate them by using simple observations to provide our children with a supplement to the authority of the teacher. Yet to point to the possibility of acquiring *some* kind of knowledge on the basis of observation is not to dispose of the worry that, properly speaking, *mathematical statements* cannot be known in this way. Hence a complete resolution of the question of the origin of mathematical knowledge should provide an account of the content of mathematical statements, showing how statements with the content which mathematical statements are taken to have can be known on the basis of perception.

The principal task of explaining the origins of mathematical knowledge thus becomes one of providing a picture of mathematical reality which will fit with the thesis that our mathematical knowledge can originate in sense perception. Philosophers of mathematics have traditionally paid considerable attention to this enterprise. As I noted in Chapter 3, consideration of the objectivity of mathematics provides strong support for Platonism: if we are to do justice to the truth of mathematical statements then it seems that we have to maintain that mathematics describes a realm of mind-independent abstract objects. Yet this conclusion is far from unproblematic. Some philosophers have viewed the Platonistic picture as unclear or incoherent; others have wondered how, given that picture, mathematical knowledge is possible. In the next chapter I shall provide a non-Platonistic view of the content of mathematical statements, which I take to accommodate some of the worries of previous anti-Platonists while also doing justice to the important ideas which inspire the acceptance of Platonism.

In this part of my project I am engaging in an enterprise whose ground rules are well understood. There are certain clear desiderata which constrain an account of mathematical reality. I shall try to elaborate the constraints, explain why I believe that traditional approaches to the problem fail to satisfy them, and advance a theory that does. The second part of my project is more problematic. Since the issue of the growth of mathematical knowledge is rarely broached in philosophical discussion, there are no standard philosophical accounts with which my own proposals can be compared. Hence I think it worthwhile to close this chapter with some methodological remarks, describing in more detail than I have given so far the type of answer to the question "How does mathematical knowledge grow?" which I take to be appropriate.

IV

The question of how to understand the growth of mathematical knowledge may appear to be very straightforward. One might be tempted to think that the appropriate strategy is to elaborate a full theory of correct inference and to show that the historical development of mathematics can be reconstructed in accordance with this theory. My decision in Chapter 1 to refrain from providing an analysis of knowledge should already have indicated that I reject this strategy. Epistemology has no Archimedean point from which it can exert leverage on the knowledge claims of those who participate in the various kinds of human inquiry. A full account of what knowledge is and of what types of inferences should be counted as correct is not to be settled in advance. Rather, it must emerge from consideration of the ways in which humans actually infer and from the knowledge claims which we actually make. Nor can we expect to come to mathematics with a theory that has already been developed from other areas of inquiry. Much of our thinking about knowledge is still dominated by the case of perceptual knowledge, and conceptions of correct inference are overshadowed by the areas in which previous investigations have been most intense: the deductive reasoning of mathematicians and the inductive and statistical inferences which play an important role in *intra*-theoretic scientific decision. If recent philosophical studies of the history of science have shown anything at all,[2] they have revealed the poverty of our detailed conceptions of rational inference in the face of the complex arguments which figure in *inter*-theoretic debate. Moreover, even if we had a detailed account of rational infer-

2. The studies I have in mind are the works of T. S. Kuhn, P. K. Feyerabend, S. Toulmin, I. Lakatos, L. Laudan, D. Shapere, and the many authors who have extended and developed the proposals of these writers.

ence drawn from study of natural scientific practice, there would still be an open question as to whether this account should be broadened to encompass types of inference which play a special role in the growth of mathematical knowledge.

My claim is that an adequate epistemology must do justice to the kinds of inferences which mathematicians make, and that, since the growth of mathematical knowledge has not hitherto been taken as a serious object of epistemological investigation, it will first be necessary to isolate the kinds of transitions which have occurred in the history of mathematics and to note some pervasive patterns of argument. Yet it should not be thought that the epistemologist is simply reduced to a cipher, a figure who simply endorses those knowledge claims of mathematicians which are unearthed by historical research. To deny that epistemology can be a critical tool fashioned in advance of any study of the nature of various types of inquiry is not to refuse any place to criticism. Our task is to systematize the inferences and claims to knowledge made and advanced by previous and contemporary investigators in a variety of fields, and in performing this task we may easily reject some popular types of inference and repudiate some knowledge claims. We do not simply describe the history of science, exclaiming triumphantly "There's human knowledge for you!" Instead, we attempt to present the history in a way which will conform to a growing account of rationality, an account which tries to expose the most general features of human knowledge and correct human inference. Both the general account and the historical narratives continue to be modified and adjusted as we endeavor to achieve a more widely encompassing theory.

This approach to epistemology may recall Nelson Goodman's comments on dissolving the classical problem of induction. In a famous passage, Goodman writes:

> The point is that rules and particular inferences alike are justified by being brought into agreement with each other. *A rule is amended if it yields an inference we are unwilling to accept; an inference is rejected if it violates a rule we are unwilling to amend.*[3]

I am not concerned here to decide whether this passage is an adequate response to the riddle of induction which descends from Hume. I would simply recommend it as excellent advice to the aspiring epistemologist who hopes to use the history of science to develop an account of correct inference. We bring to the history a view of human rationality, itself the product of prior reflection on our past and present practices, and that view can be used to criticize the inferences and claims made by those we study or their inferences and claims can be used to amend the view. The balance, as Goodman goes on to note, is delicate. The epistemologist's role is neither that of an autocrat who assesses the perfor-

---

3. *Fact, Fiction and Forecast*, p. 64.

mances of inquirers by laws that he lays down, nor that of a petty bureaucrat whose task is only to approve whatever others do.

I shall tackle the question of how mathematical knowledge grows by starting with some very general considerations about the evolution of science. Previous philosophical investigations of the kinds of inferences which figure in acceptance of new scientific theories provide us with a general framework within which mathematical change can be discussed. I shall then consider the various types of changes which occur in mathematics that are, from this general epistemological perspective, epistemologically significant, and I shall attempt to expose pervasive patterns of inference in a way which will make them recognizably rational. What will emerge from this will be an account of inference and theory change in mathematics which is analogous to discussions of theory choice in the philosophy of science. Writers of different persuasions have supposed that choice among scientific theories is motivated by a desire to achieve certain "virtues": simplicity, explanatory power, theoretical coherence, problem-solving efficacy, falsifiability, and so forth.[4] I shall aim to specify the epistemic desiderata which are relevant to mathematics. Finally, I shall use the account of mathematical inference which I have developed to explain one important historical episode, arguing that, in the terms which I have articulated, we can understand the historical evolution of the calculus as a rational process.

To engage in this project it is necessary to take the history of mathematics seriously. I shall try to develop enough historical examples to allay worries that my treatment of mathematical change is biased toward one type of case. However, it is also important not to lose sight of the general epistemological themes in a welter of historical studies. For this reason, my treatment of historical materials in Chapters 8 and 9 will be relatively brief. Kinds of mathematical change and patterns of mathematical inference will be illustrated by appeal to short snippets of history. The main confirmation for my picture must await Chapter 10, in which I shall consider historical material in much more detail. I should also note that, because the questions I am asking have not previously received much attention from professional historians, my treatment of my primary examples draws on my own historical research, rather than relying on the narratives of others. In particular, in the case of the calculus, my discussions are based on an analysis of the historical texts, which diverges at some points from traditional histories, histories that, either explicitly or implicitly, are committed to mathematical apriorism.

Let me conclude by responding to an obvious worry about my program. It may seem that I am simply begging the question, assuming in advance that it is possible to reconstruct the history of mathematics as a sequence of rational transitions. That is not so. It is conceivable that study of the history of mathe-

4. See C. G. Hempel, *Philosophy of Natural Science,* chapter 4; W. V. Quine and J. Ullian, *The Web of Belief,* chapter 5; T. S. Kuhn, "Objectivity, Value Judgment and Theory Choice."

matics should reveal no patterns of inference or principles of theory choice which could be integrated into a general account of the growth of knowledge. Any reconstruction of the historical development of mathematics might violate views of correct inference well confirmed in other areas. This is conceivable, but unlikely. Although philosophers of mathematics have typically ignored the history of mathematics [5]—and perhaps some have even thought of that history as a sequence of benighted blunders—it would be surprising if two millennia of haphazard development had bequeathed to us a corpus of knowledge. Indeed, on apriorist grounds, the history of mathematics would be almost a miracle. It surely strains our credulity to suppose that a process which was insusceptible of rational reconstruction could produce a body of statements which someone (Frege, Brouwer, or Gödel, for example) could transform into an a priori science! Hence I suggest that even the staunchest apriorist should believe that there must be some method to the mathematicians' madness—and that the task of clearly explaining that method is a philosophically important one.

5. As I pointed out in the introduction, Lakatos is the most prominent example of a philosopher who takes the history of mathematics seriously. Two other writers who have used the history of mathematics to make important philosophical points are Ernest Nagel and Mark Steiner. See Nagel's articles "Impossible Numbers" and "The Formation of Modern Conceptions of Formal Logic in the Development of Geometry," and Steiner, *Mathematical Knowledge*, chapter 3.

6
Mathematical Reality

I

If it is correct to suppose that we have some mathematical knowledge then some mathematical statements must be true. So, for example, if almost everyone knows that $2 + 2 = 4$ and if the *cognoscenti* know that $\int_{-\infty}^{+\infty} e^{-x^2} dx = \sqrt{\pi}$, then the statements "$2 + 2 = 4$" and "$\int_{-\infty}^{+\infty} e^{-x^2} dx = \sqrt{\pi}$" are both true. Yet to advance that conclusion is immediately to raise the question of what makes those statements true. In this chapter, I shall try to answer the question.

The most obvious conception of mathematical truth is Platonism. We begin with the observation that statements like those considered above contain what appear to be singular terms—'2,' '4,' 'e,' '$\sqrt{\pi}$.' If these statements are to be true, and if we are to accept the standard (Tarskian) account of truth,[1] then the terms in question must refer. More generally, any singular term which occurs in a true mathematical statement must refer and any variable which occurs in a true mathematical statement must range over a set of values. What are the

---

1. Some writers have been prepared to give up standard Tarskian semantics for first-order language in order to find an alternative to Platonism. So, for example, in "The Truth about Arithmetic," Dale Gottlieb uses substitutional quantification to defend a nominalist (or, at least, non-Platonist) account of arithmetic. Proposals of this type need to be supplemented with an epistemology and to be articulated to handle real analysis, set theory, and so forth. But perhaps the most obvious worry about them is the bifurcated semantics which they bring. There are two difficulties here. First, we give up a uniform semantics which will cover both mathematical and non-mathematical language. Second, the semantics has to be integrated at least to the extent of coping with sentences which *mix* mathematical and extra-mathematical language. Interestingly, one writer who proposes a non-Tarskian semantics for mathematics is able to avoid both difficulties. In "Myth and Math," Leslie Tharp campaigns for a non-Tarskian approach to the quantifiers of mathematical and extra-mathematical discourse. Since Tharp's interesting ideas were still in the process of development at the time of his death, I shall not try to evaluate them in the present work. (I regret that I did not have the chance to meet Tharp and to discuss with him the issues about which we corresponded. I very much hope that his manuscript "Myth and Math" will be edited and published).

referents of the singular terms and the values of the variables? Apparently, they
have to be objects which do not exist in space-time, and which exist independently of our mental activity. For there are probably not enough spatio-
temporal objects to go round, and, in any case, the truth of mathematical statements does not depend on the fate of any spatio-temporal object. Nor can we
take the objects with which mathematics is concerned to be dependent on our
mental activity, for we believe that the truths of mathematics were true prior
to the time at which humans first indulged in any such activity, that they would
have been true if humans had never existed (or never engaged in mental mathematical activity), and that there are far more truths of mathematics than mental
acts that humans have performed, or ever will perform. Hence mathematical
objects must exist, but they can neither be spatio-temporal objects nor mental
constructs. They therefore deserve the title of abstract objects. So we arrive at
the Platonist thesis: true mathematical statements are true in virtue of the properties of abstract objects.

In Chapter 3, we looked at an abbreviated version of this argument, noting
that Platonist epistemology tends to be subordinate to the concerns of providing
an adequate ontology for mathematics. Platonists standardly argue that mathematics *must* be about abstract objects—and then look around for a means of
knowing about them. We found that the Platonist conception of mathematical
intuition is indefinite and that, in consequence, Platonistic apriorism is doomed.
Of course, this does not mean that we must repudiate Platonism. Perhaps we
can do justice to mathematical knowledge by abandoning mathematical apriorism but retaining the thesis that mathematics describes a realm of abstract objects. Indeed, the argument rehearsed above appears to be very powerful, and
the ways of resisting it do not look initially attractive.[2] Nevertheless, there are
troubles with the position to which it leads, troubles which are sufficiently
serious to justify us in exploring a non-Platonistic approach to mathematical
truth.

The first difficulty, forcefully presented in a paper by Paul Benacerraf, challenges us to square a Platonistic theory of mathematical truth with a causal
theory of mathematical knowledge, without sacrificing the truism that we know
some mathematics.[3] Benacerraf's basic point is very simple. According to the
Platonist, mathematics is concerned with mind-independent abstract objects,
and such objects do not causally interact with other objects; in particular, they
do not interact with human subjects. Yet if we adopt an enlightened theory of
knowledge, we should hold that when a person knows something about an
object there must be some causal connection between the object and the person.

2. One of the most thorough and interesting attempts to take the argument for Platonism seriously
and to resist it is Charles Chihara's *Ontology and the Vicious Circle Principle*.
3. See P. Benacerraf, "Mathematical Truth." A related difficulty, concerning how, on the Platonist's account, reference to mathematical objects is possible, is presented by Jonathan Lear in "Sets
and Semantics."

Given that we know some mathematics, it follows that either our best theory of mathematical truth (Platonism) or our best theory of knowledge (a causal theory of knowledge) is mistaken.

Plainly, there is a preferred strategy for the Platonist to adopt when faced with this argument. Platonism can be defended by showing that, when our "best theory of knowledge" is probed more carefully, it will be found to contain no constraint which precludes knowledge of abstract objects. Many ingenious arguments have been advanced to implement this strategy.[4] To consider them in the detail they deserve would lead us into many byways, and I shall content myself with a brief, dogmatic, evaluation. Benacerraf's original point does depend on an oversimplification of issues about knowledge and causation; but the intuition behind that point is deep enough to enable it to be reformulated so as to cause difficulty for the Platonist responses which have been offered to the original version.

Yet, even if the Platonist were completely successful in turning back Benacerraf's challenge, that would be only a small step towards achieving an adequate position. To show that it is in principle possible for us to have knowledge of mathematical reality, Platonistically construed, is not to explain how we *do* have such knowledge. Here the Platonist has two options. One is to claim that there must be some process, perhaps called 'intuition,' about which we know very little but which does give us access to abstract objects. The other is to propose that some well-understood source of knowledge, such as sense perception, can provide us with the requisite access. The former tactic seems to me a desperate measure, tantamount to abandoning the enterprise of explaining our knowledge. The latter requires the Platonist to explain the nature of abstract objects in a way which will enable us to appreciate how standard perceptual processes could furnish information about them. Anti-Platonists are worried by the picture of ethereal entities lurking behind ordinary things, and they wonder how it is possible for the scattering of light from the surfaces of ordinary things to engender knowledge of those entities. The Platonist's task is to provide a better picture.[5]

4. See, for example, chapter 4 of Mark Steiner's *Mathematical Knowledge* and Penelope Maddy, "Perception and Mathematical Intuition."

5. Maddy's defense of Platonism, which I take to be the best that has so far been given, shows that there may be certain possibilities for constructing a better picture but it does not provide a complete answer to the worry just raised. Moreover, it seems to me that, while Maddy's account of perceptual mathematical knowledge presupposes that perception is direct, it does not fit well with the best available theory of direct perception, namely that advanced by the ecological realists. Even if we abstract from the distinctive doctrine that we perceive the affordances of objects, the more general claim that the information which we gain in perception concerns transformations of the sensory array caused by events in which perceived objects participate seems to be at odds with the idea that we can acquire perceptual information about unchanging abstract objects. (See Michaels and Carello, *Direct Perception*, chapter 2.) Hence I am not sure that Maddy's version of Platonism escapes the basic objection raised by Benacerraf.

I do not regard this criticism as a decisive argument against Platonism. Proponents of Platonism would be rational to admit that their theory is incomplete in various respects, that there are "research problems" to be solved. My aim is simply to note some areas of "incompleteness," to indicate why the "research problems" might be difficult, and to use this survey to motivate a rival theory.

Let us turn to a second difficulty for Platonism which has been discussed in recent years.[6] Suppose that we admit that mathematics describes a realm of abstract objects. What kinds of mathematical objects are there? At first sight, mathematicians seem to discuss an assortment of entities—numbers (natural, rational, complex, and so forth), functions, spaces, groups, and a host of other things. Set-theoretic investigations show us that all of these entities can be identified as sets. Canons of parsimony and explanatory unification seem to press the identification upon us. Surely we ought to admit no more entities than are necessary, and to carry out our mathematical research from the perspective of a single, all-encompassing theory if we can? But set theory gives us an *embarras de richesses*. Consider, for example, the natural numbers. Apparently, our ancestors discussed them for generations, and, on the view under present discussion, they were talking about sets. But *which* sets? What is the referent of '2' as used by our predecessors (and by those of our contemporaries who do mathematics without explicitly invoking set theory)? Or, to bypass possible worries about the existence of a common referent for many tokens, what is the referent of the token of '2' which occurs on a particular page of a particular mathematical manuscript? There are too many ways for us to reduce arithmetic to set theory for us to give straightforward answers to these questions. We might be happy to assert that $2 = \{\{\emptyset\}\}$—until we realize that our arithmetical purposes could be served equally well by claiming that $2 = \{\emptyset, \{\emptyset\}\}$ (or any of a large number of other identities). Thus the Platonist is torn between the methodological directive to identify numbers as sets and the difficulty of saying what sets the numbers are.

As with the first worry which we considered, there is room to manoeuvre. Platonists may try to argue that the alleged methodological directives to identify numbers and sets need not be obeyed, or that it is possible to maintain that numbers are sets without there being any particular sets which the numbers are. I have tried to show elsewhere that these efforts come to naught, and I shall not repeat my arguments here.[7] Consideration of a further problem about Platonism will lead us to understand the fundamental difficulty which is behind the trouble caused by multiple possibilities of set-theoretic reductions.

Mathematical truths are useful to us. But why are they so useful? An answer

6. See Paul Benacerraf, "What Numbers Could Not Be." For a more extensive discussion of these issues, which parallels the treatment of the next pages but adds some technical details, see my paper "The Plight of the Platonist."

7. In "The Plight of the Platonist."

to this question need not be a direct consequence of a theory of mathematical reality. However, a good theory of mathematical reality ought not to make this sensible question look like an unfathomable mystery. But that, I maintain, is what Platonism does.

One of the primary motivations for treating mathematical statements as having truth values is that, by doing so, we can account for the role which these statements play in our commonsense and scientific investigations. The fictionalist proposal, which avers that what we standardly regard as mathematical statements with truth values are merely marks produced in playing elaborate games, can be countered by pointing to the value of mathematics in advancing our understanding of the world. If mathematics is just a sequence of recreational scratchings, then why do the games we engage in prove so useful? When we draw conclusions from a mixture of scientific and mathematical premises, what accounts for our success? Platonism gains its initial plausibility by recognizing that these questions can be answered if we are prepared to return to the idea that mathematical sentences are what they appear to be, to wit, statements with truth values. Yet to move from this point to Platonism is to assume that Platonism has a monopoly on accounts of the truth of mathematical statements. I now want to suggest that the reasons which incline us to take the first step with the Platonist should also make us suspicious of the thesis that mathematics describes a realm of abstract objects. If we are seriously persuaded that the usefulness of mathematical statements pays tribute to their truth, then we should ask whether the account of mathematical truth which the Platonist offers helps us to understand the utility of mathematics.

We juxtapose a commonplace about mathematics, the thesis that mathematics is useful in explaining and predicting the behavior of ordinary physical things, with the Platonist's contention that mathematical statements describe a realm of abstract objects. How well do these fit together? One obvious question is why these abstract objects should be so important to us. Why is it that, by studying them, we improve our ability to describe and explain the behavior of more mundane things? On the Platonist's account, the world is bifurcated. There are ordinary physical objects, and there are the abstract objects which mathematics characterizes. Somehow, by investigating the second realm we learn truths which can be used to give us greater understanding of the first. If Platonism is to be fully intelligible, we need an account of why this should be so.

There is an old explanation of the utility of mathematics. Mathematics describes the structural features of our world, features which are manifested in the behavior of all the world's inhabitants. This line is common to many writers before the twentieth century, and, in our century, it finds expression in Russell's remark that arithmetic is concerned with "the more abstract and general features" of the world. I do not wish to pretend that such remarks are precise, but I do claim that they are suggestive. The challenge for the philosopher of mathematics is to construct a picture of mathematical reality which will give them a clear sense. What I hope to accomplish in later sections is an adequate response to the challenge. My present goal is to use the inchoate view of mathematics as describing "the structure of reality" to isolate what is fundamentally wrong with Platonism.

Intuitively, the Platonist's mistake is to replace the picture of mathematics as descriptive of structural properties which are manifested in a host of concrete instances with the picture of mathematics as describing abstract entities which manifest the structure. Platonists can be construed as espousing the vague thesis of the last paragraph, and elaborating it as the claim that mathematics owes its truth to some abstract instantiation of mathematical structure. But, by making this move, they destroy the original intuition about the utility of mathematics. Assuming that there *are* abstract instantiations of mathematical structure, they are no more of interest to mathematics than any other instantiation. We are equally concerned with all the instantiations, and equally unconcerned about any of them. More exactly, we are interested in the structure they share, and it is misleading to formulate the contents of mathematics by identifying one instantiation, even an "abstract" instantiation, as privileged.

This point casts some light on the issue (at which we looked briefly above) of how to relate numbers to sets. One very natural suggestion is to see arithmetic as articulating what is common to the various ways of identifying numbers as sets. Arithmetic would be rewritten as the theory of ω-sequences.[8] Unfortunately, there is a technical difficulty. Pursuit of the suggested strategy would require us to use the notion of function or sequence or some other notion such as relation or ordered pair. But there are problems in giving set-theoretic identifications of these notions, problems which are exactly parallel to those concerned with the numbers. Moreover, the line of attack which was used to resolve the question of the identity of the numbers cannot be replicated to deal with functions, ordered pairs, and so forth. The strategy was to replace Frege's *explicandum* '$x = n$' with 'x is an n in p,' where the variable 'p' ranges over ω-sequences. If we try to adapt this strategy to the case of the ordered pairs, we must replace the usual *explicandum* '$x = <y, z>$' with something like 'x is a $<y, z>$ in w,' where 'w' ranges over a set (more precisely, a *class*) of ordered pair *explicata* (such as the Wiener ordered pairs or the Kuratowski ordered pairs). But it is relatively trivial to show that no particular correspondence between the ordered pair *explicata* and the *explicanda* is privileged. We cannot claim that a certain set in a class is a particular ordered pair unless we have fixed an assignment function. However, relativizing to assignment functions vitiates the enterprise, since our goal was to develop the theory of ordered

---

8. This proposal was made by Nicholas White in "What Numbers Are," and, in a somewhat different way, by Hartry Field in "Quine and the Correspondence Theory." Both Field and White give modern versions of an idea which is present in Dedekind's *The Nature and Meaning of Numbers*. I criticize their proposals in section II of "The Plight of the Platonist." That paper contains a more detailed version of the argument given in the rest of this paragraph.

pairs, relations and functions from scratch without presupposing any of the usual (arbitrary) set-theoretic identifications.

The breakdown of a very natural Platonist strategy is significant. The Platonist's attempt to use abstract objects to articulate the idea that mathematics is about abstract structure founders on the case of the ordered pairs, precisely because the original introduction of abstract objects was a bad way of doing justice to the insight that mathematics is concerned with structure. We should apply the suggestion of the last paragraph in a more thoroughgoing way. Instead of supposing initially that mathematics is about abstract objects and then, when we find multiple instances of a common structure, reinterpreting statements as descriptive of the structure exemplified in those objects, why do we not begin from the thesis that mathematics is descriptive of structure without making the initial move to Platonistic objects?

I shall attempt to work out an interpretation which will give sense to the thesis that mathematics is about structure. In doing so, I aim to overcome the difficulties I have uncovered in Platonism. By taking mathematical structure to be reflected in the properties of ordinary things, we can begin to dissolve epistemological perplexities. Perception can be viewed as a process in which our causal interaction with ordinary objects leads us to discern the structure which they exemplify. There is no suggestion of a gap between these ordinary objects and other, more ethereal, entities which lurk behind them. The great utility of mathematics will be explained by reference to its delineation of a structure exemplified by all physical objects. Finally, the avoidance of abstract objects will free us from those troublesome questions of identity which Platonists seem forced to answer.

Although the thesis that mathematics is about structures present in physical reality is, at present, vague and programmatic, the substitution of that thesis for the Platonist account enables us at least to glimpse answers to questions which arise for Platonism—questions which, I suggest, the Platonist has trouble answering. The challenge is to remedy the vagueness, and to present a defensible picture of mathematical reality.

II

I begin with an elementary phenomenon. A young child is shuffling blocks on the floor. A group of his blocks is segregated and inspected, and then merged with a previously scrutinized group of three blocks. The event displays a small part of the mathematical structure of reality, and it may even serve for the apprehension of mathematical structure. I shall try to find a way of construing mathematical structure which will enable us to see clearly why this is so.

Children come to learn the meanings of 'set,' 'number,' 'addition' and to accept basic truths of arithmetic by engaging in *activities* of collecting and segregating.[9] Rather than interpreting these activities as an avenue to knowledge of abstract objects, we can think of the rudimentary arithmetical truths as true in virtue of the operations themselves. By having experiences like that described in the last paragraph, we learn that particular types of collective operations have particular properties: we recognize, for example, that if one performs the collective operation called 'making two,' then performs on different objects the collective operation called 'making three,' then performs the collective operation of combining, the total operation is an operation of 'making five.' Knowledge of such properties of such operations is relevant to arithmetic because arithmetic is concerned with collective operations.

As a first approximation, we might think of my proposal as a peculiar form of constructivism. Like the constructivists I hold that arithmetical truths owe their truth (at one level) to the operations we perform. (I shall later qualify this thesis and explain more carefully what it amounts to.) Unlike most constructivists, I do not think of the relevant operations as private transactions in some inner medium. Instead, I take as paradigms of constructive activity those familiar manipulations of physical objects in which we engage from childhood on. Or, to present my thesis in a way which will bring out its realist character, we might consider arithmetic to be true in virtue not of what *we can do* to the world but rather of what *the world* will let us do *to it*. To coin a Millian phrase, arithmetic is about 'permanent possibilities of manipulation.' More straightforwardly, arithmetic describes those structural features of the world in virtue of which we are able to segregate and recombine objects: the operations of segregation and recombination bring about the manifestation of underlying dispositional traits.[10]

I have now sketched my main thesis. My next task is to explain it and to add qualifications. Let us begin with the notion of 'truth in virtue of' which is casually employed in my previous discussions (and in many discussions in the philosophy of mathematics). I want to suggest both that arithmetic owes its truth to the structure of the world and that arithmetic is true in virtue of our

9. I do not mean to deny that their learning is aided by teachers and parents. As I have emphasized in Chapter 5, most items of mathematical knowledge are to be explained by reference to authority. However, we can view the activities of contemporary children as indicating the ways in which our ancestors, unaided by authority, began the mathematical tradition.

10. The account of mathematics indicated here and articulated below develops the view I suggested in ''Arithmetic for the Millian.'' In that paper, I also tried to show how it related to Mill's claims about arithmetic in *A System of Logic*. Here I am not concerned with the historical pedigree of the view, and I use the term 'Millian' simply as an apt label.

My Millian phraseology might easily give way to the technical terminology of ecological realism. One way to gloss Gibson's thesis that we (and other animals) perceive affordances is to maintain that perceivers perceive the possibilities of interaction with the environment. In the mathematical case, humans perceive possibilities which are afforded by *any* environment. Hence, in the Introduction, I translated my view of mathematical reality into the language of ecological realism by suggesting that mathematics is an ideal science of universal affordances.

constructive activity. How can this be? Consider an analogy with geometry. A pre-Lobatschevskian survivor in our century might maintain that the Euclidean theorem "In any triangle the sum of the angles is 180°" is true in virtue of the properties of triangles *and* that it is true in virtue of the structure of space. Set aside the fact that we do not regard the theorem as true of physical space. What concerns me here is the issue of whether the desire to maintain two accounts of the truth of geometry is confused or inconsistent. I suggest that it is not. For someone may reasonably contend that the ontological thesis that a geometrical statement owes its truth to the properties of triangles is simply an articulation of the ontological thesis that geometry owes its truth to the structure of space. Because space has the structure it does, triangles have the properties they do; conversely, what spatial structure amounts to is, *inter alia,* the fact that triangles have those properties. In other words, we are not being offered two separate answers to the question "What makes the geometrical theorem true?" but two versions of the same answer. The moral is that we can sometimes simultaneously defend two different claims of the form ⌜S is true in virtue of . . . ,⌝ and this is exactly what I want to do in the case of arithmetic. The slogan that arithmetic is true in virtue of human operations should not be treated as an account to rival the thesis that arithmetic is true in virtue of the structural features of reality. Once we understand the 'true in virtue of' locution, we can allow that these are compatible, and that taking arithmetic to be about operations is simply a way of developing the general idea that arithmetic describes the structure of reality.

Next, let us note explicitly that construing the structure of reality to be manifested in the operations we *actually* perform is obviously inadequate. Given our biological limitations, the operations in which we actually engage are limited. Thus the fact that we do not do certain things—and that, in the span of human lifetime, we *cannot* do certain things—should not be taken as setting forth some structural trait of reality. Arithmetic owes its truth not to the actual operations of actual human agents, but to the ideal operations performed by ideal agents. In other words, I construe arithmetic as an *idealizing theory:* the relation between arithmetic and the actual operations of human agents parallels that between the laws of ideal gases and the actual gases which exist in our world. We may personify the idealization, by thinking of arithmetic as describing the constructive output of an ideal subject, whose status as an ideal subject resides in her freedom from certain accidental limitations imposed on us. There is obvious kinship here with some developments of constructivism, most notably with Brouwer's doctrine of the creative subject.[11] I should emphasize, however, that the position I advocate does not endorse the epistemological and methodological views traditionally associated with constructivism. To say that

11. See L. E. J. Brouwer, "Consciousness, Philosophy and Mathematics," and A. Troelstra, *Lectures on Intuitionism.*

arithmetic in particular, or mathematics in general, is true in virtue of the constructive output of an ideal subject, does not commit me to the thesis that we can have intuitive knowledge of mathematical truths or to the thesis that there are (real or apparent) violations of the law of the excluded middle.[12] I suggest that we have no way of knowing in advance what powers should be attributed to our ideal subject. Rather the description of that ideal subject and the conditions of her performance must be tested against our actual manipulations of reality. From Kant on, constructivist philosophies of mathematics have supposed that we can know a priori what constructions we can and cannot perform, or, to put it another way, what powers should be given to the ideal constructive subject. But there is no reason to bind this epistemological claim to the basic ontological thesis of constructivism. Instead, we can adopt a more pragmatic attitude to the question of which mathematical operations are possible or what powers the ideal subject has, adjusting our treatment of these issues to the manipulations of the world which we actually perform.[13]

At this point, it is important to forestall a possible misunderstanding. In regarding mathematics as an idealizing theory of our actual operations, I shall sometimes talk about the ideal operations of an ideal subject. That is not to suppose that there *is* a mysterious being with superhuman powers. Rather, as I shall explain in the next section, mathematical truths are true in virtue of stipulations which we set down, specifying conditions on the extensions of predicates *which actually are satisfied by nothing at all but are approximately satisfied by operations we perform (including physical operations)*. This approach to idealizing theories will be very important to my account.[14]

A final clarification will prepare the way for a more definite statement of my thesis. One central ideal of my proposal is to replace the notions of abstract mathematical objects, notions like that of a colle*ction,* with the notion of a kind of mathematical activity, colle*cting.* I have introduced the notion of collecting by using a crude physical paradigm. In its most rudimentary form, collecting is tied to physical manipulation of objects. One way of collecting all the red objects on a table is to segregate them from the rest of the objects, and to assign them a special place. We learn how to collect by engaging in this type of activity. However, our collecting does not stop there. Later we can

12. For further discussion, see the final section of this chapter (Objection 5).

13. Here there is some kinship between the separating of Kant's ontological and epistemological claims about mathematics and the "Kantian constructivism," developed by John Rawls. The present formulation of my view of mathematical reality owes a debt to his three lectures on "Kantian Constructivism in Moral Theory."

14. It may help to point out that the position I defend has some affinities with the position which Chihara calls "Mythological Platonism" in chapter 2 of *Ontology and the Vicious Circle Principle.* The common theme is the idea that mathematical statements owe their truth to the stipulations on mathematical vocabulary which are laid down. The principal differences are that, on my account, the stipulations are not arbitrary but approximately characterize actual entities, and that the relevant entities are human operations.

collect the objects in thought without moving them about. We become accustomed to collecting objects by running through a list of their names, or by producing predicates which apply to them. Naïvely, we may assume that the production of any predicate serves to collect the objects to which it applies. (This naïve assumption is implicit in nineteenth-century analysis, and it was made explicit by Cantor.) Thus our collecting becomes highly abstract. We may even achieve a hierarchy of collectings by introducing symbols to represent our former collective activity and repeating collective operations by manipulating these symbols.[15] So, for example, corresponding to the set $\{\{a,b\},\{c,d\}\}$, we have a sequence of collective operations: first we collect a and b, then we collect c and d, and, finally, we perform a higher level operation on these collectings, an operation which is mediated by the use of symbols to record our prior collective activity. As I construe it, the notation '$\{. . .\}$' obtains its initial significance by representing first-level collecting of objects, and iteration of this notation is itself a form of collective activity.

Collecting is not the only elementary form of mathematical activity. In addition we must recognize the role of *correlating*. Here again we begin from crude physical paradigms. Initially, correlation is achieved by matching some objects with others, placing them alongside one another, below one another, or whatever. As we become familiar with the activity we no longer need the physical props. We become able to relate objects in thought. Once again, the development of a language for describing our correlational activity itself enables us to perform higher level operations of correlating: notation makes it possible for us not only to talk (e.g.) about functions from objects to objects (which correspond to certain first-level correlations) but also about functions from functions to functions, and so forth.

I promised a few paragraphs back to provide a more definite statement of my central thesis. It is time to redeem that promise. I propose that the view that mathematics describes the structure of reality should be articulated as the claim that mathematics describes the operational activity of an ideal subject. In other words, to say that mathematics is true in virtue of ideal operations is to explicate the thesis that mathematics describes the structure of the world. Obviously, the ideal subject is an idealization of ourselves, but I explicitly reject the epistemological view that we can know a priori the ways in which the idealization should be made. Finally, I interpret the actual operations, for which mathematics provides an idealized description, as comprising both collective and correlative operations. With respect to both types of operation, further distinctions must be drawn: not only are there very crude collectings and correlatings which consist in the rearrangement of physical objects, but there are also mathematical operations whose performance consists in the inscription of pieces of notation. Although I shall propose that the physical manipulations which

15. This point will be elaborated in Section IV.

constitute the crude paradigm of mathematical activity are epistemologically fundamental, I want to forestall the interpretation which takes these to exhaust our mathematical performances. My next task will be to explain how the ontological account I have sketched in this section can be developed in the case of arithmetic, and how it can be integrated into my empiricist epistemology.

III

Platonism can take the standard first-order versions of the Peano postulates at face value, construing the variables as ranging over abstract mathematical objects. One way to defeat the Platonist argument with which I began this chapter is to deny that the surface form of mathematical statements reveals their true logical form. So, we may try to avoid the task of developing a non-standard semantical theory for the case of arithmetic—a project which might raise uncomfortable questions about the relationship between arithmetical and nonarithmetical language—by rejecting the view of the logical form of arithmetical statements which has been standard since Frege. (I shall return, below, to the issue of whether the general account I offer is consistent with standard semantics for first-order language.)

I shall rewrite statements of first-order additive arithmetic in a first-order language, the language of *Mill Arithmetic*. The primitive notions to be used are those of a *one-operation*, of one operation being a *successor* to another, of an operation being an *addition* on other operations, and of the *matchability* of operations. (I ignore multiplication solely for reasons of simplicity; the account I propose can easily be extended to multiplicative arithmetic.) These notions are readily comprehensible, either in terms of our crude physical paradigm or in more abstract terms. We perform a one-operation when we perform a segregative operation in which a single object is segregated. An operation is a successor of another operation if we perform the former by segregating all of the objects segregated in performing the latter, together with a single extra object. When we combine the objects collected in two segregative operations on distinct objects we perform an addition on those operations.

We now turn to the important notion of matchability. This notion will play in our theory a central role akin to that of identity in the standard presentation of arithmetic. Individual operations are of little interest to us. We are concerned with *types* of operations, where operations which are matchable belong to the same type. Two segregative operations will be said to be matchable if the objects they segregate can be made to correspond with one another. (The notion of matchability will thus be an equivalence relation.) Arithmetic is concerned with those properties of segregative operations which are invariant under matchability.

It may already be clear that the arithmetical notions taken as primitive here can be related to more general notions—the notions of a collective operation and a correlative operation—which will be the concern of a reformulated set theory. I shall develop this point below. For the time being, my aim is simply to provide a formal system which will recapitulate the work of the standard systems of first-order arithmetic. We take a first-order theory with identity with (nonlogical) primitive predicates 'Ux,' 'Sxy,' 'Axyz,' 'Mxy' ('x is a one-operation,' 'x is a successor operation of y,' 'x is an addition on y and z,' 'x and y are matchable'). Clearly, we need to provide some axioms about matchability. Prominent among these will be assertions that matchability is reflexive, symmetric, and transitive. But we know much more than this about our intended concept of matchability. Anything matchable with a one-operation is a one-operation and, conversely, any two one-operations are matchable. If two operations are successors of matchable operations then they are matchable. If an operation a is matchable with a successor of some operation b then there is an operation matchable with b of which a is a successor. So we already arrive at the following axioms of *Mill Arithmetic*.

(1) $(x)Mxx$
(2) $(x)(y)(Mxy \rightarrow Myx)$
(3) $(x)(y)(z)(Mxy \rightarrow (Myz \rightarrow Mxz))$
(4) $(x)(y)((Ux \ \& \ Mxy) \rightarrow Uy)$
(5) $(x)(y)((Ux \ \& \ Uy) \rightarrow Mxy)$
(6) $(x)(y)(z)(w)((Sxy \ \& \ Szw \ \& \ Myw) \rightarrow Mxz)$
(7) $(x)(y)(z)((Sxy \ \& \ Mxz) \rightarrow (\exists w)(Myw \ \& \ Szw))$.

The first-order Peano postulates need to be embodied within our system. The principle that no two distinct numbers have the same successor will be reformulated as the statement that if two operations are successor operations and are matchable then the operations of which they are successors are matchable. The statement that one is not the successor of any number is analyzed as the claim that no one-operation is a successor operation. Finally, the induction principle is glossed as the assertion that whatever property is shared by all one-operations and which is such that if an operation has the property then all successor operations of that operation have the property is a property which holds universally. So we add to our axioms:

(8) $(x)(y)(z)(w)((Sxy \ \& \ Szw \ \& \ Mxz) \rightarrow Myw)$
(9) $(x)(y) \sim (Ux \ \& \ Sxy)$
(10) $((x)(Ux \rightarrow \Phi x) \ \& \ (x)(y)((\Phi y \ \& \ Sxy) \rightarrow \Phi x)) \rightarrow (x)\Phi x$
    for all open sentences 'Φx' of the language.

The recursive definition of addition is often introduced by adding two further axioms: the result of adding one to the number n is the successor of n; and the result of adding the successor of m to n is the successor of the result of adding m to n. These specifications can easily be reformulated in our language. We add the axioms:

(11) $(x)(y)(z)(w)((Axyz \ \& \ Uz \ \& \ Swy) \rightarrow Mxw)$

(12) $(x)(y)(z)(u)(v)(w)((Axyz \ \& \ Szu \ \& \ Svw \ \& \ Awyu) \rightarrow Mxv)$

Do these axioms suffice for the development of arithmetic in the usual way? Unfortunately not. The reason is not hard to discern. A standard system of first-order arithmetic must be interpreted by assigning an object to '1' and functions to the symbols for successor and addition. These constraints, together with those imposed by the Peano postulates, require us to choose an infinite domain of interpretation. Mill Arithmetic, as presented so far, is much more liberal. The twelve axioms given allow for interpretations in finite domains. As an example, let D be a set of arbitrary finite cardinality; assign to 'U' the entire set D, to 'M' the Cartesian product $D \times D$, to 'S' and to 'P' the null set.

The inadequacy of the reformulation appears as soon as we try to prove some of the usual theorems of elementary (formal) arithmetic. Standard proofs of such results as the commutativity of addition require us to assume the existence of numbers, which we can establish by using the fact that successor and addition are functions.[16] (Thus we can conclude that each number has a unique successor and that there is a unique number which is the sum of two other numbers.) When we try to mimic these proofs in Mill Arithmetic, as I have so far presented it, our attempted derivations break down because we have no right to the analogous existence assumptions. We can see, *in general,* how the proof would go, and the transcription of standard derivations into Mill Arithmetic would be easy *if* we could provide for the existence of operations corresponding to the commitments introduced in the standard Peano postulates.

We can easily remedy the situation by adding principles which ensure the existence of enough operations. We have only to add the following:

(13) $(\exists x)Ux$

(14) $(x)(\exists y)Syx$

(15) $(x)(y)(\exists z)Azxy.$

These principles are obvious analogs of the commitments involved in taking '1' as a constant and the symbols for successor and addition as function symbols. By adding them, we can replicate in Mill Arithmetic the standard development of elementary arithmatic. The usual proofs of theorems of formal arithmetic can be transformed, by a routine procedure, into proofs of theorems of Mill Arithmetic.

16. For such standard proofs see chapter 3 of E. Mendelson, *Introduction to Mathematical Logic.* A brief explanation of the problem in recasting them in Mill Arithmetic is given in the appendix to "Arithmetic for the Millian."

Yet, if the addition of (13) through (15) makes Mill Arithmetic adequate, it may also appear to render our recasting of standard arithmetic entirely pointless. We are driven to include the existence assumptions because we need enough substitutes for the numbers of standard arithmetic. Once we include them, we seem compelled to travel the road to Platonism which I described at the beginning of this chapter. We require infinitely many objects as values of our variables; there are not sufficiently many physical objects to go round; hence we shall need, it seems, to posit abstract objects. Or, to look at the issue from another angle, in exposing the need for existence assumptions, we appear to have uncovered the Achilles heel of the suggested treatment. We have used the crude physical paradigm of operations of spatial segregation as a means of explaining the primitive predicates and we have suggested that the occasions on which we perform such operations might serve as the basis for our arithmetical knowledge. The axioms (1) through (12) accord perfectly with this scheme of interpretation. Yet when we try to construe the existence assumptions (13) through (15) by taking the predicates to apply to physical operations we run into trouble. For when the predicates are read in this way, (14), for example, is straightforwardly false. Given that the existence of an operation consists in its performance, it is just not true that for any physical operation of segregation there is a successor operation.

Mill Arithmetic thus appears to founder on the same difficulty which besets traditional nominalist proposals. For example, if one regards arithmetic as having an ontology of stroke-symbols, construed as concrete physical inscriptions, there is the familiar problem that there are not enough physical inscriptions to do duty for the natural numbers. My response is that this objection is not fatal to traditional nominalism, and is not fatal to the account of mathematical reality I favor. I shall try to counter the objection shortly, and if my rebuttal is correct then an exactly parallel defense will be available to the traditional nominalist. However, before I tackle this issue, it is worth noting my reasons for departing from traditional nominalism. The Millian approach I favor has the advantage over standard nominalist programs that it offers a direct explanation of the applicability of arithmetic. Arithmetical truths are useful because they describe operations which we can perform on any objects. The classical nominalist, by contrast, faces the puzzle of why studying the properties of physical inscriptions should be of interest and of service to us in coping with nature (or, alternatively, the puzzle of why the features of extralinguistic entities are reflected in the properties of signs). Hence I reject nominalism for one of the same reasons that I reject Platonism: in neither case do we receive an adequate account of the usefulness of arithmetic.

Suppose we concede that the existence assumptions (14) and (15), when interpreted by means of our crude physical paradigm are false. This means that, when we construe the primitive predicates of Mill Arithmetic according to the crude physical paradigm, Mill Arithmetic will either contain false statements or else it will only provide an attenuated version of standard arithmetic. Trivially, knowledge requires truth. Hence, if we are to know arithmetical principles, then either it will be necessary to give up the crude physical paradigm for understanding the primitive notions of Mill Arithmetic or our knowledge will only cover a fragment of what has traditionally been taken as its domain. This conclusion is forced on us, and I shall not try to oppose it. Instead I shall suggest that it is by no means so disastrous as it seems. I consider two interpretations of arithmetical statements. The first cleaves to the crude physical paradigm, construes arithmetic straightforwardly as a science of physical operations of segregation, and embraces the conclusion that our arithmetical knowledge is considerably less rich than we had thought. The second substitutes for the crude physical paradigm a more sophisticated scheme of interpretation. To understand how this can be achieved, while retaining the epistemological advantages of the first interpretation, we need to develop some of the ideas introduced in the last section.

Arithmetic, I said, is an *idealizing theory*. The relation between arithmetical statements and our actual operations is like that between the laws of ideal gases and actual gases.[17] Let us fix our ideas by considering the latter example in more detail. When we write out the Boyle-Charles Law in full—making the quantifier *implicit* in the usual equation "$PV = RT$" explicit—we obtain a statement of the form:

(16) $(x)(Gx \rightarrow P(x) \cdot V(x) = R \cdot T(x))$.

Is (16) true? The answer depends on how we interpret the predicate 'G.' If we take the extension of 'G' to comprise the actual samples of actual gases then (16) is false: the pressures, volumes, and temperatures of actual gases are not related in the way (16) takes them to be. Alternatively, we can read 'G' as applying to *ideal* gases, viewing (16) as true in virtue of the *definition* of an ideal gas. We might regard (16) itself as part of an implicit specification of the properties which ideal gases must possess, or we might take the definition of an ideal gas to be furnished by higher level principles (such as those of the kinetic theory) from which (16) can be derived. (I shall exploit this analogy later when we consider the relation between arithmetic and set theory.) In Chapter 4, I have argued that stipulation does not automatically provide us with a priori knowledge. The account offered there allows me to claim that *appropriately grounded* stipulations *can* engender knowledge, knowledge which is dependent on the "appropriate grounding." In the present case, our knowledge that ideal gases satisfy (16) is not entirely explained by pointing out that (16) is part of the implicit definition of 'ideal gas' or that we can deduce (16) from a definition of 'ideal gas' in kinetic-theoretic terms. To complete the explana-

---

17. For a clear treatment of the theory of ideal gases and its relation to the kinetic theory of gases, see P. Morse, *Thermal Physics*.

tion, we must explain what makes these stipulations reasonable. Why are we entitled to claim that (16) is an item of our knowledge on the grounds that the reference of the predicate 'ideal gas' is fixed so as to ensure that (16) will be true, while we would not be entitled to claim that (17)

(17) $(x)(Hx \rightarrow [P(x) \cdot T(x)]^{\frac{3}{4}} = R \log T)$

is an item of our knowledge on the grounds that the reference of the predicate 'horrible gas' is (hereby) fixed so as to ensure the truth of (17)? The answer is not hard to find. The stipulation of the characteristics of an ideal gas is warranted by the experience we have of the properties of actual gases. We find that actual gases approximately satisfy the condition set down in the Boyle-Charles Law, and, when we develop kinetic theory, we understand why that is so. The kinetic theory of gases tells us that if the molecules of gases were of negligible size and if there were no intermolecular forces then gases would obey the Boyle-Charles Law. This discovery prompts us to abstract from certain features of the actual situation, introducing the notion of an ideal gas to describe how actual gases would behave if complicating factors were removed. The fruits of this stipulation are simple and elegant laws, which, in many contexts, we can treat as applicable to actual samples of actual gases.

I want to take the same view of Mill Arithmetic. *Although Mill Arithmetic cannot accurately be applied to the description of the physical operations of segregation, spatial rearrangement, and so forth, that is not fatal to the applicability of Mill Arithmetic.* We can conceive of the principles of Mill Arithmetic as implicit definitions of an *ideal agent*. An ideal agent is a being whose physical operations of segregation do satisfy the principles (1) through (15). (As I shall suggest in the next section, there is a parallel to the kinetic-theoretic specification of an ideal gas in the set-theoretic specification of the collective activity of the ideal agent.) Just as we abstract from some of the accidental and complicating properties of actual gases to frame the notion of an ideal gas, so too we specify the capacities of the ideal agent by abstracting from the incidental limitations on our own collective practice.[18]

At this point, let me face directly the question of how we obtain perceptually based mathematical knowledge. We observe ourselves and others performing particular operations of collection, correlation, and so forth, and thereby come to know that such operations exist. This provides us with rudimentary knowledge—proto-mathematical knowledge, if you like. I take it that our knowledge of the existence of these operations is epistemologically unproblematic: we per-

18. At this point it should be clear why, on my account of arithmetic, there is no commitment to the existence of an ideal agent or to ideal operations. Statements of arithmetic, like statements of ideal gas theory, turn out to be vacuously true. They are distinguished from the host of thoroughly uninteresting and pointless vacuously true statements (like (17)) by the fact that the stipulations on ideal gases abstract from accidental complications of actual gases and the fact that the stipulations on the ideal agent abstract from accidental limitations of human agents.

ceive what our fellows do in the same way that we perceive other types of events, and, if we have a different perspective on our own actions, then that is an epistemic advantage rather than a source of obstacles to knowledge.[19] The main task is to understand how this proto-mathematical knowledge provides the basis for arithmetical knowledge.

Here we face two possibilities, corresponding to the two interpretations I distinguished several paragraphs back. The first is to contend that, strictly speaking, our arithmetical knowledge is limited to the collective proto-mathematical knowledge of the human race. What we know is restricted to a jumble of propositions about the collective operations which have actually been performed.[20] This construal of our arithmetical knowledge is hardly attractive. It is comparable to the thesis that, strictly speaking, our knowledge of the relations among pressure, volume, and temperature of gases is restricted to the propositions which record exactly the observed results. More satisfactory is the second interpretation of arithmetical statements and the account of arithmetical knowledge which it engenders. We envisage an ideal agent and stipulate that the powers of this agent are to be characterized by the principles (1) through (15). This stipulation is warranted by our recognition that the ideal operations which we attribute to the ideal agent abstract from the accidental limitations of our own performances. We do not perform successor operations for all the collective operations we perform. But, given any one of our collective operations, we regard ourselves as able to perform a successor to it. This conclusion is based on simple inductive generalization from our past practice. We find that, when we have tried to perform a successor operation, we have always succeeded. From this finding, we project that we could have succeeded in cases where, for one reason or another, we did not try. Hence we come to view a principle like (14) as descriptive of our collective practice, as it would be if accidental limitations on it were removed. Once we have adopted this perspective, we are justified in introducing the stipulations I take to constitute Mill Arithmetic, on the grounds that these stipulations idealize our actual collective activity.

Alternatively, we can approach the question of how to theorize about our collective practice by asking how best to systematize the proto-mathematical knowledge we glean directly from experience. The principles (1) through (12) could be supported as generalizations of features we have discerned in our experiences of segregating and matching objects. Plainly, there is no problem

19. If ecological realism is correct, then perceptual knowledge of basic arithmetical truths does not even require us to engage in the operations. Independently of our performance of collective operations, we would recognize that our environment affords them. For details, see Michaels and Carello, *Direct Perception*, chapters 2–4, and, in particular, chapter 3.

20. Occasionally, constructivists offer remarks which, if read at face value, would commit them to the view that our mathematical knowledge is restricted to this kind of jumble. See, for example, A. Heyting, *Intuitionism*, pp. 8–12.

in justifying (13), for it is easy for us to observe a one-operation. The trouble will come when we try to support other existential assumptions, parallel to (14) and (15). We shall want to add *some* such assumptions, for some successor operations and some additions are performed and any theory which ignores this fact will give an inadequate picture of our collective activity. But what form should these assumptions take? Let us imagine, first, that we define, successively, the predicates 'x is a 2-operation,' 'x is a 3-operation,' and so forth, in the obvious way. This would enable us to idealize our practice slightly differently, assuming the existence of successor operations of "small number operations" but not of all number operations. So we might add a finite number of axioms of the form

$$(14') \ (x) \ (x \text{ is an } n\text{-operation} \rightarrow (\exists y)Syx).$$

But now, we must ask, why should we stop adding such axioms at any particular point? What distinguishes the choice of n such that we allow for successors of n-operations but not for successors of those successors? Perhaps it will be maintained that there are some number operations which humans could not perform. It is certainly true that humans, given our psycho-physical constitution and life span, cannot perform very large collective operations. Yet we would be reluctant to admit that there is some definite point at which the ability to perform the successor operation disappears. When we envisage the collective activity of a human agent who is bent on iterated performance of successor operations, we recognize that the iteration must eventually stop—at the agent's death, if not before. Despite this, we conceive of the performance as one that could easily have been continued. It was only an accident that it stopped when it did. Thus any limit to the addition of axioms of form (14') appears arbitrary, and we accept (14) as part of the best idealization of our collective practice.

I have described the basis of our knowledge of Mill Arithmetic somewhat fancifully, as if the route just traced were our way to arithmetical knowledge. However, as I have suggested in Chapter 5, the arithmetical knowledge *we* possess is the product of our early educational experiences. My problem here has been to provide an account of arithmetical reality which would allow for perceptual knowledge of arithmetical truths. This problem does not arise because I regard our knowledge of arithmetic as perceptually based. Instead, I take current mathematical knowledge in general, and current arithmetical knowledge in particular, to be explained by the transmission of that knowledge from contemporary society to the contemporary individual, and by the transmission of knowledge from one society to its successor. Since I claim that the knowledge of the mathematical tradition is grounded in the experiences of those who initiated the tradition, what I have offered can best be regarded as an attempt to explain how the arithmetical knowledge of our remote ancestors might have been obtained.

In presenting my answers to the question of how arithmetical knowledge can be based on elementary experiences, I have used the analogy with the theory of ideal gases to explain my thesis that arithmetic is an idealizing theory. The advantage of this analogy is that it has enabled me to stay neutral on the question of how to analyze the notion of an idealizing theory. Let me conclude this section by outlining an analysis which I am inclined to favor. Since I am not convinced that this analysis is correct, I have not built it in to my account of arithmetical knowledge. Hence deficiencies in the analysis should not impugn the picture of arithmetical knowledge painted above.

One can naturally think of an idealizing theory as describing a close possible world, one which is similar to the actual world but has the advantage of lending itself to simpler description.[21] We accept idealizing theories when we believe that these theories describe the world not as it is, but as it would be if accidental or complicating features were removed. Thus we can conceive of idealization as a process in which we abandon the attempt to describe our world exactly in favor of describing a close possible world that lends itself to much simpler description. (Interestingly, some scientific cases of idealization, such as the theory of gases, seem to involve several levels of idealization: by moving from the actual world to a different possible world, we obtain the van der Waals's idealization; by moving further from actuality, the treatment of ideal gases provides further simplicity at the cost of accuracy.) Although this conception of idealization has some advantages, it can easily cause epistemological worries. When we begin to think of possible worlds as planets in space, some of which are relatively close to the actual world (the home planet) and others of which are remote, then we may wonder how experience which is tied to one planet can give us knowledge of others. Aficionados of possible worlds naturally protest that the picture misleads us and that epistemological anxieties are unfounded.[22] However, unless we are given an account of how to construe the talk of possible worlds which makes it clear to us how to achieve the benefits in clarity that the possible worlds semantics brings to complex modal statements while at the same time explaining how we have knowledge of modal statements, there is room to wonder whether the possible worlds interpretation of idealization will not engender difficult epistemological problems. Hence, since my aim is to *resolve* epistemological puzzles about mathematics, I have not used the possible worlds approach to articulate the thesis that Mill Arithmetic is an idealizing theory.

Nevertheless, there are some people who would be happy to concede that modal statements, construed as on the possible worlds picture, are epistemologically unproblematic.[23] To make this concession is to maintain that there is an

21. One type of difficulty for the possible worlds approach to idealizations, which I shall not consider here, concerns the problem of integrating various types of idealizations that are often made. See, for example, Dudley Shapere, "Notes Towards a Post-Positivistic Interpretation of Science," and Geoffrey Joseph, "The Many Sciences and the One World."

22. See Kripke, *Naming and Necessity*, pp. 43ff., and Stalnaker, "Possible Worlds."

23. Putnam appears to take this position. See "What is Mathematical Truth?" p. 71.

adequate way to interpret the possible worlds picture which will allow for the possibility that we can obtain modal knowledge in the ways we standardly take ourselves to do so. Sometimes it seems that we know that particular things are possible by simple generalization from our perceptual experience. Proponents of possible worlds semantics may insist that our understanding of what a possible world is must conform to this elementary epistemological observation: however we conceive of possible worlds, they must be regarded as entities to which we have access *via* perception and inductive generalization. For someone who is content to leave the epistemological issues there, there is a straightforward way in which to develop the account of arithmetic which I have proposed. Regard the principles (1) through (12) as (partial) implicit definitions of the primitive notions of Mill Arithmetic. These principles (perhaps together with others) pin down the concepts of one-operation, successor operation, addition, and matchability. Once the concepts have been fixed we can consider worlds in which there exist a stock of operations which fall under them, a stock which is rich enough to satisfy the principles (13) through (15). Call such worlds *M-worlds*. Let the operations which fall under the concepts of one-operation, successor operation, and addition in an M-world be the *arithmetical operations* of that M-world. Mill Arithmetic is applicable to our world not because our world is an M-world whose arithmetical operations are our physical operations of segregation, but because, if accidental, complicating factors were removed, our world *would be* such a world.

If we take (1) through (12) to provide (partial) definitions of the notions of the various arithmetical operations and (13) through (15) to characterize the notion of an M-world, then the usual theorems of arithmetic can be reinterpreted as sentences which are implicitly relativized to the notion of an M-world. The analogs of statements of ordinary arithmetic will be sentences describing the properties of operations in M-worlds. ('$2 + 2 = 4$' will be translated as 'In any M-world if x is a 2-operation and y is a 2-operation and z is an addition on x and y then z is a 4-operation.') These sentences will be logical consequences of the definitions of the terms they contain. Hence arithmetic will explore the consequences of a system of definitions. We are justified in accepting those definitions because we can regard operations which we perform (for example, operations of physical segregation as well as less crude operations) as if they conformed to the notions of the arithmetical operations as they are defined. Having made this idealization, we can regard our world as an M-world.

Let me make one final point about this development of my proposal. Other writers have suggested a modal approach to mathematical statements. There are several ways to try to work this out. For example, one can rewrite mathematical statements so as to introduce modal operators into their surface form. So, for example, the axiom which seems to assert that every number has a successor would be recast as "$(x)\Diamond(\exists y)Sxy$" or as "$\Diamond(x)(\exists y)Sxy$." Alternatively, and more simply, one can continue to read arithmetical statements at face value, construe the variables as ranging over concrete objects (inscription tokens, let us say), and reinterpret the notion of arithmetical "truth" as "truth in some possible world."[24] Finally, it is possible to achieve the same effect by giving a modal reinterpretation of the quantifiers: '$(\exists x) \ldots x \ldots$' is read as "there is a possible object x such that $\ldots x \ldots$."[25] My proposal is distinguished from these approaches by its deliberate linkage of arithmetical truth to the applicability of arithmetic. On my account, the stipulations of arithmetical notions ensure that the principles of Mill Arithmetic are true of any M-world. But to advance from this point to the claim that in learning Mill Arithmetic we acquire knowledge, we must understand why the study of M-worlds is important to us. In my account, the gap is bridged by my thesis that an M-world is a world from which the accidental complications of the actual world are removed.

IV

I hope to give an account of mathematical reality which will enable us to see how perception can lead us to elementary mathematical knowledge. I have attempted to present some doubts about the Platonist picture of mathematical reality, and to articulate a rival picture for one part of mathematics—arithmetic—which initially appears to require a Platonistic approach. But it would still be reasonable to wonder if the task I have set myself can be completed. Will it be possible to provide a similar account of the content of other mathematical statements? And will the account allow for us to have perceptual knowledge of those mathematical truths which I take to be elementary? In this section I shall try to address these concerns.

What kinds of knowledge are relevant? I take the basic parts of mathematical knowledge to comprise arithmetic and geometry. The account of arithmetic provided in the last section shows how perception can yield proto-mathematical knowledge. I still face the task of providing a parallel account of our geometrical knowledge, including an account of that knowledge of real numbers which is embedded in ancient geometry. In addition, I need to sketch a picture of that reality whose nature is revealed in the later endeavors of mathematicians, as they build upon the knowledge they have inherited.

It is not difficult to understand the perceptual basis of parts of our geometrical knowledge. We can easily conceive how our ancestors could have introduced the concept of congruent figures, the concepts of particular shapes, and how their observations of ordinary physical objects could have acquainted them with some of the propositions which Euclid systematized. The notions of congruence as sameness of size and shape, sameness of length, sameness of area, and so forth, can be introduced by reference to the operations we can perform

24. For this type of approach—and a clear treatment of the issue of the ontological commitments it brings—see Charles Parsons, "Ontology and Mathematics," section 3.

25. This is the line favored by Tharp in "Myth and Math."

on different figures. Crudely, two pieces of a surface have the same area if, by performing certain operations on one, you can fit it exactly into the other. (The allowable operations would include cutting into parts and changing the relative positions of the parts.) [26] Names of kinds of figures are introduced to formulate generalizations about congruence relations. From the perspective of Euclid's axiomatization of geometry, we can view the situation as follows. The first principles of geometry can be formulated as a set of definitions (the definitions of the names of the various types of figures) supplemented with a set of existence assumptions (assertions that circles exist with any center and radius etc.). The existence assumptions, in turn, can be regarded as characterizing the notion of a Euclidean space. We may then reformulate the standard theorems of geometry as statements which are implicitly relativized to the notion of a Euclidean space. The relativization can be made explicit: so, for example, we can rewrite the usual claim that the sum of the angles of any triangle make two right angles as "In any Euclidean space, the sum of the angles of any triangle is two right angles." When they are rewritten in this way, the theorems of geometry are logical consequences of the definitions of the terms they contain. We are justified in accepting those definitions because we can regard objects in our world as conforming to the definitions of the names of geometrical figures (we can regard rigid rods as covering straight paths in space, coins as covering circular regions, and so forth), and, having made this idealization, we can regard our world as if it were a Euclidean space. The stipulations made in geometry are warranted by their systematization of the manipulations of objects which we actually perform.

Let us look more closely at the particular theorem cited in the last paragraph. What we *observe* is an episode in which an actual object, whose rough shape is

is disassembled so as to yield the configuration

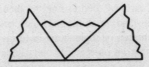

26. But they would not be exhausted by such operations. I conjecture that the concept of congruence was originally introduced by reference to paradigms: someone declared that figures are congruent in area if you can cut up one and reassemble it to match the other *or if you can do something of that kind.* Obviously, coming to identify the kind of operation involved here took highly sophisticated mathematical developments. For discussion of the introduction of mathematical terms by reference to paradigms, see Chapter 7, Section V.

We systematize this observation, and kindred observations, by introducing the notions of Euclidean geometry. Our geometrical statements can finally be understood as describing the performances of an ideal agent on ideal objects in an ideal space. However, these performances are sufficiently similar to what we do with actual objects in actual space to justify us in the attempt to unfold the character of those performances, an attempt which constitutes Euclidean geometry.

As I mentioned above, I believe that *part* of our geometrical knowledge allows of this relatively straightforward treatment. Many of the statements of elementary geometry can easily be interpreted by taking them to be part of an idealization which systematizes facts about ordinary physical objects which are accessible to perception. Trouble begins to threaten when we turn to that part of ancient geometry in which statements about real numbers are formulated. Since the importance of this area of mathematics in the growth of mathematical knowledge can hardly be denied, it is incumbent on me to provide an explanation of statements which are ostensibly about real numbers, an explanation which will allow for the genesis of our knowledge of those statements. To focus the issue, let us consider the explanation which a Platonist who favored my general approach to mathematical knowledge might propose. Conceiving of real numbers as abstract objects, the Platonist would regard the ancients as discovering, perceptually, truths about real numbers, arriving at the disturbing result that there are real numbers which are irrational, and continuing to develop the theory of real numbers without a clear arithmetical description of what they were talking about. Only after two millennia of referring to these objects did mathematicians arrive at a position in which they could adequately identify them. Now I want to preserve some features of this account. I believe that the theory of real numbers, like so many areas of research in the natural sciences, began from an inadequate characterization of the reality it attempted to describe and only achieved a clear specification of its subject matter at a relatively late stage in its development. What I want to deny is the Platonistic thesis that the subject matter of real number arithmetic is to be understood by taking it to be about certain abstract objects, to wit, the real numbers.

The real numbers stand to measurement as the natural numbers stand to counting. Hence I shall attempt to link the theory of real numbers to the practice of measuring just as earlier I tried to link the theory of natural numbers to the activity of segregating and combining. Let us begin by noting that *some* operations of measurement can be understood in terms of elementary arithmetical operations. To say that an object is n units long is to say that a certain kind of decomposition can be performed on it: we can split it into segments which match the unit and when we collect these segments we perform an n-collection. To say that an object is $\frac{m}{n}$ units long is to say that it can be decomposed into matchable parts such that (i) an assembly of an n-collecting on these parts matches the unit and (ii) an assembly of an m-collecting on these parts

125

matches the object.[27] The genesis of the notion of measuring thus lies in the ordinary arithmetical notions of which Mill Arithmetic treats. We can also define, in these terms, the notion of *commensurability:* two objects x, y are commensurable if and only if there is an object z and number collectings on objects congruent to z such that x is congruent with the assembly of the objects collected in one number collecting and y is congruent with the assembly of the objects collected in the other number collecting.

Unfortunately, as the ancient Greeks discovered, not all pairs of objects are commensurable. Now one can think of the operation of decomposing an object into units as exposing a relation between the object and the unit. More generally, one can think of those kinds of measurement discussed in the last paragraph as cases in which the performance of the collective operations I have indicated makes manifest a relationship between the objects. Alternatively, we may think of the relationship between the objects as a disposition to admit those kinds of collectings. What now of those objects which are incommensurable? Here, I suggest, the Greeks attempted to generalize the notion of a measurement operation to those cases in which it could not be understood in terms of collecting and assembly. They introduced a language for describing the relations between objects—articulated in a restricted domain by the restricted operation of measurement as collecting—and proposed that this language should apply even in those cases in which no concrete construal in terms of collecting and assembly is possible. What goes on here is parallel to what scientists frequently do. Language which is well understood is extended beyond its limits of antecedently available specification because the extended language can be used to help answer questions which arise in the limited language.[28] Expressions of the form "AB measures r with respect to CD" could be unpacked in cases of rational commensurability by employing the strategy described above. The relationship between commensurable segments can be articulated in terms of operations of collecting, assembling, and matching. Greek geometry provides rules for deploying expressions of the form "AB measures r with respect to CD" even when AB and CD are incommensurable, gaining thereby a host of results which can be corroborated through the performance of operations of collecting, assembling, and matching. The classical problem of the continuum is to provide an interpretation of these useful statements.

On the natural Platonist account, the Greek geometers were referring to real numbers even though an adequate specification of the reals had to await the work of Weierstrass, Dedekind, and Cantor at the end of the nineteenth century. Thanks to the profound investigations of these three men, we are now in a position to explain what the Greeks were talking about. For the Platonist, the referents of the expressions which the Greeks introduced are sets: thus, follow-

27. Strictly speaking, what we rely on here are elementary arithmetical operations together with geometrical operations (assembly, covering) which are readily interpretable.

28. For exploration of this theme, see Chapter 9, Section II.

ing Dedekind (or, more exactly, a contemporary Dedekindian approach), one might identify the referent of '$\sqrt{2}$' as {r: r is rational and $r^2 < 2$}. Now there is much of this explanation with which I want to agree. Dedekind *et al.* did solve the classical problem of the continuum by enabling us to specify the referents of expressions which had been in use since antiquity. What I want to reject is the thesis that the discourse we have inherited from the Greeks employs expressions which refer to abstract objects. Hence my task is to indicate how we can interpret the set-theoretic language in which the classical problem of the continuum is solved without committing ourselves to the existence of abstract objects. The Platonist is right to insist that Dedekind *et al.* told us what our ancestors had been talking about, but may be wrong in thinking that the Platonistic interpretation of Dedekind's language is forced upon us.

Our reformulation of set theory will be similar to our reformulation of arithmetic. Language in which the variables appear to range over abstract objects (sets) will give way to language in which we discuss the properties of mathematical operations.[29] The primitive notions which will be used are those of an operation's *being a collecting on* certain objects, an operation's being an *ordering of a pair of objects*, and of the *equivalence* of operations. In terms of our crude physical paradigm we can easily explain these notions. Operation x is a collecting on object y if y is one of the objects segregated in performing x. Operation x is a pairwise ordering of y and z if one performs x by giving y some priority over z, for example, by touching y and z in sequence or drawing an arrow from y to z. As I shall explain below, I take pairwise ordering to be the root notion in correlating. The notion of equivalence, like the notion of matchability in our treatment of arithmetic, is used to group together operations which we do not want to distinguish within our theory. Different individuals can collect (order) the same objects, and one individual can collect (order) the same objects on different occasions, but the differences in the operations of collecting are not significant. So we treat such operations as equivalent.

I think it is relatively easy to see how the reformulation of set theory will go. We take 'Cxy,' '$Pxyz$,' and 'Exy' as primitive predicates ('x is a collecting on y,' 'x is a pairwise ordering of y and z,' 'x and y are equivalent'). Our intention to treat equivalent operations as indistinguishable will be captured by the principles:

(18) $(x)xEx$
(19) $(x)(y)(xEy \rightarrow (\Phi x \rightarrow \Phi y))$
      for all open sentences 'Φx' of the language.

---

29. However, as in the case of Mill Arithmetic, I do not suppose the existence of an ideal subject who performs these ideal operations. Following the approach of Section III, the axioms laid down are construed as stipulating the conditions on an ideal subject and our theory unfolds the consequences of these conditions.

The unrestricted comprehension principle (for sets which are not relations) becomes:

(20) $(\exists x)(y)(\Phi \leftrightarrow Cxy)$

for any open sentence 'Φ' of the language with 'y' as its sole free variable.

We also need a principle corresponding to the standard assertion of the existence of ordered pairs: [30]

(21) $(y)(z)(\exists x)Pxyz$.

Finally, the axiom of extensionality becomes:

(22a) $(x)(y)((z)(Cxz \leftrightarrow Cyz) \rightarrow xEy)$
(22b) $(x)(y)(z)(w)(u)(v)((Pxzw \,\&\, Pyuv \,\&\, zEu \,\&\, wEv) \rightarrow xEy)$.

Plainly, the scheme of translation employed here consists in replacing occurrences of '$=$' with occurrences of 'E,' occurrences of '$x\epsilon y$' with 'Cyx.' Standard statements about the ordered pair of two objects will go over into universally quantified statements about pairwise orderings of those objects.

Is the theory I have just sketched inconsistent in the way that ordinary naïve set theory is? Well, nothing I have said debars the use of expressions of the form 'Cyy.' Hence we can obtain from (20):

(23) $(\exists x)(y)(\sim Cyy \leftrightarrow Cxy)$

and, from (23), we can derive the usual Russellian contradiction. Yet, once the paradox has been exposed within our reformulated set theory, the solution which motivates the theories of both Russell and Zermelo immediately suggests itself. Sentences of the form 'Caa' appear pathological because we are troubled to make sense of a collective operation which collects itself. Yet this observation provokes a more fundamental worry. In presenting (18) through (22), I have ignored the question of whether a single universe of discourse can serve for the variables. Someone may reasonably deny that the variables can be taken to range over one domain, on the grounds that the relation C relates two different types of entities. It makes no sense to assert 'Cab' if a is an object which is not an operation or if b is an operation. Now the first disjunct of the complaint need not concern us much. Little harm would be done to our theory by declaring that no sequence which assigns an object to the variable 'x' satisfies 'Cxy.' The second disjunct is more problematic. To debar the possibility of satisfying 'Cxy' by assigning an operation to 'y' would restrict our embryonic set theory

30. Of course, in the usual versions of set theory, ordered pairs are identified with sets. Part of my program is to treat the notions of ordering and collecting separately, so that my transcribed set theory is analogous to that of the Bourbaki (see *Elements of Mathematics: Theory of Sets*) in which the existence of ordered pairs is asserted separately. Another consequence of my treatment is the need for an explicit principle of extensionality for ordered pairs, (22b).

to an analog of "first-rank set theory": we would reconstruct only that fragment of the subject which deals with individuals and sets containing them. The
challenge is to provide an interpretation of the open sentence 'Cxy' which will
allow us to regard it as satisfiable by sequences which assign an operation
to 'y.'

What is troublesome here is the thought that we cannot make sense of higher-
order collecting unless we envisage ourselves as collecting *objects* which have
been brought into being by prior acts of collecting. But why should this be? Is
it not possible for us to conceive of higher-order collectings as operations which
operate on the previous operations themselves? I suggest that we do not need
any intermediate entities—products of collecting—to make sense of the notion
of iterated collecting.[31] For a sequence which assigns an operation to 'y' to
satisfy 'Cxy' it is necessary and sufficient that the sequence should assign to
'x' an operation which collects the operation assigned to 'y.' But what is it like
for one operation to collect another? Or, to put the question more exactly,
under what conditions is it correct to say that a collective operation has been
performed on a prior operation?

Once again, let us recall the way in which we explain the notion of a set to
someone who is not already familiar with it. We begin with operations in which
physical objects are segregated. At this point, many students who are satisfied
that they understand the idea of collecting objects find the notion of a set whose
members are sets much more mysterious. To respond to their concerns, we find
ourselves drawing pictures. We represent the activity of collecting objects by
enclosing the objects in circles, and we depict many such collectings:

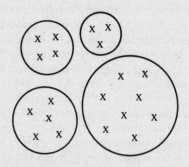

31. Notice also that we shall need to make sense of iterating pairwise ordering to make sense of
the general notion of a finite ordering (the analog of the finite sequence $<a_1, \ldots , a_n>$, typically
defined, of course, as $<<a_1, \ldots , a_{n-1}>, a_n>$). Furthermore, we shall require the idea of collectings on orderings to generate the analog of relations. In the text I shall simply discuss the notion
of iterated collecting, but what I say will apply, *mutatis mutandis,* to these other cases of operations
on operations.

Then we suggest that the neophyte forget about the "interiors of our circles," and form the representation:

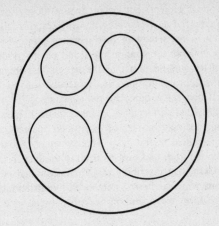

This, we suggest, is a way to represent sets whose members are sets. I propose that we can use this crude heuristic device to understand iterated collective activity. In general, to collect some objects is to represent them together. For the beginner, an appropriate form of representation is to introduce symbols standing for the objects and to draw a closed curve around them. Higher-level collecting uses the representations generated in prior collectings as the starting point for further collecting. Thus the closed curves ("circles") already introduced are seen as representations of the products of prior collective activity, representations which can themselves be enclosed in order to generate a higher-level representation. The trained mathematician is able to forego such crude types of representation (although, as anyone who has ever struggled with certain parts of linear algebra, topology, and analysis will appreciate, closed curves linked by arrows are often a valuable aid to mathematical understanding), but it nonetheless remains true that the activity of collecting involves achieving a representation and that, in higher-level collecting, prior representations are themselves elements out of which the further representation is fashioned.

Let me note a possible misinterpretation. I am not suggesting that higher-order collective activity consists in collecting *symbols*. As I have explicitly claimed, we collect prior operations. But in performing these operations we use symbols. *To collect is to achieve a certain type of representation, and, when we perform higher-order collectings, representations achieved in previous collecting may be used as materials out of which a new representation is generated.*

It is worth pursuing this point further, for, if I am correct, mathematical language plays a dual role. Not only do the sentences which occur in mathematics books describe ideal mathematical operations (more exactly, the ideal operations of an ideal subject) but, in producing those sentences, the mathematician may be engaged in *performing* those operations. In inscribing the token '$\{a, \{b\}\}$' I may be achieving that representation which constitutes the performance of a collective operation on a and on the prior operation which collects just b. (The qualification that the act of inscribing *may* count as a performance of the appropriate operation is needed for the simple reason that one may carry out that act without the right intentions, beliefs, attention, and so forth. My point is that performance of mathematical operations may involve acts of inscription, not that the acts of inscription automatically count as performing the mathematical operations.) Thus a sentence-token inscribed by a mathematician may be true, and owe its truth to the fact that some mathematical operation performed in inscribing that sentence-token belongs to a family of actual operations which is best systematized by attributing to the ideal subject a power to perform the operation described (or asserted to exist) by the sentence-token.

This double functioning of mathematical language—its use as a vehicle for the performance of mathematical operations as well as its reporting on those operations—has figured in some constructivist discussions of mathematics.[32] It can, I think, be used to account for the emphasis which mathematicians frequently place on finding *perspicuous* notation: to have a perspicuous notation rather than an unperspicuous one may make no difference to the reporting function of one's mathematical language, but it can be highly significant in facilitating the performance of those operations for which the language is a vehicle. Moreover, the account I have offered enables us to sharpen a criticism which constructivists and nominalists have often directed against Platonism. This criticism begins by noting that the mathematician engaged in problem-solving or theorem-proving does not seem to engage in any activity which is plausibly construed as "observing" or "investigating" abstract objects. If a Platonistic account of mathematical reality were to be correct, then we might wonder at this phenomenon. When mathematicians get stuck why do they not renew their acquaintance with the objects which, on the Platonist account, they are attempting to characterize? Why, instead, do they so often engage in symbolic manipulation? And why, when the problem is finally cracked, does the insight so often seem to depend on recognizing analogies which are present in the notation? I do not pretend to have complete answers to these questions. Such answers would require more study of the psychology of mathematical research than has yet been undertaken. However, I do think that the point about the

32. I think that it is present, for example, in Kant's discussion of algebra in the *Transcendental Methodology*. See *Critique of Pure Reason* A717, B745.

double functioning of mathematical notation indicates that the mathematician's problem-solving behavior is less mysterious than Platonists would make it appear. To solve a problem is to discover a truth about mathematical operations, and to fiddle with the notation or to discern analogies in it is, on my account, to engage in those mathematical operations which one is attempting to characterize.

Let us now return from the speculations of the last paragraph to the central issue. The challenge which confronted us was to interpret the notion of collective operation in a way that would allow for the performance of higher-level collectings. I have tried to explain how we might construe the iterated collective activity of the mathematician. Given this construal the route to a reformulation of set theory—rather than some fragment of set theory—is open. At the same time, we inherit the problem posed by the paradoxes, for nothing in our treatment so far debars the move to (23) and the consequent contradiction. Thus the task confronting us is to use the scheme of translation outlined above to reformulate one of the set theories which have been developed in response to the paradoxes. In what follows, I shall continue to focus on the theory of collecting and temporarily ignore the extra complications imposed by pairwise ordering. This will simplify the discussion. After formulating an analog of standard set theory, it will not be difficult to introduce a counterpart to the theory of ordered pairs and relations.

Currently, one set theory, ZF, enjoys the favor of most mathematicians. There is good reason for this. ZF encapsulates a unifying idea, the idea of the "iterative conception of set." In a lucid article,[33] George Boolos has presented this conception in detail and has shown how the core axioms of ZF derive from it. Intuitively, the conception takes sets to be formed in stages: at the first stage, all sets of individuals are formed; at the second stage, all collections of individuals and sets formed at the first stage are formed; . . . ; at the $(n + 1)$th stage, all collections of individuals and sets formed at stages up to and including the nth stage are formed; . . . . So we regard the ZF hierarchy as generated from the iterated formation of sets out of "materials" which have previously been produced. (Of course, the iteration may be taken to proceed into the transfinite.) Boolos describes ZF from three perspectives. First he offers an informal presentation of the generation of sets in stages, similar to (but more detailed than) the characterization I have just indicated. Next he provides a formalization of this "stage theory," which spells out precisely the ideas given in the informal description. Finally, he demonstrates how the main axioms of ZF can be obtained from the formal stage theory. Now it is not important, for my purposes, to review the details of Boolos's conception. What is central to our concerns is to recognize the possibility of a translation which will reformulate *each* of Boolos's three descriptions of ZF in our preferred idiom. At

33. "The Iterative Conception of Set."

the first level, we replace the talk of the successive formation of collec*tions* with talk of the iterated collec*ting* of an ideal mathematical subject. Instead of saying that, at the $(n + 1)$th stage, all collections of individuals and sets formed at stages up to and including the nth stage are formed, we declare that, at the $(n + 1)$th stage of his collective activity, the ideal subject performs all collectings on individuals and the prior collectings performed up to and including the nth stage. When we proceed to the formal presentation of the stage theory, we simply use the correspondence between ordinary set theoretic notions and our preferred idiom which has already been outlined in the discussion of naïve set theory. Thus Boolos's axiom (VIII), which states that every member of a set is formed *before*, i.e. at an earlier stage than, the set,

    (24) $(x)(y)(s)(t)((y \epsilon x \ \& \ xFs \ \& \ yFt) \rightarrow tEs),$

becomes

    (25) $(x)(y)(s)(t)((Cxy \ \& \ xOs \ \& \ yOt) \rightarrow tEs).$

In (25), we replace Boolos's locution '*xFs*' ("*x* is formed at *s*") with the locution '*xOs*' ("*x* occurs at *s*"), to provide a natural transition from the language of collec*tions* to the language of collec*tings*. (Boolos's principles about the sequence of stages go over untransformed. The sequence of stages can be thought of as the collective life of the ideal subject.) Finally, the principles of set theory which are standardly taken as axioms—and which Boolos derives from the stage theory—obtain exactly the same type of reformulation as the axioms of naïve set theory. Thus, in place of the standard axiom of unions

    (26) $(z)(\exists y)(x)(x \epsilon y \leftrightarrow (\exists w)(x \epsilon w \ \& \ w \epsilon z))$

we have

    (27) $(z)(\exists y)(x)(Cyx \leftrightarrow (\exists w)(Cwx \ \& \ Czw)).$

Hence I claim that once we have learned how to interpret the notion of iterated çollecting, statements which the Platonist construes as asserting the existence of abstract objects, sets, can be recast as statements asserting the existence of operations performed by the ideal subject. Moreover, the translation will even preserve that motivating idea which singles out ZF as preferable to other set theories. We can reformulate, in our preferred idiom, the intuitive conception which Boolos takes to underlie ZF, and we can parallel Boolos's construction of ZF from it. Indeed, I think it is arguable that our reformulation expresses more clearly the "thought" which Boolos—correctly, I think—takes to lie behind ZF. Consider Boolos's own attempt to motivate the contention that there is something pathological about sets containing themselves as members (and the larger "circles" of form $x \epsilon y \ \& \ y \epsilon z \ \& \ . \ . \ . \ \& \ w \epsilon x$, which naïve set theory seems to let us envisage).

. . . when one is told that a set is a collection into a whole of definite elements of our thought, one thinks: Here are some things. Now we bind them up into a whole. *Now* we have a set. We don't suppose that what we come up with after combining some elements into a whole could have been one of the very things we combined (not, at least, if we are combining two or more elements).[34]

Note that this passage, with its emphasis on *our* activity, accords quite well with the language of operations which I have suggested. The agreement is even more striking, in a footnote which Boolos appends, in which he cites a figure of Kripke's to illustrate the notion of binding into a whole: we put a "lasso" around the objects. Yet, as Boolos later points out, the language which he uses in presenting his initial description of the iterative conception of set is at odds with a central theme of Platonism. "From the rough description it sounds as if sets were continually being created, which is not the case."[35] This disclaimer is somewhat embarrassing. If the language of "binding objects together" to "form" sets is misleading, then can we really use the passage I have quoted above to explain the pathological character of the idea that sets can be elements of themselves? On the approach I have recommended, no such tension is present. The "rough description" of the stage theory may be viewed as a literal account of the iterated constructive activity of the ideal mathematical subject.[36]

At this point, let me explain how the account offered by Boolos should be modified to yield the independent treatment of ordered pairs and relations which I favor. On Boolos's picture the hierarchy of sets arises from the iterated performance of one fundamental operation, that of collecting. I propose to ascribe to the ideal subject *two* basic powers: the power to perform collectings and the power to perform pairwise orderings. Hence the informal description would begin with the claim that, at the first stage, all collectings on individuals and all pairwise orderings of individuals are performed. At the $(n + 1)$th stage, all collectings and pairwise orderings on individuals and previously performed operations are performed. In this scheme, the analogs of finite sequences (ordered n-tuples) arise by iterated ordering. The analogs of relations arise by collectings on previously performed orderings (pairwise orderings in the case of binary relations, iterated pairwise orderings in the case of other relations).

It is easy to provide an informal argument for principle (21), which is the counterpart of the standard axiom, or theorem, asserting the existence of ordered pairs:

(21) $(y)(z)(\exists x)Pxyz$:

34. *Ibid.*, p. 220.

35. *Ibid.*

36. For an account of the iterative conception of set which is closer to the constructivist approach I favor than that given by Boolos, see chapter 6 of Hao Wang, *From Mathematics to Philosophy*. I only became aware of Wang's treatment after I had arrived at the account given in the text.

Let y, z be any individuals, collectings or orderings. Then there will be a first stage at which y exists and a first stage at which z exists. Let the later of these two stages be the nth stage. Then both y and z exist at the nth stage. By hypothesis, all possible pairwise orderings on individuals or operations existing at the nth stage are performed at the $(n + 1)$th stage. Hence, at the $(n + 1)$th stage there exists a pairwise ordering of y and z. So (21) is true. It is relatively trivial to formalize the "stage theory" and to argue formally for (21) in a similar way to that in which Boolos derives the standard axioms.

I shall conclude my discussion of the theory of collecting by pointing out that it differs from standard set theory in three respects. First, I replace references to objects (sets) with references to operations. Second, I view the principles of the resultant theory as true in virtue of the stipulative conditions laid down on an ideal mathematical subject. Like Mill Arithmetic, my revised set theory is not committed to the existence of an ideal subject and of ideal operations. Rather, it works out the consequences of specifications of the powers of a nonexistent being, powers which abstract from the actual limitations of our collectings and orderings. Third, like a minority of set theorists (the Bourbaki, Dedekind), I develop the theory of ordering and of relations separately from the theory of sets. Thus I ascribe to the ideal subject the power to perform two fundamental operations.

Once we have available to us a scheme for reformulating statements of set theory, it is relatively easy to resolve the difficulty about real numbers from which our excursion into set theory began. Consider, once again, the Platonist's account of what our ancestors were talking about. Real numbers are identified with sets, by taking them to be (lower) Dedekind cuts or Cauchy sequences of rationals. There are infinitely many ways to identify the real numbers as sets—besides the identifications we owe to Dedekind, Cantor, and Weierstrass, it is relatively trivial to see how more artificial identifications can be contrived. A Platonist who wishes to avoid the arbitrariness of taking one identification to be privileged—to pick out the *real* reals—can use the natural strategy which we considered briefly at the end of Section I.[37] Each of the adequate schemes of identification can be viewed as identifying a "real number structure," and the theory of real numbers may then be taken to articulate what is common to the real number structures. Thus, instead of formulating the density of the reals as

$$(28) \quad (x)(y)((x\epsilon\mathbf{R} \ \& \ y\epsilon\mathbf{R} \ \& \ x<y)\rightarrow(\exists z)(z\epsilon\mathbf{R} \ \& \ x<z \ \& \ z<y))$$

which would then raise the issue of how to provide a set theoretic identification of $\mathbf{R}$ (the real numbers) and the relation $<$, the Platonist just envisaged will prefer the characterization

---

37. This is the strategy of talking about structure by talking about all instances. The primary example of the strategy is the development of arithmetic as the theory of ω-sequences suggested by White and Field. See the references cited in note 8 above.

(29) If $\langle \mathbf{R}, < \rangle$ is a real number structure then
$$(x)(y)((x\epsilon\mathbf{R} \ \& \ y\epsilon\mathbf{R} \ \& \ x<y)\rightarrow(\exists z)(z\epsilon\mathbf{R} \ \& \ x<z \ \& \ z<y))$$

where the problem will be to formalize the antecedent in the language of set theory by finding a set-theoretic identification of the property of being a real number structure.[38] The resulting presentation of real number theory in explicitly set-theoretic terms will then expose the objects which the Platonist takes us to have been talking about since antiquity. To achieve a parallel account in the terms I prefer, we simply need to translate the Platonist's set-theoretic formulation of the theory of the reals, using the scheme of translation indicated above. The result would be to view the Greek extension of the theory of rational measuring to consist in the introduction of language standing for measuring operations which could not initially be characterized in the available terms of the language of arithmetical operations, but which we have learned to characterize by formulating an account of iterative collective activity. More concretely, one *kind* of real measuring operation—that which corresponds, on the Platonist's account, to the identification of reals with lower Dedekind cuts—would be the operation of collecting those rational measurings which correspond to intervals smaller than the interval to be measured.[39]

Similarly, the various set-theoretic approaches to arithmetic can be recast in our language of collective operations. Let me begin by pointing out that *any* account—whether it be Platonistic or of the type I prefer—can adopt the strategy of construing arithmetic as the theory of ω-sequences, so long as it can solve the problem of the reduction of the theory of ordered pairs.[40] I shall suggest below that my own approach yields a resolution of this latter difficulty. Before I do so, I want to argue that it has the additional advantage of providing a very natural account of the arithmetical notions.

Recall my characterization of the primitive notions of Mill Arithmetic. We perform a one-operation when we perform a segregative operation in which a single object is segregated. An operation is the successor of another operation if we perform the former by segregating all of the objects segregated in performing the latter, together with a single extra object. When we combine the objects collected in two segregative operations on distinct objects we perform an addition on those operations. These explanations are readily formalized as definitions of the arithmetical vocabulary in terms of the theory of collecting.

---

38. The obvious way to do this is by using either Dedekind's Principle of Continuity (see *Continuity and Irrational Numbers*, p. 11) or the Bolzano-Weierstrass Theorem.

39. Here the notion of an interval's being smaller than another would have to be given a concrete geometrical interpretation (e.g. in terms of insertion), and one would also have to extend the scheme of interpretation to allow for "negative measurements" (for example, by taking the direction of the intervals into account).

40. See above Section I and the references given in note 8.

(30) $(x)(Ux \leftrightarrow (\exists y)(z)(Cxz \leftrightarrow z = y))$

(31) $(x)(y)(Sxy \leftrightarrow ((z)(Cyz \rightarrow Cxz) \ \& \ (\exists u)(v)((Cxv \ \& \sim Cyv) \leftrightarrow v = u)))$

(32) $(x)(y)(z)(Axyz \leftrightarrow (\sim (\exists w)(Cyw \ \& \ Czw) \ \& \ (u)(Cxu \leftrightarrow (Cyu \lor Czu))))$

Moreover, the notion of matchability is straightforwardly definable. Two seg-regative operations are said to be matchable if the objects they segregate can be made to correspond with one another. We can formalize this idea as follows.

(33) $(x)(y)(Mxy \leftrightarrow (\exists z)((w)(Czw \leftrightarrow (\exists u)(\exists v)(Pwuv \ \& \ Cxu \ \& \ Cyv))$
         $\& \ (u)(Cxu \rightarrow (\exists w)(\exists v)(s)(t)((Czs \ \& \ Psut) \leftrightarrow (s = w \ \& \ t = v)))$
         $\& \ (v)(Cyv \rightarrow (\exists w)(\exists u)(s)(t)((Czs \ \& \ Pstv) \leftrightarrow (s = w \ \& \ t = u)))))$.

(The first clause of the biconditional tells us that there is a correlating between the objects collected by x and y; the two further conjuncts tell us that it is one to one.)

Now in order to generate the axioms of Mill Arithmetic within our theory of collecting it will be necessary to amend the formulations (1) through (15) of Section III. The reason for this is that (1) through (15) do not contain any predicate which restricts the variables to range over *number operations*. No trouble threatens until we include within our domain of interpretation opera-tions which are not number operations. However, to obtain a derivation of Mill Arithmetic from the general theory of collecting and ordering we shall need to recognize that the axioms of Mill Arithmetic are restricted to number opera-tions. So, for example, instead of (14), we shall have:

(14*) $(x)(Nx \rightarrow (\exists y)(Ny \ \& \ Syx))$.

Of course, to arrive at the reformulated principles of Mill Arithmetic from the principles of the general theory of collecting and ordering we shall need a characterization of the property of being a number operation. (30) tells us what a one-collecting is, and (31) describes the relation of successor collecting. In-tuitively, one-collectings are number operations and any operation related to a one-collecting by a finite chain of applications of successor is a number oper-ation. Thanks to the labors of Frege and Dedekind, we know how to make this intuitive idea formally precise. We can characterize the ω-sequence with a par-ticular first member a and a particular generating relation T, as the intersection of all those sets which contain a and are such that, for any x, if they contain x they contain the T-successor of x. We can adapt this idea to our present needs as follows. First we introduce the idea of a collecting which collects a unique one-collecting and a unique successor for any collecting it collects. Call this a hereditary collecting. Formally:

(34) $(x)(x$ is a hereditary collecting $\leftrightarrow ((\exists y)(z)((Cxz \ \& \ Uz) \leftrightarrow z = y)$
         $\& \ (y)(Cxy \rightarrow (\exists z)(w)((Cxw \ \& \ Swy) \leftrightarrow w = z))))$

Each hereditary collecting is associated with a particular "choice" of a one-collecting and of the successor relation. We can define the one-collecting of a hereditary collecting and the succession of a hereditary collecting as follows:

(35) $(x)(x$ is a hereditary collecting $\rightarrow (y)(y$ is the one-collecting of $x \leftrightarrow (Uy$ & $Cxy)))$

(36) $(x)(x$ is a hereditary collecting $\rightarrow (y)(y$ is the succession of $x \leftrightarrow (z)(Cyz \leftrightarrow (\exists u)(\exists v)(Cxu$ & Cxv & Suv & $Pzvu))))$

We can now introduce the analog of the ω-sequences whose initial members are one-collectings and which are generated by successions. A minimal hereditary collecting is one which collects any object that would be collected by every hereditary collecting with its one collecting and a restriction of its succession.

(37) $(x)(x$ is a minimal hereditary collecting $\leftrightarrow (y)((z)(u)(v)((u$ is the one-collecting of x & v is the succession of x & Czu & $(w)(Czw \rightarrow (s)(t)((Cvs$ & $Pswt) \rightarrow Czt))) \rightarrow Czy) \rightarrow Cxy))$

Finally, number operations are just those operations collected in minimal hereditary collectings. That is:

(38) $(x)(Nx \leftrightarrow (\exists y)(y$ is a minimal hereditary collecting & $Cyx)).$

Lest the complications of the formalism should detract from the basic idea, it may help to present a picture. On my account we can view the universe of number collectings as a multitude of tree-like structures.

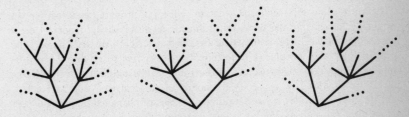

Here the nodes at the bottom represent one-collectings, and the lines represent various ways to perform successor operations. The notion of a minimal hereditary collecting corresponds to an infinitely ascending path through the structure, beginning from some bottom node. The definitions (34) through (37) pick out such paths by starting with the notion of structures containing a path of the type in question and filtering out any unwanted "extras" which these structures may include.

Using (30) through (38), we can effect a derivation of Mill Arithmetic from the general theory of collecting and ordering. This derivation does not depend on a particular identification of number collectings, but reflects the natural idea that a one-collecting is *any* collecting in which a single object is segregated and a successor collecting is *any* collecting in which *any* single extra object is segregated.

Let me now return to my diagnosis of the Platonist's problems with the multiple reductions of arithmetic to set theory. The strategy I have implemented is available to the Platonist if he can define the notion of an ω-sequence in a non-arbitrary way. However, I claim that this cannot be done. From the Platonist's perspective, there is no basis for distinguishing between sets and ordered pairs. Hence considerations of economy and explanatory unification compel Platonists to adopt a reduction of ordered pairs to sets, and this can only be done by making an arbitrary choice. (The strategy of trying to construe the theory of ordered pairs as treating of what is common to the various *explicata* is unworkable, because one must appeal to the notion of relation to develop it.) However, when we switch from thinking of mathematics as descriptive of a realm of abstract objects to construing it as an idealized science of operations, there *is* a basis for distinguishing between *collecting* and *ordering*. These notions are to be kept separate, because they are to idealize operations we actually perform, and the operations we perform fall into two distinct types. From our first crude collectings and orderings to the more sophisticated operations we may eventually learn to perform, the activities of collecting and ordering are different. Hence I suggest that my interpretation of mathematics supports the minority tradition of set theorists (such as the Bourbaki), and the intuitive views of many mathematicians who have wanted to treat ordering as an irreducible mathematical notion.

Before closing this section I want to redeem a promise made earlier. In originally introducing the theme of arithmetic as an idealizing theory, I drew an analogy between the arithmetical case and the theory of ideal gases. The principles of Mill Arithmetic were initially regarded as specifying the performance of the ideal mathematical subject, just as one might use such laws as the Boyle-Charles law to specify the notion of an ideal gas. But, as I noted, kinetic theory provides us with a different means of characterizing the concept of an ideal gas and, having given this characterization, we can show that ideal gases have the properties attributed to them by the phenomenological laws (such as the Boyle-Charles law). Similarly, in developing the general theory of collecting and ordering, we provide a different way of characterizing the performance of the ideal mathematical subject and, given this characterization, it is possible to specify that part of the performance which consists in carrying out the arithmetical operations, demonstrating that the arithmetical performance of the ideal subject meets the conditions attributed by the principles of Mill Arithmetic. Thus, as I promised above, the analogy between arithmetic and the theory of ideal gases can be sustained.

V

I now want to respond to some objections which may have suggested themselves. The criticisms that I shall address are those which I have had to overcome in arriving at the position articulated above, and it is, of course, quite
possible that the blinkers of prejudice have prevented me from recognizing
problems which others will feel to be both obvious and devastating.

Objection 1. "You have offered an account of arithmetic in which arithmetical sentences are recast in a first-order language, and you have also claimed
to provide a semantical account of arithmetic which will not divorce arithmetical language from the rest of our discourse. So, for example, you reject some
approaches to arithmetic—such as those based on substitutional quantification—because they bifurcate the semantics of English. However, the standard
Tarski semantics for first-order languages uses set-theoretic notions. Hence it
seems that the reformulation of arithmetic is pointless. Although you struggle
to avoid commitment to abstract objects in the object language, reference to
these objects will be necessary in giving a semantics for that language."

Reply. It is true that I have cited as an advantage of my view that it enables me to provide a uniform semantical account of mathematical and extra-
mathematical discourse. But I see no reason to believe that that uniform account requires commitment to sets. Given the reformulation of set theory which
I have sketched above, I shall reinterpret the language in which Tarski semantics is given, replacing the references to sets by references to collectings. Tarski
semantics will itself be translated into my preferred idiom, thus allowing for a
uniform semantics for our discourse which does not bring commitment to abstract objects.

Objection 2. "The account of arithmetic and of set theory assumes that
mathematical truth can be identified with derivability. If we take the statements
of Mill Arithmetic or of the reformulated set theory to be true it is because
they are consequences of stipulations made in specifying the ideal subject. Yet
we know from Gödel's Theorems that the notion of mathematical truth outruns
derivability in any formal system. For any formal system which is rich enough
to be a candidate for stating all mathematical truths, there will be a true sentence in the language of the system which is not a theorem of the system.
Hence no specification of the powers of the ideal subject will generate all mathematical truths."

Reply. This type of objection has often been launched against views which
take mathematical truth to flow from stipulation. Such objections work quite
well against anyone who believes that the stipulations which characterize mathematical truth can be completed. Suppose someone were to assert that there is
a formal system F, containing first-order arithmetic, such that mathematical

truth is identifiable with theoremhood in F. This person would be vulnerable to the charge that Gödel's first theorem holds for F, so that there is a closed sentence G in the language of F, such that neither G nor its negation is provable in F. Given the principle of bivalence, one of G and its negation is true, so that, contrary to the assertion, there is a true mathematical statement which is not a theorem of F.

I am not, however, committed to the problematic assertion. It is perfectly consistent to hold that the stipulations from which mathematical truth flows can never be completed (at least not by ordinary humans). One standard way to handle the semantical paradoxes is to consider ordinary languages, like English, as a hierarchy of languages: first-level English contains a truth predicate for sentences not containing 'true,' second-level English contains a truth predicate for sentences containing the first-level truth predicate, and so forth. Similarly, we can consider a hierarchy of stipulations. There is no point in the hierarchy at which the stipulations and their consequences exhaust mathematical truth (just as there is no stratum in the hierarchy of languages for English at which we find a truth predicate for all English sentences). However, as we take any truth of English to belong to the extension of the truth predicate of some stage of the language hierarchy, so too we can regard each mathematical truth as flowing from the stipulations at some stage of the stipulative hierarchy.[41]

Objection 3. "Stipulative theories of truth do not introduce any genuine notion of *truth*. To have a notion of truth is to employ the concepts of reference and satisfaction. '. . . implicit definition, conventional postulation, and their cousins are incapable of bringing truth. They are not only morally but practically deficient as well.' "[42]

Reply. I have already responded to something akin to this objection.[43] The main point to emphasize is that an account taking truth to flow from stipulation need not bypass the concepts of reference and satisfaction. Rather, the stipulations are construed as fixing the referents of the expressions employed so that the right referential relations obtain. Thus, if we specify the notion of an ideal gas in the standard kinetic-theoretic way, we fix as true the statement that the temperature, volume, and pressure of an ideal gas are related by the equation $PV = RT$. This statement has the logical form (16), and we can provide a per-

---

41. Replies somewhat similar to this have been offered to similar Gödelian objections by Hilary Putnam ("The Thesis that Mathematics Is Logic") and Michael Resnik (*Frege and the Philosophy of Mathematics*, pp. 124–25). I should point out that my reply is not tied to one specific method for handling the semantical paradoxes. I conjecture that alternative approaches to the semantical paradoxes—say along the lines of Kripke's "Outline of a Theory of Truth"—could offer different ways of working out the stipulative approach to mathematical truth.

42. Benacerraf, "Mathematical Truth," p. 679.

43. See Chapter 4, Section V.

fectly good referential explanation of its truth by pointing out that, because there are no ideal gases, no sequence satisfies the antecedent of the conditional whose closure is (16). It is wrong to think of stipulational approaches to truth as at odds with referential explanations. The stipulations are better thought of as deepening the referential explanations, showing why the referential relations hold.

Yet there is an important point behind the objection. Stipulation is practically deficient if construed as an automatic means to knowledge. To engage in a stipulation that brings the consequence that p is not necessarily to provide a route to knowledge that p. Acts of stipulation which are to engender knowledge must be well grounded. Thus my account of mathematical reality stresses the fact that the stipulations which specify the powers of the ideal subject are intended to systematize the practical activities in which we engage and about which we gain empirical knowledge. In this way, I think my account brings out the important insight which underlies the objection.

Objection 4. "The account of arithmetical statements proposed interprets such statements as having an unobvious logical form. Platonistic approaches to arithmetical truth have an advantage in that they take the surface form of arithmetical statements to be their logical form, thus enabling us to read at face value the sentences which mathematicians write down."

Reply. Certainly, when other things are equal, an account of a particular body of discourse which reads that discourse at face value is to be preferred to one which suggests a complicated reformulation of it. In the first section of this chapter I have tried to show that other things are *not* equal, that there are reasons to be dissatisfied with Platonism. I now want to add to that a suggestion about how the apparently Platonistic language of contemporary mathematics might have arisen, a suggestion which will let us treat references to abstract objects as a harmless *façon de parler*.

When we think of the operations which, on my account, are the subject matter of mathematics, it is sometimes convenient to think of them as having a product. So, for example, we might conceive of our collecting of some objects as an activity in which we produce something, bringing into being a new object, the collection, or set, of those objects. Yet this picture is unstable, for it suggests that sets are impermanent entities which are brought into being by our efforts. So we shift the picture, arriving at the view that the set is abstract and permanent and that our operations just bring us into relation to it. Then we construct a language for talking about sets, as so conceived. The language is simple and we are able to use it to make concise and elegant statements. But we forget the route through which we arrived at it, and, thereby, come to inherit the problems which I discussed in Section I. The remedy is to remind ourselves of the underlying subject matter which our mathematical language is attempting to describe. I do not suggest that we abandon the standard language

of contemporary mathematics—any more than someone with different ontological views would recommend that all mathematics be done in the primitive
notation of set theory. Nor is it even necessary to forego the claim that mathematics studies abstract objects—*so long as we regard that claim as ultimately
interpreted in terms of ideal operations.* What is central to my account is a
scheme for recasting mathematical language so that we can dissolve the mysteries which Platonism spawns, and this, I suggest, is consistent with viewing
Platonism as a convenient *façon de parler,* a position which errs by adopting a
picture of mathematical reality without recognizing the route through which
that picture emerged.

Objection 5. "The position adopted shares some of the central tenets of
constructivism, but it explicitly ignores the limitations which modern constructivists have imposed on themselves. For example, you have assumed classical
logic, disregarding the objections ʻ. ʻle law of the excluded middle which have
been offered by the intuitionists and others. What justifies this selective attitude
towards constructivist doctrine?"

Reply. As I have indicated above, I think that discussions of constructivist
views about mathematics have been confused by a failure to distinguish between ontological and epistemological claims. The constructivist ontological
thesis is that true mathematical statements owe their truth to the constructive
activity of an actual or ideal subject. The constructivist epistemological thesis
is that we can have a priori knowledge of this constructive activity, and so, in
particular, recognize that it is limited in certain respects. My position develops
the ontological thesis, while repudiating the epistemological thesis. I claim that
the ideal subject is an idealization of ourselves, and that the powers rightly
attributed to this subject are determined by the possibility of giving a simple
account of our own constructive activity. Thus my use of classical logic rests
on the assumption that the best idealization of our practice construes the activity of the ideal subject as complete in a certain respect.

To articulate this point will require me to offer an interpretation of intuitionism, which I take to be illuminating independently of its significance for my
project here. Let me begin by acknowledging a prevalent interpretation which
differs from my own. Recently it has become fashionable to understand the
intuitionistic repudiation of classical logic as based upon a rejection of the
classical conception of a theory of meaning. Thus Michael Dummett has argued
that classical logic depends upon the choice of the concept of truth as the
central concept of the theory of meaning, while, for the intuitionist, the concept
of assertability occupies this position. Instead of explaining the meanings of
the connectives by specifying the *truth* conditions of sentences containing them,
the intuitionist specifies the assertability conditions of those sentences: instead
of declaring that ⌜P v Q⌝ is true iff. at least one of P, Q is true, one declares
that ⌜P v Q⌝ is assertable iff. at least one of P, Q is assertable. Dummett goes
on to contend that systematic constraints on theories of meaning should lead us

to prefer a theory of meaning which takes assertability rather than truth as its central concept.[44]

This is not the place to respond in detail to Dummett's elaborate arguments. Dummett's account has the merit of offering a philosophical explanation for the intuitionistic rejection of classical logic. However, I think that a simpler explanation is available and that Dummett's thesis that assertability is the root concept of an intuitionistic theory of meaning is ultimately unwarranted. Consider the latter issue first. We must begin by asking how the notion of assertability is to be unpacked. Assertability conditions are taken to be *proof* conditions. But, in its turn, the notion of an intuitionistic proof is, deliberately, open-ended. Intuitionists are quick to deny that the notion of proof is to be identified with the concept of proof in some formal system. Rather something counts as a proof because it bears a special relation to the constructions of an idealized mathematician. A sequence of symbols counts as a proof if it correctly describes a sequence of constructions performed by an ideal subject. Indeed, in many intuitionist writings, we find the suggestion that a sequence of statements is a proof because it provides a means of *verifying* that certain properties hold of constructions. So, although it may *appear* that the intuitionist is providing an account of the connectives which is couched in terms of assertability conditions, the notion of assertability is a derivative one, ultimately cashed out by appealing to the concept of truth. Hence I am puzzled by Dummett's claim that assertability is the central concept of an intuitionistic theory of meaning.

Let us now turn to the simpler explanation of the intuitionistic rejection of classical logic which I promised in the last paragraph. The most obvious way to understand intuitionistic statements is to regard the surface form of *atomic* statements as deceptive. Thus, when an intuitionist inscribes a simple arithmetical statement, such as "$2 + 3 = 5$," she does not intend to describe the properties of certain mind-independent abstract objects but to record the performance of certain constructions. Heyting's spokesman makes the point forthrightly. "Every mathematical assertion can be expressed in the form: 'I have effected the construction A in my mind.' "[45] This approach can be articulated using Brouwer's theory of the creative subject. We begin by supposing that what we initially take to be mathematical statements correspond to rules for construction. The distribution of quantifiers and connectives in the surface forms of mathematical statements corresponds to a particular structure among the associated rules of construction. This structure is delineated in what Dummett (and others) identify as the semantics of the connectives and quantifiers. Thus, for example, if the statement A corresponds to a rule of construction R

44. The most detailed treatment of this theme is in "What Is a Theory of Meaning? (II)." See also *Elements of Intuitionism* and several essays in *Truth and Other Enigmas*.
45. *Intuitionism*, p. 19.

then the statement $\ulcorner -A \urcorner$ corresponds to a rule of construction which directs the effecting of a construction showing that the supposition of a construction according to R engenders a contradiction. Let us now generalize by taking R(p) to be the rule corresponding to p. Then the intuitionistic account of an instance of the law of the excluded middle $\ulcorner p \vee -p \urcorner$ should be as follows:

> At some stage in the life of the creative subject, a construction according with R(p) has been effected or a construction according with R($-$p) has been effected.

In this formulation, the quantifier and the disjunction are purely classical. Viewed from this perspective, intuitionistic rejection of arithmetical statements (or other mathematical statements) which *look* as though they take the form $\ulcorner p \vee -p \urcorner$ is readily comprehensible. The underlying form of these statements is revealed in the transcription I have suggested, and denial of them stems from nothing more than the thesis that the operations of the creative subject may be incomplete in an obvious sense. Suppose that we take p to be some statement involving what would be construed (classically) as quantification over an infinite domain. Then the intuitionist may assert that the powers of the creative subject are inadequate either to effect a construction according with R(p) or to effect a construction according with R($-p$). This assertion would then take the *surface* form $\ulcorner p \vee -p \urcorner$.

I now want to suggest that this approach represents *one* way of idealizing the constructive activity in which we actually engage. There are certain types of constructions which we are not able to perform: we cannot, for example, check universally quantified statements in arithmetic by generating representations of each number and verifying that, in all cases, the property alleged to hold genuinely does. Intuitionistic mathematics results from building in to the notion of the creative subject this limitation and others which are akin to it. Classical mathematics will be generated if we are more generous to the creative subject. In my reconstruction of mathematics I have been extremely liberal in specifying the powers of the ideal subject: witness the reformulation of set theory offered in the last section. My motivation in this has been of a piece with the practice of idealization generally. To idealize is to trade accuracy in describing the actual for simplicity of description, and the compromise can sometimes be struck in different ways. Recall the analogy which I used extensively in developing my account: just as there are different ways to idealize the findings of actual gases, so too there are different ways to develop an idealized treatment of the operations we actually perform. Intuitionism plays to classical mathematics the role of the theory of van der Waals's gases to the theory of ideal gases: it stays closer to actuality at the cost of simplicity.

I hope that this somewhat lengthy reply clarifies my position with respect to contemporary constructivism. Obviously, a far more detailed analysis of the intuitionist program could be given, but I think I have said enough to indicate the way in which my analysis would develop. If the general perspective developed here is correct, then my picture of mathematical reality can sustain the doctrine which classical mathematicians have wanted to uphold. Intuitionism is a part of mathematics, for there is room for many different idealizations of our constructive practice, but, by the same token, it cannot lay claim to being the *only* legitimate form of mathematics. And, on grounds of the simplicity it brings, the classical idealization seems preferable.

Objection 6. "You have claimed to be able to reconstruct classical mathematics using a liberal idealization from our actual constructive practice. But to give a faithful reconstruction one must show how to cope with the impredicative definitions which are used by classical mathematicians. Since the use of impredicative definition is at odds with the basic idea of a constructivist set theory, the project will fail."

Reply. I believe that certain kinds of impredicative definition can be sanctioned by the approach I have recommended. (If this were not so, then I should have to argue that something resembling classical mathematics can be developed without using impredicative definitions. Charles Chihara has made an impressive attempt to do this.)[46] The kinds of impredicative definition I hope to allow can be described using the language of Boolos's stage theory. These definitions specify sets formed at a particular stage s out of entities formed prior to s but making reference in the specification to sets formed at stages later than s. Here is an example in the language of ordinary set theory:

$$N = \{x : (y)((1\epsilon y \ \& \ (z)(z\epsilon y \rightarrow Sz\epsilon y)) \rightarrow x\epsilon y\}.$$

This, of course, is the Fregean identification of the set of natural numbers as the intersection of all sets containing 1 and closed under successor.

Now the standard method of motivating impredicative definition is to use the Platonist's picture of mathematical reality. Platonists tell us that the purpose of a definition is not to enable us to construct a set out of materials that are already available but to identify a set from a pre-existent universe of sets. When we adopt the picture I have recommended it seems that we are doomed to forfeit this motivation. However, I think that we can achieve something similar. If we imagine the hierarchy of collectings as generated by the iterated collective activity of the ideal subject, we can consistently hold the following principles: (i) collectings performed at any stage must be performed on entities available at that stage (e.g., individuals, prior collectings); (ii) the subject can use references to future collectings (collectings performed at later stages) to single out available entities and to collect them. Intuitively, in the more familiar language employed by Boolos, sets can only be formed out of materials already available but you can use references to higher-level sets to specify a property which will select the available individuals you want to form into a set. It is as if my ideal

46. See *Ontology and the Vicious Circle Principle,* chapter 5.

subject could talk about subsequent collecting and use that talk in performing collectings on the entities so far produced. Given this construal of the collective activity of the ideal subject, we can allow for some forms of definition which might seem, on my account, to be debarred, including the types of definition which are central to classical analysis.

The reply just offered attempts to establish a stronger conclusion than is strictly necessary. For impredicative specifications of collectings may be justified even if we concede that the ideal subject herself cannot use that specification to pick out the objects collected. What is crucial is that the powers of the subject should be taken to be determinate and as full as possible. Thus we may claim that, at each stage, the subject performs all possible collectings on the available entities and that one of these collectings will turn out to satisfy the impredicative specification, whether or not she conceives of it in this way. Hence, even if the proposal of the last paragraph is not adopted, there should be no more objection to impredicative specifications on my account than there is on standard versions of ZF set theory.

Objection 7. "Even granting that it is appropriate to attribute to the ideal mathematical subject abilities which allow for the retention of classical logic and of impredicative definitions, it is still not clear that one can allow for the full set-theoretic hierarchy. For example, if one is to achieve a set theory which is equivalent to that used in recent investigations (such as those which consider the possibility of adopting various kinds of axioms about inaccessible cardinals), one must assume that the "stages" at which the ideal subject carries out constructive operations are highly superdenumerable. At this point, the idea that the iterated constructive activity represents the "life" of the subject no longer seems justifiable, and one may even wonder if it is coherent."[47]

Reply. It is perfectly correct to point out that if we conceive of the stages of the constructive activity as instants in the life of the ideal subject and if we take the structure of time for the subject to be that of ordinary time, then we shall not be able to ascribe to the subject constructions which correspond to the entire "Zermelo-Frankel paradise." What this shows is that, if we are to obtain a comparably rich set theory, we must take some further abstractions. Two questions then arise. Can we specify a coherent idealization? Can we justify that idealization?

I believe that the answer to the first question is "Yes." I see no bar to the supposition that the sequence of stages at which sets are formed is highly superdenumerable, that each of the stages corresponds to an instant in the life of the constructive subject, and that the subject's activity is carried out in a medium *analogous* to time, but far richer than time. (Call it "supertime.") Plainly,

---

47. A similar objection is made by Charles Parsons against Wang's version of the iterative conception. See "What Is the Iterative Conception of Set?"

to make this supposition is to idealize still further from our own thoroughly finite performances. But the move is no different in principle from earlier idealizations in which we abstract from our own mortality or from our inability to survey infinite domains. The view of the ideal subject as an idealization of ourselves does not lapse when we release the subject from the constraints of our time.

The second question is more tricky, and it leads inevitably into issues which I shall take up in subsequent chapters. I suggest that we would have to justify the introduction of supertime by appealing to the methodological directive of generalizing the mathematical results which have already been achieved. In order to systematize the results of analysis we need to ascribe to the ideal subject an ability to perform iterative collective activity through an infinite sequence of stages. To do this is to introduce *two* principles governing the sequence of stages: one which asserts that each stage is followed by another, and another which allows for the existence of stages, besides the initial stage, which do not have immediate predecessors. The first of these allows for stages by *succession;* the other allows for what we may call "limit stages." (The ωth stage is, of course, the first limit stage.) Full ZF set theory allows for the unrestricted use of *both* principles to generate further stages in the life of the creative subject. We can justify our view of the ideal subject as operating in supertime by regarding it as the result of allowing for general application of a principle of stage-introduction—introduction of limit stages—which we must use at least once to allow for classical analysis.

In the spirit of my response to Objection 5, we can regard ourselves as faced with a progression of idealizations which take us further from our actual performances. To secure a set theory which suffices for classical analysis, we shall need to suppose that the life of the ideal mathematical subject contains more than an indefinitely proceeding sequence of stages, or, in other words, that it contains at least one limit stage. Mathematical theories which permit general application of the principle of introducing limit stages depart further from actuality, obtaining their justification from their claim to generalize what was artificially restricted in previous practice. Whether such claims to generalization are sufficiently strong to support the attributions of such striking powers to the ideal subject is a delicate issue which I shall not try to decide.[48]

In this chapter, I have tried to review some shortcomings of the traditional, Platonist conception of mathematical reality, to suggest that we might overcome these difficulties by viewing mathematics as describing the structure of the world, and to show how that view can be articulated. Plainly, my major

---

48. I shall give an account of the rationality of generalization in Section IV of Chapter 9. That account will describe the criteria to which one would appeal in making a decision.

concern has been to develop a picture of mathematical reality which will conform to the general epistemological position advanced in this book.[49] Hence,
although I believe that my picture of mathematical reality can be used to illuminate many metaphysical issues about mathematics (such as the topic of the
modal status of mathematical truths) I have not pursued those issues here.

My central claim is that proto-mathematical knowledge can be obtained by
manipulating the world and observing the manipulations. From these humble
beginnings, mathematical knowledge develops into the impressive corpus of
contemporary theory. *How* it does so is for me to explain in subsequent chapters. I hope to describe the ways in which the historical development of mathematics has disclosed the mathematical structure of reality, beginning from crude
physical manipulations and erecting on that basis an ever more refined theory
of the constructive activity of the ideal subject. My task will be to understand
the methodological principles which have directed the advancement and acceptance of successive parts of that theory.

49. It is worth pointing out that the most sophisticated attempts to save Platonism from epistemological difficulties—such as those of Mark Steiner (*Mathematical Knowledge,* chapter 4), Michael
Resnik ("Mathematical Knowledge and Pattern Cognition"), and Penelope Maddy ("Perception
and Mathematical Intuition")—would allow for perceptual knowledge of elementary mathematical
truths. Hence, if a Platonist account should prove workable (and if it should prove superior to that
which I have offered), I suspect that it will be able to be assimilated into the general epistemological framework of this book.

7
Mathematical Change and Scientific Change

I

The existence of mathematical change is obvious enough. Contemporary mathematicians accept as true statements which our predecessors did not accept. In 1400, the members of the mathematical community did not believe that every polynomial equation with rational coefficients has roots; their nineteenth-century descendants did. Conversely, later writers sometimes abandon claims which have been espoused earlier. Leibniz and some of his followers believed that $1 - 1 + 1 - 1 + 1 \ldots = \frac{1}{2}$. Cauchy and Abel scornfully rejected this and kindred statements. Yet the shifting allegiance to some statements is only one facet of mathematical change. Equally evident are alterations in mathematical language, variations in style and standards of reasoning, changes of emphasis on kinds of problems, even modifications of views about the scope of mathematics. The fact of mathematical change provokes a series of questions. Why do mathematicians propound different statements at different times? Why do they abandon certain forms of language? Why do certain questions wax and wane in importance? Why are standards and styles of proof modified? In short, what kinds of changes occur in the development of mathematics, and what general considerations motivate them?

To raise these questions is to begin to investigate the methodology of mathematics, in a way which is parallel to recent and contemporary inquiries about the methodology of the natural sciences. Neglect of the methodology of mathematics stems from distrust of the parallel. In turn, that distrust gains powerful support from mathematical apriorism. Yet, even if we reject the apriorist conception of mathematical knowledge, we may still wonder whether the development of mathematical knowledge is analogous to that of natural scientific

knowledge. My goal in this chapter is to investigate the similarities and differences between mathematical change and scientific change. By doing so, I hope
to dispose of some myths about mathematical change and to use the comparison
with natural science to formulate more sharply the enterprise of investigating
the methodology of mathematics.

Suspicion about the kinship of mathematical change and scientific change,
when it is not simply a by-product of apriorist doctrine, is prompted by two
important observations. One apparent major difference between the growth of
scientific knowledge and the growth of mathematical knowledge is that the
natural sciences seem to evolve in response to experience. As observations and
experiments accumulate, we find ourselves forced to extend and modify our
corpus of beliefs. In mathematics, however, the observation of previously
unobserved phenomena and the contrivance of experiments seem to play no
important role in stimulating change of belief. So we are easily led to conclude
that the springs of change are different in the two cases. A second feature of
the growth of mathematical knowledge is the appearance of cumulative development in mathematics in ways which seem absent in the natural sciences.
Because contemporary mathematics appears to preserve so much more of what
was accepted by the mathematicians of the past, it is tempting to suppose that
the manner in which mathematical knowledge evolves must be fundamentally
different from that in which scientific knowledge grows. Mathematical methods
must be more sure-footed than those used by natural scientists.

In this section, I want to consider the first of these apparent disanalogies. I
shall consider the issue of the cumulative character of mathematical knowledge
in Section II. Our first task will be to uncover the picture of scientific change
which underlies the complaint that, unlike the natural sciences, mathematics
does not grow by responding to observation and experiment.

Consider the simplest empiricist view of the growth of scientific knowledge.[1]
According to this picture, the statements accepted by the scientists of a given
period can be divided into two classes: there are observation statements (*O-
statements*) and theoretical statements (*T-statements*); the former are accepted
on the basis of observation and are unrevisable; the latter are adopted on the
basis of inference from the accepted O-statements, indeed on the basis of inferences which accord with principles of the "logic of scientific inquiry," principles which hold for all scientists at all times.[2] As science develops, the change

1. The view I shall present appears to accord with the central ideas of such thinkers as Carnap,
Hempel, and Feigl. Since these thinkers do not consider the question of providing a philosophical
reconstruction of the historical development of natural sciences, it is no surprise that their writings
contain no explicit endorsement of the view.

2. It should be clear from this characterization of them that T-statements are not necessarily couched
in a special ("theoretical") vocabulary. The distinction I am drawing here is that between the
alleged foundations of scientific knowledge and the theoretical superstructure erected upon them.
The latter includes what are sometimes called "empirical laws" as well as the principles which
are expressed in the technical language of theories.

in the corpus of O-statements is by accumulation. New O-statements are added, but old O-statements are never deleted. However, amendment of the class of T-statements is not by accumulation. Even though a particular set of T-statements may have been justified in the light of the limited set of O-statements adopted at an earlier stage, extension of the corpus of O-statements can force us to retract what we formerly believed, substituting a quite different set of T-statements in its place. There are two features of this picture of scientific change to which I wish to draw attention: (i) the match between observation and theory at any stage in the history of science is assumed to be perfect (the adopted O-statements justify the accepted T-statements in the light of the universal principles of the "logic of scientific inquiry"); (ii) addition of new O-statements can disrupt the match, forcing the modification of the corpus of T-statements to accommodate the broader class of O-statements. Together, these features combine to distinguish observation as the source of scientific change. Without new observations, science would be static.

I do not know whether anyone has held exactly this picture of scientific change, but something very close to it seems to be implicit in the writings of many logical empiricist philosophers of science. A variety of considerations makes it clear that this simple empiricist picture of scientific change cannot be sustained.

In the first place, there have been severe (and, to my mind, conclusive) attacks on the thesis that there is a class of unrevisable reports of observation, with consequent denial that the history of science can be viewed as a series of responses to an observational corpus which develops cumulatively.[3] Yet this critique, in and of itself, does not compel us to abandon those features of the simple empiricist picture which generate the view that observation is the source of scientific change, and thereby foster our suspicion that mathematical change is importantly different from scientific change. We may continue to suppose that the science of an epoch is a collection of statements determined jointly by the stimuli which have so far impinged upon those who adopt it and the canons of scientific inquiry. New stimuli can still be viewed as the sole inducers of modification of the corpus of beliefs, even though we agree that there is no level at which modification must be cumulative.

A second major assault on the simple empiricist picture challenges us to understand the large upheavals in science—such "revolutions" as the transition from Aristotelian cosmology to Copernician cosmology, the overthrow of the phlogiston theory, and the replacement of Newtonian physics with the special and general theories of relativity—using the terms which simple empiricism

---

3. The *loci classici* of the attacks are W. V. Quine, "Two Dogmas of Empiricism" (sections 5 and 6), and W. Sellars, "Empiricism and the Philosophy of Mind." For earlier doubts about the observational foundations of scientific knowledge, see Karl Popper, *The Logic of Scientific Discovery,* chapter 5 (especially p. 111), and, for a clear recent presentation of the major criticisms, Michael Williams, *Groundless Belief.*

supplies.[4] Can we account for these episodes as consisting in the modification of a corpus of statements in the light of new stimuli and a set of universal canons of scientific inquiry? A number of writers, most notably Paul Feyerabend, Stephen Toulmin, and Thomas Kuhn, have argued that we cannot, and their writings have provoked several attempts to offer a view of scientific change which will do justice to scientific revolutions. Among these writers I shall take Kuhn as the most important representative, since his views are at once most systematic and most sensitive to the history of science. Kuhn's seminal book, *The Structure of Scientific Revolutions*, argues for a conception of scientific revolutions which is at odds with simple empiricism and which has been much discussed by philosophers. On Kuhn's account, scientific revolutions involve: *conceptual changes*, which can render impossible the formulation of prerevolutionary and postrevolutionary theories in a common language; *perceptual changes*, which produce new ways of seeing familiar phenomena; and, perhaps most important, *methodological changes*, which, by amending the rules of justification for scientific theories, make the rational resolution of the differences between earlier and later theories impossible. The simple empiricist picture of science as developing by rational adjustment to observation is completely undermined if this account of revolutions is accurate. Scientists engaged in revolutionary debate do not share enough rules of justification to reach agreement, even if they could begin from shared observations. But they do *not* begin from shared observations. Moreover, their rival claims cannot be formulated in a common language. Small wonder, then, that, in one of the most cited discussions in his much-quoted book, Kuhn talks of scientific decision in terms of "conversion experience" and "faith."[5]

Despite the fact that Kuhn's account of revolutions is obviously important, what concerns me is not the correctness of the view of revolutions just sketched, but whether that view alters our previous estimate of the distinction between mathematical change and scientific change. I think it does not. For, as I have so far presented it, the central thrust of the view is that observation does not rationally compel us to modify our scientific beliefs. Unless we yearn for a change of fashion, faith in the old corpus can be maintained. To accept this thesis is not to abandon the claim that observation is the source of scientific change, but only to contend that not even new observation need provoke us to amend our old ways.

Yet my presentation of the historically inspired attack on the simple empiricist picture of scientific change has been deliberately one-sided. In the last paragraph I have briefly rehearsed the view which most philosophers have found

4. See, for example, T. S. Kuhn, *The Structure of Scientific Revolutions;* P. K. Feyerabend, "Explanation, Reduction and Empiricism," "Problems of Empiricism," *Against Method,* and *Science in a Free Society;* N. R. Hanson, *Patterns of Discovery;* S. Toulmin, *Human Understanding.*
5. *The Structure of Scientific Revolutions,* pp. 150–59.

in *The Structure of Scientific Revolutions*.[6] However, besides its apparent commitment to the thesis that scientific revolutions can only be resolved "by faith," Kuhn's book contains another very important claim, which not only controverts the simple empiricist picture but is also relevant to our project here. To put the point in its simplest terms, Kuhn contends that almost all theories are falsified at almost all times. Thus, contrary to feature (i) which we distilled from the simple empiricist picture, the match between theory and observation is *not* perfect. In the discrepancy between theory and observation, or, more generally, between different *parts* of theory, Kuhn finds the source of the problems which occupy scientists for most of their careers. On this account, scientists (justifiably) accept a general form for theory-construction in a particular field, adopting particular pieces of work as paradigmatic, selecting certain questions as important, choosing rules for answering those questions, and so forth. Given this set of background views, they put forward proposals, modifying and articulating them so as to achieve, insofar as possible, successful conformity both to the canons which govern all scientific activity and to the rules of their own particular enterprise. Discrepancies are always with them, presenting challenges even in the absence of new observations.[7] The problems may be more or less empirical (for example, puzzles about unanticipated experimental data) or they may be highly theoretical. The latter are of especial concern to us. Scientists are frequently challenged to answer a question posed by existing theory. Newton struggled with the issue of whether his theory of gravitation could be reconciled with the thesis that all action is by immediate contact. Darwin was confronted with the difficulty of resolving conflicts between his account of rates of evolution and geophysical estimates of the age of the Earth. Wegener and his early adherents were challenged to propose a mechanism which could move the continents. Contemporary evolutionary theorists have exhibited considerable ingenuity in devising theoretical models to show how apparently maladaptive traits may become fixed in a population. Molecular biology still faces the problem of reconciling our knowledge of the differential development of the cells of an embryo with our understanding of the synthesis of intracellular products. The examples could be multiplied almost indefinitely. They show that the simple empiricist picture of scientific change is badly mistaken. Even without the provocation of new observations, factors to stimulate scientific change are always present.

We are now in a position to become clearer about the complaint from which we began. It would be futile to deny that observation is *one* source of scientific change. The burden of the last paragraph is that observation is not the *only*

6. In particular, this interpretation of Kuhn's work is advanced by Dudley Shapere, Israel Scheffler, and Carl Kordig. See Dudley Shapere, "Meaning and Scientific Change"; Israel Scheffler, *Science and Subjectivity*; Carl Kordig, *The Justification of Scientific Change*.

7. *The Structure of Scientific Revolutions*, chapters 3–5.

such source. There are always "internal stresses" in scientific theory, and these provide a spur to modification of the corpus of beliefs. I propose to think of mathematical change as akin to this latter type of modification.[8] Just as the natural scientist struggles to resolve the puzzles generated by the current set of theoretical beliefs, so too mathematical changes are motivated by analogous conflicts, tensions, and mismatches.

To oversimplify, we can think of mathematical change as a skewed case of scientific change: all the relevant observations are easily collected at the beginning of inquiry; mathematical theories develop in respone to these and *all* the subsequent problems and modifications are theoretical. This is an oversimplification because new observations are sometimes important even in mathematics. The efforts of the inhabitants of Königsberg to cross all of the famous seven bridges without retracing their steps suggested to Euler a mathematical problem, for which he found a solution, integrated by later mathematicians into a new branch of mathematics. Nor is this an isolated case. Pascal's investigations in probability theory, the study of possibilities of map coloring, and the recent work in catastrophe theory (whatever its merits) can all be viewed as mathematical responses to observable features of everyday situations. Moreover, as with the natural sciences, the "new" observation is often concerned with some familiar phenomenon whose significance has not hitherto been appreciated.

Before leaving the issue of the relation between observation and mathematical change, we should take note of the *indirect* ways in which experiment and observation may affect the development of mathematics. Sometimes difficulties in mathematical concepts or principles are first recognized when trouble arises in applying them in scientific cases. Thus in the eighteenth- and nineteenth-century study of functions, variational problems, and differential equations, modification both of physical theory and the mathematics presupposed by it go hand in hand. We shall examine one example of this interplay in Chapter 10.

Our initial concern was that an account of mathematical change must be very different from an account of scientific change in that the main force of scientific change is the pressure of new observations. I have responded to this in two different ways. The last two paragraphs indicate that new observations may be relevant (directly or indirectly) to the evolution of mathematical knowledge. But my principal point is that the concern thrives on a misunderstanding of scientific change. Many important episodes in the evolution of scientific knowl-

---

8. The type of view presented here has some kinship with that advanced by R. L. Wilder in his *Evolution of Mathematical Concepts*. Wilder is one of the few people to have considered seriously the question of mathematical change, and, though he modestly disclaims all intentions to philosophize, I think that his work is more relevant to philosophical understanding of *mathematics* than many of the books and papers to which philosophers of mathematics give their attention. Some of Wilder's ideas are extended further in Michael Crowe's "Ten 'Laws' Concerning the History of Mathematics." I hope that the account I shall advance in this and the ensuing chapters will provide a general framework within which the suggestive observations of Crowe and Wilder can be embedded.

edge are best viewed not as responses to new observations but as attempts to resolve pre-existing intra-theoretic tensions. The same applies to mathematics—and applies with a vengeance. Later in this chapter, I shall try to explain how this idea of intra-theoretic stress can be conveniently represented. Before I do so, I want to examine the second concern voiced above, the worry that mathematical change is cumulative in ways that scientific change is not.

II

In what sense is the development of mathematics cumulative and the development of science not? The idea that there is a difference here can receive a number of formulations: (a) there are no "revolutionary debates" in the history of mathematics; when mathematicians engage in dispute at least one party is being irrational or stubborn;[9] (b) many mathematical truths have been accepted since antiquity; (c) when mathematical statements are accepted at one time and rejected at a later time, those who originally accepted the statements were unjustified in doing so. In each case the formulation suggests a contrast with the natural sciences. Since reading Kuhn, Feyerabend, and others, philosophers have recognized that those episodes during which the natural sciences seem to make their greatest advances are marked by disputes in which the conservative protagonists cannot simply be labelled as "prejudiced," "irrational," or "stubborn." Moreover, increasing understanding of the history of science has enabled us to see that many of the scientific concepts and principles of our predecessors have been discarded or modified. Finally, our study of science finds room for the notion of a justifiable mistake. We are prepared to admit that the scientists of earlier ages held justified false beliefs. Hence each of the theses (a), (b), (c) can serve to expose a contrast between the cumulative development of mathematics and the non-cumulative development of natural science.

These ideas of an important contrast stem from the available historical studies. Hence an appropriate first response to them is to suggest that the appearance of harmony and straightforward progress may be an artifact of the histories of mathematics which have so far been written. Until the history of natural science came of age, it was easy to believe that the course of true science ever had run smooth. Unfortunately the history of mathematics is underdeveloped, even by comparison with the history of science.[10] Only in the last few years

9. This conception of revolutionary debates stems from the works of the writers cited in note 4—particularly Kuhn and Feyerabend.

10. This remark needs a little qualification. Excellent work on Greek mathematics and pre-Greek mathematics has been done by Heath, Neugebauer, and others. But, with the exception of a few insightful essays by Philip Jourdain and Ernest Nagel, the history of mathematics from the seventeenth century on has been much less sophisticated than the general history of science until quite recently.

have there appeared studies which advance beyond biographical details and
accounts of names, dates, and major achievements. One difficulty for the historian has been the prevailing philosophical view of the nature of mathematics,
with its emphasis on mathematics as a body of a priori knowledge. That emphasis has diverted attention from the rejected theories, the plausible but unrigorous pieces of reasoning, the intertheoretical struggles.

Even the most cursory look at some primary sources will dispose of a *very*
naïve conception of the cumulative character of mathematics, the idea that
mathematics literally proceeds by accumulation, that new claims are added but
old claims are never abandoned. Eighteenth-century analysis abounds with
statements that we have rejected. The history of the investigation of the distribution of prime numbers contains many false starts and blind alleys. Other
cases are more subtle. If one compares a contemporary text in analysis with a
classic text of the early part of the century (say Whittaker and Watson's *Course
of Modern Analysis*) it is impossible to regard the later work as a simple extension of the former. True, there is significant overlap in material, but the modern
text approaches the subject from a different perspective, generalizing the treatment of some theorems and omitting other topics altogether. *In some sense,*
most of nineteenth-century analysis survives in the contemporary treatment, but
it does not do so in any straightforward way: we no longer care for the systematic exploration of special functions which our Weierstrassian predecessors loved
so well.

The formulations I have given to the idea that mathematics is cumulative in
a way that natural science is not are more sophisticated than the position just
considered, and less easy to dismiss. Nevertheless, we can point to episodes
from the history of mathematics which call each of them into question. Just as
there are protracted disputes in the history of science in which we are reluctant
to characterize any of the protagonists as stupid or wrongheaded, so too in
mathematics there are parallel controversies. Consider, for example, some of
the debates which surround the early calculus. Newtonians and Leibnizians
each proclaimed the superiority of their method to that practiced by the rival
tradition. The Leibnizians pointed proudly to their problem-solving efficiency;
Newtonians emphasized their ability to preserve important features of previous
mathematics. We should no more castigate Newton and his successors for
clinging to a style of mathematics which the calculus was eventually to transform than we should condemn Priestley for his attempt to salvage the phlogiston theory and to use it to account for his own experimental results. As a
further illustration, we can turn to the late nineteenth-century dispute about the
legitimacy of various construals of the real numbers and of Cantor's transfinite
set theory. We disagree with those, like Kronecker, who insisted on a literal
application of the slogan that analysis should be arithmetized. Yet we would
find it just as hard to convict Kronecker of irrationality and dogmatism as to
press the same charges on the more subtle of the Aristotelians who debated

Galileo. Hence I conclude that we should not articulate the contrast between mathematics and natural science along the lines suggested by (a).

Let us now examine (b). Even if we grant that standard presentations of the history of mathematics conceal the existence of genuine disputes and noncumulative changes, it appears at first that vastly more of ancient mathematics than of ancient science has survived intact into the present. We have not abandoned the truths of arithmetic, or Euclid's theorems, or the solutions to quadratic equations obtained by the Babylonians. Does this not indicate an important difference between the development of mathematics and the development of science? It is crucial here to find the right scientific analogs for these mathematical results. Let us recognize that many statements have in fact persisted through the history of science. We continue to share with our ancestors a wealth of beliefs about the ordinary properties of ordinary things. To claim that there is no privileged level of observational reporting, that all our observation statements are revisable, is quite consistent with the admission that many of the claims we make on the basis of observation coincide with judgments that have been made for centuries. I anticipate an objection. When we say, for example, that feathers float on water or that the sun rises in the east, can we really be taken to agree with our predecessors? Perhaps the translation of their utterances by these sentences of ours blurs important conceptual differences which separate us from them. I believe that such worries are unfounded. When the notion of conceptual change in science is properly understood, we see that it is possible to allow for the existence of conceptual differences between ourselves and our ancestors while claiming that we can record some of their beliefs in sentences of contemporary language to which we would assent. However, even if this were not so, the objection would not be pertinent to our present discussion. For any argument for shifts in our concepts of the ordinary things around us and of their ordinary properties could be mirrored by an argument for parallel shifts in our concept of number. If, for example, we suppose that our concept of water has been transmuted by the discovery that matter is discontinuous, so too we may take our concept of number to have been altered by the introduction of negative, rational, real, complex, and transfinite numbers. Hence it would be wrong to claim that our arithmetical beliefs have been preserved through the centuries, while our everyday physical beliefs have not.

Finally, we must address the suggestion that mathematicians, unlike natural scientists, cannot justifiably hold false beliefs (the suggestion offered by (c)). Were we to adopt this suggestion we would be forced to some harsh judgments concerning those mathematicians who have advanced inductively based conjectures about formulas for generating prime numbers. More importantly, we would fail to do justice to the numerous occasions on which acceptance of a simplified principle paves the way for the development of concepts which can be used to correct that principle. Euler and Cauchy justifiably believed, for example, that trigonometric series representations of arbitrary functions could not be given.

Only in the wake of Cauchy's attempt to articulate the reasons which he drew from Euler could it become apparent how the claim was incorrect. To develop the concepts required to correct Cauchy's mistake took approximately a quarter of a century. Here, and in many other cases, we find mathematicians making the best use of their epistemic situations to advance false claims, whose falsity only becomes understood through the efforts of those very mathematicians to articulate their reasons. If we accept (c) we shall not only divorce the notion of justification in mathematics from justification in other fields, but also make the progressive uncovering of subtle errors look like a sequence of blunders which culminates, miraculously, in apprehension of the truth.

So far, then, we have failed to discover a sense in which the growth of mathematical knowledge is cumulative and the growth of scientific knowledge is not. However, I believe that there is something to the suggestion that we have so far failed to credit. Mathematical *theories* seem to have a far higher rate of survival than scientific theories. Newton's "method of fluxions" is very different from contemporary calculus, and Hamilton's theory of quaternions is by no means identical with modern linear algebra; yet, in some sense, both Newton's and Hamilton's ideas live on in modern mathematics. Obviously, similar remarks can be made about some past scientific theories. What we do not seem to find in mathematics are the analogs of the discarded theories of past science: there appear to be no counterparts of Aristotle's theory of motion, the phlogiston theory of combustion, or theories of blending inheritance. I shall now try to explain why this is so.

Consider the difference between the development of non-Euclidean geometry and the (roughly contemporary) development of the oxygen theory of combustion. In the former case, after nearly two millennia of attempts to prove Euclid's fifth postulate (which is *equivalent* to the statement that, given a line in a plane and a point of the plane which does not lie on the line, there is a unique line through the point which is parallel to the given line), three mathematicians, Lobatschevsky, Bolyai, and Gauss, decided to investigate the consequences of adding to the first four postulates a statement asserting the existence of many parallels. Their efforts produced the non-Euclidean geometry we call "Lobatschevskian." Once they became convinced that the new geometry was consistent, mathematicians accepted it as part of mathematics, and they set about proving Lobatschevskian theorems, trying to find characteristics which would distinguish Lobatschevskian geometry from Euclidean geometry, attempting to generalize geometrical theories, and so forth. As far as mathematics is concerned, there was no need to choose between Lobatschevsky and Euclid (although tradition credits Gauss with an investigation designed to determine if space is Euclidean). Contrast this course of events with the debate over theories of combustion. The phlogiston theory claimed that something—phlogiston—is emitted from substances when they burn. Lavoisier's oxygen theory contends that combustion involves not emission but absorption of a constituent of the air. By 1800, the scientific community had decided in favor of the oxygen theory, and, after Priestley's death in 1804, no major scientist explored further consequences of the phlogiston theory.

What appears at first to be mathematical competition issues in peaceful coexistence. By contrast, scientific competition ends in the death of one theory. Lobatschevsky's geometry sits alongside Euclid's in the pantheon of mathematical theories, because for the mathematician both theories are correct descriptions of different things; Lobatschevsky, Bolyai, and Gauss provided an accurate account of a particular kind of non-Euclidean space; Euclid's geometry remains the correct theory of Euclidean space; the question of which kind of geometrical space is realized in physical space is given to the physicists (or, if the apocryphal story about Gauss is true, to mathematicians moonlighting as physicists). Yet we should appreciate that this distinction of questions is a consequence of the construction of non-Euclidean geometry. Both geometries survive because both are interpreted differently from the way in which geometry had previously been construed. Between the time of Descartes and the investigations of Lobatschevsky, Bolyai, and Gauss, mathematicians did not distinguish geometrical space from physical space. Euclid's geometry was, at once, part of mathematics and part of physical science. The mathematical investigation showed that there was (apparently) a rival theory of physical space.[11] The mathematicians equipped both the old and the new geometry with a new style of interpretation, and left the physicists to determine which theory was true on the old construal.

The move is typical of mathematics, especially of the recent history of mathematics. Yet the root idea is readily comprehensible in terms of a division of labor which began in ancient science.[12] Initially, mathematics included optics, astronomy, and harmonics as well as arithmetic and geometry: our contemporary division of fields does little justice to the classificatory system of the ancient world. What has occurred since is a continued process of dividing questions among specialists. The old mathematical investigations of light, sound, and space are partitioned into explorations of the *possibilities* of theory construction (the province of the mathematician) and determinations of the correct theory (the province of the natural scientist). This division of labor accounts for the fact that mathematics often resolves threats of competition by reinterpretation, thus giving a greater impression of cumulative development than the natural sciences.

Consider this practice in light of the picture of mathematical reality advanced

11. Here, and in what follows, I ignore the issues raised by the apparent "conventionality" of geometry as a theory of physical space. For classic discussion of these issues, see H. Reichenbach, *The Philosophy of Space and Time*. Excellent recent treatments are available in L. Sklar, *Space, Time and Space-Time*, chapter 1, and C. Glymour, "The Epistemology of Geometry."

12. See T. S. Kuhn, "Mathematical versus Experimental Traditions in the Development of Physical Science," especially p. 37.

in the last chapter. Mathematics begins from studying physical phenomena, but its aim is to delineate the structural features of those phenomena. Our early attempts to produce mathematical theories generate theories which, we later discover, can be amended to yield theories of comparable richness and articulation. When this occurs, we regard both the original theory and its recent rival as concerned with different structures, handing over to our scientific colleagues the problem of deciding which structure is instantiated in the phenomena we set out to investigate. Our consideration of "neighboring" structures is scientifically fruitful both for enabling us to formulate and test scientific hypotheses about which structures are instantiated in the actual world, and for advancing our understanding of those structures which are instantiated.

The case of Lobatschevskian geometry is worth examining at slightly greater length, for it may appear that the status of that geometry is problematic. After all, someone may complain, Lobatschevskian geometry does *not* apply to the world, and so how can it be claimed that, in developing that geometry, Lobatschevsky, Bolyai, and Gauss were unfolding part of the mathematical structure of reality? My answer draws on the interpretation of the thesis that mathematics describes the structure of the world which I gave in the last chapter. Mathematics consists in a series of specifications of the constructive powers of an ideal subject. These specifications must be well grounded, that is, they must be successful in enabling us to understand the physical operations which we can in fact perform upon nature. What makes an idealization appropriate is its relation to prior idealizations and, ultimately, to the concrete manipulations in which we engage. We attribute to the ideal mathematical subject a power to perform Lobatschevskian as well as Euclidean operations because, by doing so, we are able to enhance our understanding of powers which have already been attributed. It is important to emphasize that, in doing this, we adopt an inclusive policy of attributing powers to the ideal subject. We extend our account of the powers of that subject in any way which is illuminating or fruitful. Thus whether or not Lobatschevskian geometry finds instances in the physical world, that geometry counts as part of mathematics because it is an appropriate idealization to introduce in our inquiries into the physical world, and what makes it an appropriate idealization is its relation to prior idealizations which were themselves properly grounded.

There is a tendency to be drawn in one of two directions. On the one hand, someone may suggest that mathematics is the investigation of the consequences of arbitrary stipulations.[13] This proposal has the advantage of accounting for those episodes in which prior mathematical theories are reinterpreted to resolve the problem of a threatened dispute. Yet, as I have already argued at some length, it fails to be epistemologically satisfactory. Moreover, one might note

---

13. Historically, this position has taken the development of non-Euclidean geometry as its primary example. For a fine discussion of the merits and shortcomings of the position, see Michael Resnik, *Frege and the Philosophy of Mathematics,* chapter 3.

that the historical development of mathematics does not reveal a *random* set of investigations of the consequences of arbitrary stipulations. The opposite pull is to anchor mathematics in what actually exists, to suggest that mathematics describes those entities (Platonic objects, structures, operations) which the world contains. I have offered what I hope is a middle course. Mathematics consists in idealized theories of ways in which we can operate on the world. To produce an idealized theory is to make some stipulations—but they are stipulations which must be appropriately related to the phenomena one is trying to idealize. I maintain *that* the idealizations which have been offered in the course of the history of mathematics satisfy this latter condition, and, in taking the methodology of mathematics seriously, I shall try to understand *in what* the satisfaction of that condition consists.

Mathematics is cumulative in a way that natural science is not, because threats of competition are often resolved by reinterpretation. Furthermore, this important role of reinterpretation does indicate the significance of stipulation in mathematics. Yet we should not conclude from this that mathematical method is simple, that all the mathematician has to do is set down his stipulations and work out the consequences. The power to stipulate is constrained by canons of mathematical method, akin to those which govern the practice of natural science. Hence my concession to the thesis that mathematics is cumulative should not be taken to invalidate the project of describing mathematical methodology. Nor, since science also proceeds by achieving idealizations, should it convince us that parallels between scientific change and mathematical change are not worth pursuing.

III

The previous sections of this chapter have attempted to clear some ground. My next step will be to use recent insights about scientific change to pose in a more precise form the question of how mathematical knowledge grows. One of the most important contributions of those philosophers of science who have been sensitive to the historical details of scientific change has been their recognition that the great clashes of opposing views involve more than a simple opposition of theoretical statements, and that, by the same token, the development of a field of science during periods of relative calm proceeds against the background of shared extratheoretical assumptions which expedite the resolution of disagreements.[14] The simple empiricist picture (as well as the most obvious re-

14. This applies not only to the work of Kuhn but also to others. For Kuhn, a revolution consists in a clash between rival paradigms, not rival theories, and "normal science" is always governed by a single paradigm, even though, during periods of normal science, the field may employ a succession of theories. Similar conceptions can be found in the writings of Toulmin, Laudan, and Imre Lakatos.

finements of it) aims to understand scientific change by finding principles which govern the modifications of sets of theoretical statements in response to observational changes. One way to reject this picture is to give up its view of the units of change. So, for example, we might replace empiricist talk of modifications of theory with Kuhnian talk about articulations and changes of "paradigms."

The concept of a paradigm is as suggestive as it is unclear.[15] It would be tangential to my main theme to offer detailed exegesis of Kuhn's discussions of paradigms. What I wish to emphasize is that the notion of a paradigm is designed to fulfil two different philosophical purposes. First, and perhaps most obviously, his references to paradigms enable Kuhn to divide the history of science into large segments. The distinction between normal and revolutionary science separates those periods in which paradigms are articulated from those in which paradigms are abandoned, and, taken at face value, Kuhn's book encourages us to apply this distinction throughout the history of science. However, in the linguistic move from the empiricist mode of discussing scientific change as theory change to the Kuhnian idiom of paradigm change, we find a second function which paradigms serve. Kuhn intends to deny that we can understand the history of science simply by talking about modifications of the set of statements which the scientists of an era accept. To chart the development of a field we need more indices of its state at any given time. Hence, Kunn introduces the richer—and vaguer—notion of a paradigm in place of the empiricist concept of a theory or corpus of beliefs.

The first point I wish to make is that the second function of the paradigm concept is independent of the first. It is quite possible for someone to be sceptical about the possibility of subsuming all episodes in the history of science under Kuhn's normal/revolutionary distinction while consistently maintaining that scientific change should be understood in terms of the modification of more than a set of accepted statements. To suppose that the science of a time is to be regarded as multi-faceted is not to endorse the idea that the history of science must reveal discontinuities, or that changes in some components of the science are so fundamental that those changes should be hailed as revolutionary. We can disregard Kuhn's doctrines about the segmentation of history, while retaining his insight that the units of change are more complicated than empiricists have traditionally supposed.

Let me elaborate on this point by drawing an analogy between an evolutionary account of human knowledge and the evolutionary theories which have been propounded in the natural sciences. With any evolutionary theory, there is a danger that one will fail to isolate the principles which govern the devel-

---

15. Kuhn's conception of paradigm (or "disciplinary matrix" as he now prefers to call it) is well known for the difficulty of analysing it. (See Margaret Masterman, "The Nature of a Paradigm," and Kuhn, "Second Thoughts on Paradigms.")

opment of the system under study because one has failed to pick out all the relevant variables. A physicist who tried to chart the changes in pressure of a gas by attending only to temperature variations, or an ecologist who studied the career of a population by considering only food supply and neglecting threats posed by predators, would be engaged in a hopeless enterprise. Evolutionary theories, whether they are concerned with the thermal behavior of gases, the modification of organic phenotypes or the development of human knowledge, hope to understand the state of the system at later times by relating it to previous states of the system by laws of development, and to achieve their goal they must provide a sufficiently detailed characterization of the states of the system.

I interpret Kuhn's challenge to simple empiricism as applying this point to the growth of scientific knowledge. Kuhn denies that we can understand scientific change by focussing simply on the shifts in allegiance to theoretical principles. Instead we must view what changes as a *scientific practice* with many components: language, theoretical principles, examples of experimental and theoretical work which are deemed worthy of emulation, approved methods of reasoning, problem-solving techniques, appraisals of the importance of questions, metascientific views about the nature of the enterprise, and so forth. Unfortunately, Kuhn fuses this important idea with a claim that certain types of changes in practice are intrinsically different from others, so that the notion of a paradigm is expected to cover those sequences of practices in which no "fundamental" transitions occur.[16]

I wish to salvage the notion of a practice and jettison the concept of a paradigm which Kuhn generates from it. One of Kuhn's major insights about scientific change is to view the history of a scientific field as a sequence of practices. I propose to adopt an analogous thesis about mathematical change. I suggest that we focus on the development of *mathematical practice,* and that we view a mathematical practice as consisting of five components: a language, a set of accepted statements, a set of accepted reasonings, a set of questions selected as important, and a set of metamathematical views (including standards for proof and definition and claims about the scope and structure of mathematics). As a convenient notation, I shall use the expression "$<L,M,Q,R,S>$" as a symbol for an arbitrary mathematical practice (where L is the language of

16. Moreover, the two theses I have distinguished here are themselves intertwined with passages in which Kuhn suggests a subjectivism about science, which has excited some readers and received most of the attention of his critics. (See the works cited in note 6.) I think it is worth pointing out that, when he is interpreted in the way I favor, Kuhn's view is not inevitably subjectivist. It is one thing to say that some of the components of scientific practice involve judgments of value, and quite another to say that such judgments are arbitrary. It would be compatible with the position I have ascribed to Kuhn to propose that the value judgments which scientific communities make about the merits of various kinds of theories, explanations, problem-solutions, and so on are rationally explicable. Moreover, in some cases, the rational explanation of these judgments could trace them to reflection upon the elements of prior practices.

the practice, M the set of metamathematical views, Q the set of accepted questions, R the set of accepted reasonings, and S the set of accepted statements). The problem of accounting for the growth of mathematical knowledge becomes that of understanding what makes a transition from a practice $<L,M,Q,R,S>$ to an immediately succeeding practice $<L',M',Q',R',S'>$ a rational transition.

In regarding a mathematical practice as a quintuple of this kind, I have selected those features of mathematical activity which seem to undergo significant change. Obviously it is possible that I may have chosen more components than I need or, conversely, that other features of mathematical activity need to be included if we are to obtain an adequate understanding of mathematical change. If I have erred in the former direction then we should find that it is possible to understand changes in some subset of the components without appealing to components which do not belong to this subset. A mistake of the latter type should be reflected in inability to reconstruct certain kinds of mathematical change. Later chapters will provide support for my analysis, both by showing how important types of mathematical change involve interconnections among all the components I have listed, and by demonstrating its capacity for handling a range of examples.

Let me conclude this section by using my reformulation of the problem of mathematical change to present more precisely the points about the similarities and differences between mathematical and scientific change which were made in Sections I and II. In the first place, scientific practices can change in response to new observations. But they can also change as the result of the existence of discrepancies among the various components of the practice. To exploit the analogy with developing systems, we may say that the movement to a new practice may result from the fact that the old practice was not in equilibrium. This type of change is the rule in mathematics. As we shall see, the components of a mathematical practice are never in complete harmony with one another, and the striving for concordance generates mathematical change. Second, we shall account for the apparently greater cumulative development of mathematics, by recognizing the existence of a particular type of linguistic change in mathematics which enables the resolution of apparent conflicts. So, where in the case of science we find the *replacement* of one theory by another (as in the case of the replacement of the phlogiston theory by the oxygen theory), in the mathematical case there is an adjustment of language and a distinction of questions, so that the erstwhile "rivals" can coexist with each other. Mathematical change is cumulative in a way that scientific change is not, because of the existence of a special kind of interpractice transition. As I have already suggested, this type of transition is found in mathematics because the task of the mathematician is to unfold the possibilities for theory construction, a task which consists in advancing appropriate stipulation of the powers of the ideal mathematical subject. We engage in this task by following an inclusive policy of attributing powers, further articulating our account of the subject in any ways which advance our understanding of the attributions already made.[17]

IV

I have used the comparison between mathematical change and scientific change to offer a very general hypothesis about the growth of mathematical knowledge. Mathematical knowledge develops through the rational modification of mathematical practices, and mathematical practices are to be understood as having five components. I now want to examine one type of interpractice transition which is especially important. To fill out the specific details of my hypothesis, we shall need to pay attention to the question of how mathematical language develops.

One of the principal obstacles to a satisfactory account of scientific knowledge has been the difficulty of understanding conceptual change in science. Any adequate study of the history of science must come to terms with the fact that the language used in the same field of science at different times seems to undergo subtle shifts. We find our predecessors using the words we use, but when we try to translate them we discover that it is difficult to record their beliefs without attributing blatant errors to them. A radical response to this predicament is to declare that the languages used in the same field at different times (at times separated by a revolution) are *incommensurable,* that statements made in one cannot be adequately translated by statements made in the other.[18] I believe that we can do justice to our predicament without making this radical response. I shall try to provide an account of conceptual change which will avoid the declaration of incommensurability, applying this account to cope with the problem as it arises in mathematics. As will become clear in later chapters, my discussion will not only help us to understand that type of interpractice transition which consists in the modification of mathematical language. It will be important in explaining other types of interpractice transition as well.

In this section, I shall investigate the general topic of conceptual change. Let us begin with the problem which leads some writers to talk of incommensurability. When we consider the language of Aristotelian physics or of phlogiston theoretic chemistry, we encounter difficulty in giving an adequate translation

17. For further discussion of the rationale of this type of transition, see Chapter 9, especially Section IV

18. This is the term favored by Kuhn and Feyeraband. In what follows, I shall adopt, without argument, a fairly straightforward reading of their claims about incommensurability. However, I want to note explicitly that there are remarks in the writings both of Kuhn and of Feyerabend which suggest that all they wish to maintain is the type of innocuous incommensurability that my account will ultimately allow.

for central expressions of the language. The standard of adequate translation invoked here is relatively straightforward: an adequate translation for an expression is one which would specify the referent of that expression. Trouble arises from the fact that we do not countenance the entities to which proponents of the old theory seem to have intended to refer, so that obvious attempts to translate their utterances construe their claims as completely false. When we reflect that the old theory seems to have been useful in developing its successor, this blanket dismissal is disconcerting. Consider, for example, the language of the phlogiston theory. We are inclined to say that there is no phlogiston, that 'phlogiston' fails to refer, and that in consequence the complex expressions 'phlogisticated air' and 'dephlogisticated air' fail to refer.[19] We are then embarrassed to find that phlogiston theorists apparently recorded many true claims about oxygen using the term 'dephlogisticated air' and that their achievements in this area were important in the development of Lavoisier's theory of combustion. How can we avoid the unfortunate conclusion that *all* phlogiston-theoretic claims are false because the phlogiston theorists were not talking about anything—without embracing the unhelpful suggestion that they were talking about the occupants of "another world" or that their theory has a different ontology from ours?

My answer is to retain the idea that adequate translation of the language of past science should specify the referents of the expressions which were formerly used, but to articulate that idea in the light of recent work on the theory of reference. I shall first review some contemporary insights about reference. This will lead me to a resolution of the problem posed in the last paragraph, and to a general account of conceptual change in science. The application to mathematics will be undertaken in the next section.

Recent studies of reference for proper names and natural kind terms have made it clear that one can refer to an object (or set) without being able to produce any description which identifies the object (or gives the condition of membership in the set) in a nontrivial way.[20] People regularly refer to Einstein without being able to say any more about him than that he was (is!) a physicist, and we can refer to aluminum without knowing any criterion which would distinguish it from molybdenum (or other metals). How is this possible? The first thing to recognize is that many of our references are parasitic on those of others. We refer to an object by intending to refer to that to which our fellows refer. Better, we acquire an ability to refer using a particular term from others who already have an ability to refer using that term. But how does the chain of reference originally start? Here, it is natural to think that the original user

19. Here, and in subsequent discussions, my use of the phlogiston theory example draws on material I have presented in more detail in "Theories, Theorists and Theoretical Change."

20. See Saul Kripke, *Naming and Necessity;* Hilary Putnam, "Meaning and Reference," "Explanation and Reference"; Keith Donnellan, "Proper Names and Identifying Descriptions," "Speaking of Nothing."

attaches the term to its referent either by providing a description of the referent or by applying it to a presented object. Thus we obtain the picture of reference as initiated by a baptismal ceremony in which the expression is fixed to its referent; thereafter, the ability to refer spreads through a community of speakers in virtue of intentions to concur in the references of other speakers (including, ultimately, the original user of the term). Baptismal ceremonies themselves divide into two types, *ostensive* and *descriptive*. In cases of the latter type, the referent is originally singled out by description, and, even though we do not assume that the description is known by all those who use the term, there are at least some members of the community who can give an identifying description of the referent. For terms introduced by ostensive baptismal ceremonies, however, it may happen that *none* of the subsequent users is able to provide an identifying description of the referent.

The history of science supplies a number of examples of terms which appear to be introduced by an ostensive baptismal ceremony. Consider, for example, terms for kinds of substances, such expressions as 'gold,' 'water,' and 'acid.' It is tempting to adopt the hypothesis that the current use of the expressions descends from occasions on which original speakers attached expressions (not necessarily the terms 'gold,' 'water,' and 'acid') to samples of gold, water, and acid respectively, intending thereby to pick out the kind of thing to which the present sample belonged. During the subsequent centuries, their successors struggled to find descriptions which would correctly characterize the kinds to which they were referring, sometimes advancing incorrect descriptions whose shortcomings were exposed by further research. Finally, we have achieved sufficient knowledge of the properties of the referents to be able to give descriptions which identify them correctly.

I think that this picture of the reference of some scientific terms has much to recommend it, but it needs to be refined in two different ways if we are to have an account which will solve the problem of conceptual change. First, we need to allow for the possibility that the links between words and the world may be constantly renewed so that, in time, a term becomes associated with a complex apparatus of referential ties, with the result that different tokens may refer differently. Second, some of the initial links between words and the world, or some of the subsequent connections, may be made by description. Recognition of the role of ostensive baptismal ceremonies should not lead us to neglect the fact that sometimes reference is fixed differently.

Both points are illustrated by the example considered above. The term 'phlogiston' was originally introduced into the language of chemistry by a declaration that phlogiston is to be the substance which is emitted in combustion. The description which is used here to fix the reference of 'phlogiston' is not satisfied by anything at all, so that, given this original establishment of its usage, the term fails to refer. As a result, insofar as the referent of 'phlogiston' is fixed through the description initially given, the term 'dephlogisticated air,' which abbreviates the phrase "the substance obtained when phlogiston is removed from the air," also fails to refer. However, when tokens of 'dephlogisticated air' occur in the writings of theorists such as Priestley and Cavendish, the best interpretation of their remarks is often to construe those tokens as referring to oxygen, a gas which Priestley was the first to isolate. For example, Priestley recounts that dephlogisticated air supports combustion better than ordinary air, that mice thrive in it, and that breathing dephlogisticated air is quite pleasant. There is a natural explanation for such remarks. Having isolated oxygen, Priestley misidentified it as "dephlogisticated air." On this occasion the referent of his token of 'dephlogisticated air' had its referent fixed in the old way: that is, dephlogisticated air is the substance remaining when the substance emitted in combustion is removed from the air (hence the token fails to refer). However, Priestley's misidentification set the stage for a new usage. Thereafter, he sometimes produced tokens of 'dephlogisticated air' whose reference is fixed *via* the misidentification (or perhaps *via* subsequent misidentifications), tokens which refer to the kind of substance which Priestley had isolated, namely to oxygen. Thus the term *type* 'dephlogisticated air' came to acquire two different modes of reference. The reference of its tokens could be fixed through the original "ceremony" in which phlogiston was picked out as the substance emitted in combustion or through encounters with oxygen. Let us call the "ceremony" through which the referent of a token is fixed the *initiating event* for that token. Then our conclusion is that tokens of 'dephlogisticated air' have initiating events of two different kinds. The fact that tokens of 'dephlogisticated air' possess different kinds of initiating events reflects the belief, explicit in Priestley's work and accepted by his fellow phlogistonians, that the different initiating events pick out the same entity. We can generalize the example by defining the *reference potential* of a term type as the set of events which can serve as initiating events for tokens of the type. The *theoretical presupposition* of a term is the thesis that all the initiating events which belong to the reference potential pick out the same entity. After Priestley's work, 'dephlogisticated air' had a heterogeneous reference potential and a false theoretical presupposition.

I want to use this general approach to account for conceptual change in the natural sciences, in general, and in mathematics, in particular. I suggest that we identify concepts as reference potentials and chart changes in concepts by following the modifications of reference potentials. If this approach is to succeed we shall need a firmer grasp on the concept of the fixing of the reference of a token through a particular event, a concept presupposed by my notion of reference potential.

What does it mean to claim that an event is the initiating event for a particular token? The question naturally arises when we try to apply the view of reference which I have offered. Attention to the case of Priestley helps us to see how to answer it. We take some of Priestley's tokens to have their referents fixed through his encounters with oxygen because, by doing so, we achieve the best explanation of why he said what he did. In this we emulate the professional historian. To understand the dicta of our predecessors, we conceive of them by analogy with ourselves, attributing to them the kinds of cognitive faculties we possess and using our knowledge of the stimuli impinging upon them to project the content of their beliefs. We do not expect them always to agree with us, for, despite the similarity of their faculties to ours, the experiences they have may be very different. What we do expect to find is a similar pattern of relationships among beliefs, desires, intentions, experience, and behavior. Claims which identify particular events as the initiating events for particular tokens should be understood in this light. We are proposing that the identification offers the best explanation of the remarks in which we are interested, where the standards for goodness of explanation are fixed by the expectation of similar psychological relations.

It will be helpful for our future discussions to recognize three main types of explanations of a speaker's token. The first is what I will call a *conformity explanation,* when we attribute to the speaker a dominant intention to agree with others and trace the referent of his token to an initiating event involving some other speaker. Although the vast majority of cases of language use require this type of explanation, many of the most interesting cases demand something different. Historical studies of mathematics and science are frequently concerned with the pioneers, those who authored new patterns of usage. Sometimes, when we attend to the utterances of a great mathematician or scientist, it is appropriate to explain her remarks by supposing that the initiating event for her tokens is an event in which she singled out a paradigm object (or paradigm objects) with the dominant intention to refer to a kind exemplified by the paradigm. I shall call these *present paradigm* explanations. They contrast with cases in which our best explanation is to take the remarks as initiated by an event in which the speaker singles out the referent by description, explanations which I shall call *stipulational* explanations. The difference between the present paradigm and stipulational types can easily be dramatized by a fictitious attribution of soliloquy. When we give a present paradigm explanation, our conception of the speaker's psychological stance is that she should say to herself, "I do not care whether or not the descriptions I am inclined to give are misdescriptions—what is important is that I am picking out a genuine kind." On the other hand, when we advance a stipulational explanation it is as if we conceived of the speaker as saying to herself, "It does not matter whether or not I am picking out a genuine kind—what is important is that the referent should satisfy these descriptions." I think that it is worth emphasizing that the attitudes manifested in these fictitious attributions are both reasonable in appropriate contexts. Among the goals of inquiry are the development of a language which will divide the world into kinds (that is, a language which will permit the formulation of simple laws) and the achievement of descriptions which will accurately characterize the referents of our terms. To sacrifice the former goal

for the latter is to risk creating cumbersome theories, while the contrary sacrifice courts the danger of vague and ill-understood language. It is sometimes
reasonable in the interests of clarity to stipulate explicitly that the referent of a
word is to satisfy a particular description.[21] On other occasions, it is equally
reasonable to allow that all of one's attempts to identify one's referent may be
premature. I conjecture that many scientific and mathematical expressions pass
through a period during which it is correct to give present paradigm explanations of the production of some tokens and stipulational explanations of the
production of others.

To sum up, conceptual change in science is to be understood as the modification of reference potentials. The reference potential of a term type is the class
of events which can initiate the production of tokens of the type. An event
counts as the initiating event for the production of a token if the hypothesis
that the speaker referred to the entity singled out in that event provides the best
explanation for her saying what she did. We can recognize a number of different forms of explanation, two of which, present paradigm explanations and
stipulational explanations, will be especially important in applying my account
to mathematics.

This approach solves the problem about conceptual change from which we
began. Proponents of incommensurability have recognized that reference potentials of terms used by former scientists need not match the reference potentials
of any terms in the language of later science. Yet it is wrong to conclude that
the referents of the individual tokens are not specifiable in the other language,
or that, in some mysterious sense, the two groups of scientists are responding
to different worlds. I shall now return from my general discussion to the specific case of mathematics, showing how to resolve some difficulties about the
development of mathematical concepts and how to make sense of an important
type of interpractice transition.

V

I propose to think of the language component of a mathematical practice as
consisting of a syntax coupled with a semantics which includes a set of reference potentials. Some kinds of changes in this component are relatively trivial.
Anyone who has attained a modest degree of sophistication in mathematics
understands the point of introducing notation to abbreviate expressions of the
existing language. To make a proof more perspicuous, or simply to avoid the

---

21. Hence I believe that the program of operationalism should not be dismissed as completely
wrongheaded. There are contexts in which the ultimate aims of science are best served by requiring
that an "operational definition" (better: a descriptive fixing of the referent of a token) should be
given. That is not, of course, to assert that all contexts are of this type, or even that a majority of
them are.

tedium of writing the same long string of symbols again and again, one decides to adopt an abbreviatory convention, and sometimes the convention spreads through the mathematical community. If such cases were the only kinds of linguistic change which occurred in the development of mathematics, we should need no elaborate account of conceptual change. For these examples are readily understood as occasions on which the syntax of the language is changed by adding a new expression and the semantics is augmented by fixing the referent of the expression through explicit stipulation in previously available terms.

Although these simple changes are very common, the history of mathematics presents us with at least two other types of linguistic change. When we look at the history of analysis, we are inclined to say that the concepts of function, continuity, integrability, and series summation change during the eighteenth and nineteenth centuries; similarly, the history of algebra seems to show the evolution of the concept of a group. I shall attempt to explain what occurs in these examples by using the approach to conceptual change developed in the last section. I shall also consider a more worrying type of case. Sometimes it appears that a new expression is introduced into mathematical language by a stipulation which violates previously accepted theorems. This seems to occur, for example, both with the initial usage of expressions for complex numbers and with Cantor's term 'ω.' Such cases are the mathematical analogs of those episodes which, like the phlogiston-theoretic example of the last section, provoke philosophers of science to appeal to incommensurability. If my approach can yield insight into them then that should count strongly in its favor.

Initially I shall discuss these cases from an ontologically neutral standpoint, without invoking the picture of mathematical reality which I presented in Chapter 6. I shall simply assume that mathematical expressions typically refer and inquire into their mode of reference without adopting my favored view of their referents. There are two advantages to this procedure. First, it will show that the thesis about linguistic change defended here can be accepted independently of any particular picture of mathematical reality. Second, discussion of the examples will be focussed more precisely by concentrating on the types of referential links between words and entities without worrying about the nature of the entities to which the words are linked. However, I shall conclude my discussion by explaining one example from the perspective of the ontological view of the last chapter.

A first approximation to an account of those conceptual changes typified by the evolution of the concepts of function, continuity, integrability, series summation, and group can be given as follows. Originally the reference of the associated terms was fixed through paradigms. Later discussions show a sequence of attempts to give a descriptive characterization of the entities which had previously been picked out. Consider, for example, the concept of function. Leibniz began from the idea that the functions of a *curve* were such things as its length or area, and that the functions of a *point of a curve* were such

things as the tangent to the curve at that point. Thus the term 'function' originally had its reference fixed through certain paradigms. As the calculus was
developed by Leibniz's successors, the set of things recognized as belonging
to the same kind came to include entities which were not obviously subject to
characterization in geometrical terms. Euler achieved a *partial* descriptive characterization of the referent of 'function.' In a famous sentence he announced
that a function is any expression however made up of variables and constants.
I call this a "*partial* descriptive characterization" to highlight the fact that
Euler's statement itself contains an expression whose reference is fixed through
paradigms. The notion of an expression's "being made up of variables and
constants" has its reference fixed by the paradigms of expression formation
used in constructing polynomial expressions. Further work was required to determine if "functions" given only by integral or infinite series representation
belong to the same kind, and to arrive at the modern general characterization
of a function.

This kind of story obviously runs parallel to the accounts we would offer
concerning the evolution of the natural scientific concepts of acid, water, and
so forth. We regard the evolution of the concepts as consisting in the replacement of reference by way of paradigms with a descriptive characterization of
the referent. Although this brings out the main features of the development of
the concept, it can be improved by drawing on some of the ideas of the last
section. Specifically, we can recognize that the referents of some tokens of the
expression under study are fixed through initiating events in which a description
which is ultimately rejected as an appropriate characterization is used to single
out the referent. In short, the reference potentials of these terms are heterogeneous, and the evolution of the concepts shows an interesting interplay between
the addition of new paradigms for use in initiating events and the discarding of
descriptions which had previously been taken to give adequate characterizations
of the referents.

Continuing the example used above, I suggest that some tokens of 'function'
which appear in the writings of eighteenth- and nineteenth-century analysts have
their reference fixed descriptively. Thus, for example, the dispute between Euler and d'Alembert concerning complete solutions to partial differential equations (in particular, the equation of motion of the vibrating string) turns in part
on their having two different conceptions of function.[22] D'Alembert offers a
specification of the referent of 'function' which excludes entities regarded by
Euler as belonging to the same kind as those which serve both men as paradigms. The eventual resolution of the dispute involved modification of the reference potential of 'function,' abandoning d'Alembert's favored description as
an appropriate means of fixing the reference of tokens of 'function.'

22. An excellent account of the dispute is provided in Ivor Grattan-Guinness, *The Development of
the Foundations of Analysis from Euler to Riemann,* chapter 1.

The cases which are most difficult are not those in which we can discern a clear pattern of development towards characterization of a previously uncharacterized referent, overlaid with occasional uses of what turn out to be inadequate preliminary descriptions, but those in which, from the beginning, there is an apparently crucial obstacle to the provision of a descriptive characterization of the referent of the newly introduced expression. Sometimes it seems that a new symbol, or complex of symbols, has its reference fixed in such a way that, from the perspective of the mathematics of the time, it can have no referent. I shall examine two examples of this type in a little more detail, since the problematic character of such cases provides the best way of exposing the strength of my account of conceptual change.

Consider first Cantor's initial introduction of symbols for transfinite ordinals in 1883. Cantor fixed the referent of his symbol 'ω' by declaring that ω is to be the first number immediately following the series 1, 2, 3, ...[23] Cantor seems to have used a description belonging to the previous language of mathematics to fix the referent of a new symbol. Yet it would be wrong to assimilate Cantor's specification to those trivial cases of abbreviation which I noted at the beginning of this section. Many of Cantor's contemporaries were puzzled—and some were outraged—by his procedure. Their response was based on an appeal to an alleged "theorem" to the effect that nothing follows the entire series 1, 2, 3, . . . , a "theorem" that depends on the *prima facie* plausible premises that the series of natural numbers does not come to an end and that it makes no sense to speak of something following an entire series unless the series comes to an end.

When we consider this example without employing the distinctions introduced in the last section we appear to have two options. If we suppose that Cantor used antecedently available language successfully to refer to a transfinite ordinal, then we shall find ourselves with the task of explaining how so many of his contemporaries viewed his specification as deeply puzzling. If, on the other hand, we credit them with a correct understanding of the old language, taking the referents of the expressions Cantor employed to be fixed so as to preclude the possibility that anything satisfies his specification, we shall have trouble seeing how he could have launched himself into the transfinite. The remedy is to recognize the expressions which figure in Cantor's characterization of ω as having heterogeneous reference potentials. Consider first the ways in which the referent of 'number' could be fixed. One way to specify the numbers is to take them to be the complex numbers. Thus we can imagine a late nineteenth-century mathematician fixing the referent of 'number' by saying: "A number is anything denoted by an expression '$a + ib$' where 'a,' 'b' are deci-

23. See Cantor, *Gesammelte Abhandlungen*, p. 195. Cantor later defined the symbols for transfinite ordinals differently, but these later definitions rested on a new analysis of the concept of number. The later definitions can also be understood by applying my approach to conceptual change.

mal expressions.'' Our mythical mathematician would obviously be puzzled by Cantor's claims, since nothing of the kind picked out by this specification follows all the natural numbers: An alternative way to fix the referent of 'number' would be to suppose that numbers are entities on which one can perform certain kinds of operations (saying, for example, that numbers are those things which can be added, subtracted, multiplied, and divided). Given this method of fixing the referent the question of the existence of transfinite numbers is an open question. To settle it, one needs to show how it is possible to define recognizable analogs of the standard operations on ordinary numbers which can be applied to transfinite numbers. The naïve methods of introducing transfinite arithmetic—such as the extension of ordinary division to allow division by zero—had been explored long before Cantor, and shown to lead to paradox. Thus Cantor had to turn back a serious challenge to the possibility of extending arithmetic into the transfinite. Transfinite arithmetic plays so large a role in his papers because Cantor's extension of arithmetical operations shows that the challenge can be met. Cantor produces an analog of ordinary arithmetic, thereby demonstrating that his transfinite numbers are indeed *numbers,* that is, entities to which arithmetical operations are applicable.[24]

Yet this is only to touch on one aspect of the problem involved in extending mathematical language to include reference to the transfinite. So far, I have addressed the worry that Cantor's specification of ω cannot characterize a *number,* but I have not examined the deeper anxiety that it fails to pick out *anything at all*. This anxiety has its source in the idea, to which I alluded above, that it only makes sense to speak of an entity as following an entire series if the series comes to an end. One way to explicate this idea would be to take the referent of '① follows all the members of the series ②' to be fixed in such a way that it only includes pairs whose second member is a series with a last member. Perhaps some of Cantor's contemporaries fixed the referent of the expression in this way, and thus concluded that Cantor's characterization of ω must inevitably be empty. Their objection could be turned back by showing that there are paradigms of succession in which all the members of a series are succeeded even though the series has no last member. In general, points of accumulation (limit points) for infinite open point sets will provide examples. (A specific instance is given by noting that 1 follows all the members of the infinite series $< \frac{1}{2}, \frac{3}{4}, \cdot \cdot \cdot, 1-\frac{1}{2^n}, \cdot \cdot \cdot >$.) Hence Cantor could appeal to paradigms of succession to rebut the complaint that I have reconstructed, and, once again, he would face the challenge of showing that the notion of succession used in his treatment of the transfinite is sufficiently similar to the paradigms in which he would anchor the reference of '① follows all the members of the series ②.'

Quite evidently, I have not given a detailed historical reconstruction of the

24. For further discussion of this example, see Section IV of Chapter 9.

way in which Cantor's extension of mathematical language generated perplexity among his contemporaries and how he was able to dissolve that perplexity. What I have tried to show is that my approach to conceptual change allows us to see how the mathematicians of Cantor's time might have seen his work as conceptually confused; how, nonetheless, Cantor could use the existing mathematical language to refer to transfinite numbers; and, finally, how he could argue for replacement of those concepts of number and succession on which opposition to his specifications is based. We may note, in passing, that Cantor's papers in fact employ the strategies I have attributed to him. The introduction of transfinite ordinals is linked to the theory of infinite point sets and to a transfinite arithmetic.[25]

Let me now consider a last example, which will reinforce the points that have just been made. Mathematicians of the late sixteenth century began to use expressions for the square roots of negative numbers. Thus expressions like '$\sqrt{-1}$' (and cognate terms) became relatively common in writings on algebraic equations, and, later, in the early integral calculus. Now, since '$\sqrt{-1}$' is an abbreviation for the expression "the number whose product with itself is -1," an expression which is a syntactically well formed expression of the old language, we face a similar dilemma to that which we encountered in the case of Cantor's 'ω.' To what does '$\sqrt{-1}$' refer? Two hypotheses present themselves: first, '$\sqrt{-1}$' refers to i; second, '$\sqrt{-1}$' fails to refer. The first hypothesis has the advantage of making it clear how reference to complex numbers became possible. The language of mathematics always had the resources to refer to these numbers. But the hypothesis fails to explain the deep and long-lasting suspicion of complex numbers and the strenuous efforts which were made to understand them. By contrast, the second hypothesis enables us to account for the resistance to complex numbers at the cost of making it mysterious how we ever came to be in a position to refer to them. '$\sqrt{-1}$' fails to refer, we might say, because, in the way in which 'number' was used at the time of the alleged introduction of complex numbers, there is no number whose product with itself is -1. The opposition to the numbers was so intense because mathematicians were all acquainted with a theorem to this effect.

Neither hypothesis is correct, but both have captured part of the story. One way to fix the referent of 'number' is to use the available paradigms—3, 1, -1, $\sqrt{2}$, π, and so forth—to restrict the referent to the reals. Given this mode of reference fixing, the theorem that there is no number whose product with itself is -1 is almost immediate. (Any number is positive, negative, or zero. The product of a positive number with itself is positive, the product of a negative number with itself is positive, the product of zero with itself is zero.) Given a different way of fixing the referent of 'number,' numbers are entities on which arithmetical operations can be performed. Here, from the point of

25. See J. Dauben, *Georg Cantor*, chapters 3–5.

view of medieval and renaissance mathematics, it is an open question whether one can find recognizable analogs of the paradigm operations which allow for the square of a "number" to be negative. In effect, Bombelli and the other mathematicians who allowed expressions of the form '$\sqrt{-\nu}$' to enter their calculations were fixing the referent of 'number' in this second way, and were referring to complex numbers. What needed to be done to show that the more restrictive mode of reference fixing should be dropped from the reference potential of 'number' was to allay fears that recognizable analogs of ordinary arithmetical operations could not be found. During the seventeenth and eighteenth centuries, algebraists, analysts, and geometers responded successfully to such fears. Gradual recognition of the parallels between complex arithmetic and real arithmetic led to repudiation of the more restrictive mode of reference fixing, so that the reference potential of '$\sqrt{-1}$' came to include only events in which i was identified as the referent.

In considering this example it is helpful to drop the stance of ontological neutrality which I have been adopting. One special feature of the concern about complex numbers was the felt need for a concrete interpretation of them. (Thus the metamathematical views of the practices of mathematicians up to the end of the eighteenth century contained a requirement that, for any kind of number, some statements about numbers of that kind must admit of concrete construal.)[26] An important episode in the acceptance of complex numbers was the development, by Wessel, Argand, and Gauss, of a geometrical model of the numbers. We can obtain a clear view about what was being demanded and how the demand was satisfied if we adopt the picture of mathematical reality given in Chapter 6. Prior to the work of Bombelli and his successors, the referent of 'number' was fixed with respect to paradigms of number operations. Each of the paradigm number operations could be given a construal in concrete physical terms: natural number operations obtained their physical interpretation in the process of counting; real number operations found theirs in the process of measurement. Bombelli can be regarded as suggesting that there is a type of number operation which had not hitherto been recognized. To eliminate from the reference potential of 'number' the restrictive mode of fixing the reference to the familiar kinds of number operation, it was not sufficient to show that the new operations would submit to recognizably arithmetical treatment. Proponents of complex numbers had ultimately to argue that the new operations shared with the original paradigms a susceptibility to construal in physical terms. The geometrical models of complex numbers answered to this need, construing complex addition in terms of the operation of vector displacement and complex multiplication in terms of the operation of rotation.

In general, of course, I want to suggest that all the examples of conceptual

26. The presence of this requirement also helps us to understand the opposition to negative numbers, apparent even in Descartes. (See Part III of the *Géométrie*.)

change in mathematics should be understood by integrating the central idea of shifting reference potentials with my picture of mathematical reality as constituted by the operations of an ideal subject. It should be easy to see how the integration is to be accomplished. At any stage in the history of mathematics, mathematical language will contain expressions referring to or qualifying the operations of the ideal subject. These expressions may have their reference fixed through paradigms of such operations or through descriptive characterizations. They may even have a heterogeneous reference potential. In modifying the reference potentials, mathematicians attempt to achieve a more adequate theory of the ideal activity of the constructive subject. Thus, to translate the point of the examples of transfinite and complex numbers, modes of fixing the referents of mathematical expressions which unnecessarily restrict that activity come to be abandoned.

8
Mathematical Changes

I

In the last chapter I tried to expose the kinship between mathematical change and scientific change, and I introduced the notion of a *mathematical practice* as an important concept for understanding the growth of mathematical knowledge. In addition, I discussed one of the five components of mathematical practices, describing what I take the language of a practice to be and the ways in which I view mathematical language as changing. My task in this chapter is to explain the other four components and the ways in which they can be modified.

I shall begin with the simplest case. The set of accepted statements belonging to a practice is simply the set of sentences, formulated in the mathematical language of the time, to which an omnivorous and alert reader of the current texts, journals, and research papers would assent. A more fine-grained analysis of this component is possible by applying two familiar distinctions from the philosophy of science. In the first place, we could discriminate among the accepted statements, as just characterized, by considering the degrees of credence which our imagined reader would give to them.[1] Second, we could allow for various *types* of assent. Recent philosophy of science has found it useful to distinguish cases in which a scientist endorses a hypothesis as true from cases in which the hypothesis is adopted as the basis of research.[2] Perhaps a similar distinction may prove useful in understanding some episodes in the history of mathematics. However, at this stage of the investigation of the methodology of mathematics, I do not think it worthwhile to adopt either of the refinements just considered, and I shall rest content with the simple characterization with which I began.

1. For a classic exposition of this type of idea, see R. Carnap, "The Aim of Inductive Logic."
2. See R. Burian, "More than a Marriage of Convenience: On the Inextricability of History and Philosophy of Science," and L. Laudan, *Progress and Its Problems,* pp. 108–14.

What kinds of changes does this component undergo? The answer seems easy: statements can be added or deleted. However, two points deserve emphasis. First, change in the corpus of accepted statements sometimes involves repudiation of what was previously firmly believed. As I remarked in the last chapter, this can occur even when the earlier acceptance was justifiable. Witness Cauchy's announcement that the sum of a convergent series of continuous functions is continuous or Fermat's claim that every number of form $2^{2^n} + 1$ is prime. In both cases not only do we have abandonment of a rationally accepted *sentence*, but there is no temptation to suggest that we should find an alternative to the homophonic translation so as to preserve what was previously believed.

The second point is one that I have not yet discussed. Frequently, modification of the set of accepted statements is intertwined with change in mathematical language. Consider one of the examples from the last chapter. Many of our predecessors would have assented to the sentence "There is no number whose square is -1." Contemporary mathematicians deny the sentence. Yet what has occurred is not simply a change in belief. Part of the content of the old assertion is preserved in our endorsement of "There is no real number whose square is -1." The reference potential of 'number' has changed, and, given *one* way in which our predecessors fixed the referent of 'number,' we agree with their assertion, even though we would reject the claim when the referent is fixed to include all those entities which submit to arithmetical operations.

In the example just considered, a change in response to a particular sentence concealed a measure of agreement with the members of the earlier community. By contrast, continued espousal of a sentence can hide disagreements. Contemporary mathematicians appear to agree with geometers since Euclid, in sometimes asserting the sentence "Given a line in a plane and a point of the plane outside it, there is a unique line in the plane through the point which is parallel to the given line." Is the agreement genuine? Well, when the mathematicians of today assert the sentence the referents of their terms 'point' and 'line' are fixed to the points and lines of Euclidean space. (The method of fixing the referents may even involve the stipulation that the sentence in question is to be true.) Perhaps in some of their utterances of the sentence, our predecessors referred in the same way so that we can endorse some of their remarks about the uniqueness of parallels. However, on at least some occasions, it is evident that they used 'point' and 'line' to refer to the points and lines of physical space. Tokens of the fifth postulate (or equivalent sentences) which involve these references are rejected in modern mathematics.

Alterations of the corpus of accepted mathematical statements go hand in hand with the modification of mathematical language. Once the point is appreciated, we can begin to see how to approach the large number of cases in which we no longer assert the sentences proclaimed by our predecessors, although we feel unhappy in describing ourselves as rejecting what they said. Consider Hamilton's quaternions. Hamilton introduced the quaternions as numbers, specifically as four-dimensional analogs of complex numbers. The theory of quaternions no longer plays any large role in mathematics (certainly it plays nothing like the role which Hamilton envisaged for it!), but it would be misleading to characterize the current view as a *rejection* of the theory of quaternions, or even as a denial that there are quaternions. Our perspective is different. Hamilton's algebra for quaternions is one among many kinds of algebra which we recognize, and we see no reason to identify the elements of one of these algebras as *numbers*. In effect, we have contracted the extension of 'number' in Hamilton's usage. The result is that many of the sentences he wrote appear to us to be wrongheaded, even though we can represent in our own idiom that part of their content with which we agree.

II

The next of the components which I shall consider, the set of accepted reasonings, requires a more complicated treatment. The accepted reasonings are the sequences of statements mathematicians advance in support of the statements they assert. It will be important for our attempts to delineate patterns of rational mathematical change to make some distinctions among these reasonings.

Prominent among the set of accepted reasonings will be the sequences of statements which are counted as proofs. Such sequences will begin from statements identified as first principles and will proceed according to accepted canons of correct inference. Both the criteria for first principles and for correct inference are set by the background metamathematical views (which I shall discuss in Section IV). Those metamathematical views are intended to specify the conditions which must be met if a sequence is to fulfil the distinctive functions of proofs.

In Chapter 2, I advanced the thesis that the concept of proof is a functional concept and I offered an apriorist construal of the function of proofs. I now wish to give a different account. The first and perhaps most obvious element of this account is that proofs sometimes yield new knowledge. There are many cases in which we know the first principles from which the proof begins and, by advancing from these principles by means of the inferences codified in the proof, we gain knowledge of the conclusion. Moreover, given the first principles, the proof provides optimal support for the conclusion, in that other ways of obtaining the conclusion from those premises would be *more* vulnerable to challenge. This is not to say that the steps in a proof are invulnerable to challenge, but merely that proofs fare better in this regard than rival forms of argument. We might sum this up by saying that proofs represent the best ways of obtaining new knowledge from old which are available to us. However, this does not exhaust the role of proofs. Our understanding of a mathematical subject is enhanced by achieving a systematic presentation of the results in the field, by finding a small number of principles from which we can infer all those results through steps which we take to expose as completely as possible the connections among statements.

On the view I propose, proofs are sequences of statements which belong to a system of such sequences, and the function of the system is twofold. In the first place, *some* of the sequences in the system represent optimal ways of obtaining new knowledge from old. Second, the system advances our understanding of the mathematical theorems we derive, by exposing clearly their connections to one another and to a small set of principles which, from the perspective of the system, are not derived. Our knowledge of the members of this set will typically be based upon our prior knowledge of *some* of the theorems we derive from them. However, *other* theorems will be known by deriving them from the principles which make up the special set. Proofs of the former theorems do not serve to generate new knowledge, but advance our understanding. Proofs of the latter theorems not only fulfil the function of increasing understanding but also yield a preferred route to knowledge of these theorems.

At a given stage in the development of mathematics, the sequences of statements accepted as proofs are those which belong to a system recognized as discharging this double function. The ways in which these functions are taken to be discharged, and hence the structure of the system of proofs, are points on which the background metamathematical views pronounce. As I shall explain in Section IV, these metamathematical views mediate between the general conception of the function of proofs just outlined—a conception which I regard as invariant through the development of mathematics—and the set of sequences actually accepted as proofs. Metamathematical views represent the judgment of the times about how the functions of proofs are achieved.

For the present, what I want to emphasize is that the set of accepted reasonings will outrun the set of accepted proofs. Some mathematical statements are asserted on the basis of inductive or analogical arguments. Such arguments standardly will not fulfil either of the functions of proofs, but they may nonetheless be used to warrant mathematical belief. Polya has presented a large variety of examples of inductive generalization from the history of mathematics.[3] The most popular field for such inferences is probably number theory. Dozens of generalizations about the natural numbers have initially been accepted on the basis of the observation that they hold for a finite sample of cases. Sometimes these generalizations come to grief—as in the case of Fer-

3. See G. Polya, *Induction and Analogy in Mathematics*.

mat's conjecture about numbers of form $2^{2^n} + 1$. However, in many cases, the acceptance of a generalization on the basis of incomplete induction is the prelude to an eventual proof.

For my purposes, the most interesting subset of the accepted reasonings of a practice will not be those which are accepted as proofs nor those which are clearly unable to discharge the functions of proofs, but those which occupy an intermediate status. I shall call these the *unrigorous reasonings* of the practice. They are analogous to proofs in promising to figure in the system of proofs, but disanalogous in that they cannot be integrated within that system. They share with accepted proofs enough structural features to make it reasonable to suppose that a proof can be given along the lines they set down. Yet it is currently impossible to reconstruct them in a way which will accord with the system of proofs.

Consider the most obvious examples of unrigorous reasonings in the history of mathematics: the reasonings which use the method of infinitesimals. Seventeenth and eighteenth century mathematicians were accustomed to calculate the derivative of the function $y = x^2$ with respect to x by employing something like the following reasoning:

$$\frac{dy}{dx} = \frac{(x+o)^2 - x^2}{o} = \frac{(x^2 + 2xo + o^2) - x^2}{o} = \frac{2xo + o^2}{o} = 2x + o.$$

Since o is small in comparison with x, $\frac{dy}{dx} = 2x$.

Although this reasoning shows superficial similarities with many algebraic proofs, it is also puzzling. Why are we allowed to "neglect" the term in o? The problem is not simply that we cannot recast the argument as a deduction from accepted premises. As the mathematicians of the seventeenth and eighteenth centuries found, it is hard to present it in any way which does not introduce premises which are obviously false.

Arguments regarded as unrigorous are those with respect to which there is a particular type of failure of understanding. They appear to be candidates for adoption within the system of accepted proofs, in that they seem to exhibit some of the features characteristic of accepted proofs. However, there is no known way of recasting them as arguments which accord with the background constraints on proofs. When mathematicians try to reformulate them as gapless arguments from acceptable premises, they fail. Thus, the mathematicians of the late seventeenth century could discover no way to recast the arguments of the methods of infinitesimals in a way which would accord with their (implicit) standards of elementary deductive inference and which would, at the same time, avoid introducing principles whose truth they had excellent reason to doubt. (Their efforts will be described in more detail in Chapter 10.) An unrigorous reasoning, we may say, is one whose fine structure is not understood.

There is another important distinction among types of accepted reasonings which is orthogonal to the divisions I have just introduced. This distinction will involve references to the set of accepted questions, which I shall consider in Section III. Among the set of accepted reasonings, we can distinguish *problem solutions*. These are reasonings which enable us to obtain an answer to a question antecedently accepted as worth asking. Now some reasonings which meet our standards for proofs will not function as problem solutions, because they do not enable us to proceed from a state of ignorance with respect to a question to a state in which we know the answer to the question. An obvious example is provided by inductive proofs of theorems about the sums of finite series. If we have already conjectured that the value of $\Sigma_{r=1}^{n} r$ is $\frac{1}{2}$ n · (n + 1), then we can use mathematical induction to prove our conjecture. But mathematical induction does not generate the answer to the question for us. Conversely, there are problem solutions which do not satisfy the conditions imposed on proofs by our metamathematical views. Some of Euler's series summations unrigorously generated answers to accepted questions (answers which were then verifiable by using techniques which could not have been used to discover the solutions). Hence we can ask two separate questions of a piece of reasoning: does it count as a problem solution? does it count as a proof? In some cases, both questions will receive affirmative answers, but we should recognize that either characteristic can occur in the absence of the other. There are examples in the history of mathematics of classes of reasonings which are explicitly recognized either as instantiating a general solution technique (but not a proof technique) or, conversely, as exemplifying a general proof technique (but not a solution technique). One interesting case is that of the Greek method of analysis. This method was identified as a means for solving problems which did not count as a proof technique. However, it was supposed that successful analyses (problem solutions), once obtained, could be transformed into successful syntheses, reasonings which would be proofs but not problem solutions. Thus, in this example, we find an association of two techniques, one which is recognized as a problem-solving technique but not a proof technique, and one which is regarded as a proof technique but not a problem-solving technique.

It is natural to ask whether any accepted reasoning must either be a problem solution or a proof. I think the answer is "No." As we shall see in Chapter 9, there are cases in which reasonings which could not be used to generate the answers to questions and which do not explain why those answers are correct are needed to verify the conclusions generated by unrigorous problem solving methods. Thus some reasonings, checking procedures, belong to the class of accepted reasonings even though they are neither problem solutions nor proofs.

We can now begin to classify the kinds of changes which can occur in the set of accepted reasonings. New reasonings may be added, old reasonings abandoned, or there may be internal shifts in the set of accepted reasonings as mathematicians change their minds about whether or not a particular reasoning should count as a proof. Additions to the set of accepted reasonings divide into three types. Most obviously, we may come to accept a new reasoning at the same time as accepting its conclusion: the common case of proving a new theorem is of this type, as is that of accepting an answer to a question and simultaneously adopting the unrigorous reasoning which generates it. Alternatively, we may adopt a new reasoning whose conclusion was already accepted but not proved. This occurs when we prove for the first time a long-standing conjecture or when we find a proof of the conclusion of an unrigorous reasoning. (It may, nevertheless, happen that the unrigorous reasoning is retained because of its status as a problem solution.) Finally, we sometimes adopt new proofs of theorems which were already proved. Many theorems of mathematics come to acquire a number of different proofs. In such cases, the provision of alternative proofs enhances our understanding of the theorem and provides greater insight into interconnections.

In a similar way, we can recognize three different kinds of changes in which reasonings are deleted. Sometimes, as with Cauchy's claim about convergent series of continuous functions, a "theorem" is rejected along with its "proof." In other cases, the conclusion survives the demise of the reasoning. The history of the four-color conjecture is full of examples of alleged proofs whose rejection has spurred mathematicians to try again. Finally, there are some theorems for which certain privileged proofs are selected while others are discarded. The original proofs of the insolubility of the quintic in radicals, which depended on laborious computations to show the impossibility of particular subgroups of groups of permutations, have given way to the elegant and general reasoning which stems from Galois. Similarly, Cantor's early proofs of the nondenumerability of the continuum are no longer needed once we have the far simpler argument by diagonalization.

Instead of simply deleting a piece of reasoning which had previously been accepted as a proof, the mathematical community may decide to retain it while recognizing it as unrigorous. So, for example, in the early days of research on infinite series there was no reason to believe that theorems about such series cannot be proved by rearranging the terms of the series. Once it was recognized that the method of term rearrangement sometimes gives unacceptable results, the method was no longer accepted as a method of proof, but reasonings exemplifying it continued to be accepted as unrigorous reasonings of the practice. The most dramatic examples of this kind of shift occur when discoveries of flaws in certain methods of reasoning generate a change in metamathematical views, producing widespread demotion of reasonings previously accepted as proofs. Thus the work of Cauchy, Weierstrass, and the early set theorists appears "revolutionary" because so much of the structure of the set of accepted reasonings is changed.

Finally, there may be shifts in the opposite direction. This does not usually occur because mathematicians decide that they were too strict in denying to a piece of reasoning, as traditionally presented, the title of a proof. (However, the acceptance of algebraic reasoning in the seventeenth and eighteenth centuries, even in cases where no arithmetical or geometrical interpretation could be given, does reveal abandonment of strictures to the effect that proofs must always be "demonstrations in Euclid.") More typically, the community accepts a proposal for *reconstructing* unrigorous reasonings which have previously been accepted. As we shall see in Chapter 9 this type of change may lead to the acceptance of new mathematical language and new principles.

III

At any stage in the history of mathematics, there is a corpus of questions which mathematicians take to be worth asking and answering. Moreover, questions have variable urgency. Some questions are regarded as extremely important and great efforts are spent in attempting to answer them. Other questions, while recognized as worthy of attention, are not given the same emphasis. Often it requires considerable mathematical talent—even genius—to formulate good questions or to draw attention to questions which have been formulated by others. One of Fermat's principal contributions to mathematics consists in his delineation of questions which have defined the field of number theory for his successors. Likewise, Hilbert's compilation of a list of mathematical problems for the twentieth century should rank among his many achievements.

The set of accepted questions belonging to a mathematical practice consists of questions, formulated in the language of the practice, which are regarded as unanswered and as worth answering. I shall distinguish two kinds of worth (or urgency) which a question may have. *Intrinsic* worth accrues to questions which are important because answers to them would advance mathematical understanding, either by illuminating entities which have previously been discussed or by bringing new entities within the purview of mathematical discussion. So, on my account, the questions "Are there real-valued functions which are continuous everywhere and differentiable nowhere?" and "Is it possible to make comparisons of the "size" of infinite sets?" are both questions which had intrinsic worth for late nineteenth-century mathematicians, although they obtained this worth in slightly different ways. By contrast, a question gains *instrumental* worth from the fact that an answer to it would enable mathematicians to tackle further questions. The further questions may be mathematical questions or questions which emerge from applications of mathematics in the sciences. Questions concerning the provision of convergence tests for infinite series had instrumental worth for Cauchy and Abel because answers could be applied to mathematical questions about the convergence of particular series. The instrumental worth of many questions posed in the study of partial differential equations obviously reflects the physical importance of solving such equations.

In drawing this distinction, I do not intend to suggest that the worth of any accepted mathematical question should be understood *either* as intrinsic *or* as instrumental. Brief reflection will reveal that many of the most prominent questions in the history of mathematics have had both intrinsic and instrumental value. My point is to draw attention to the possibility of explaining the emergence of mathematical questions in two slightly different ways.

Let us now turn our attention to the changes that occur in this component of a mathematical practice. The most obvious type of change is that by which a question exits from the set of accepted questions because mathematicians believe that they have answered it. To answer a question is to produce a statement which is true and which stands in the correct relation to the question posed. (What counts as the correct relation depends on the form of the question, and, as logicians have discovered, it is a tricky matter to say in general what the forms of answers to various forms of questions must be.)[4] When questions are apparently answered, and thus deleted from the set of accepted questions, a mathematician will have produced what seems to be a true statement standing in the correct relation to the original question and the appearance will typically be produced by providing a reasoning which supports the answer offered. Hence contraction of the set of accepted questions in this way standardly involves expansion both of the set of accepted statements and of the set of accepted reasonings.

In talking of "appearance," I do not mean to claim that accepted answers to questions are *false* statements. Typically, mathematicians are right, and what they take for a correct answer is a correct answer. However, it is possible for a question to leave the set of accepted questions, because mathematicians mistakenly think that it has been answered, and for it to re-enter when the mistake is exposed. On the basis of his faulty "theorem" about the sum of convergent series of continuous functions, Cauchy thought he had answered the question of whether an arbitrary function has a Fourier series representation (instrumentally a very important question). When Cauchy's "theorem" was discredited, the problem which he thought he had solved resumed its former place.

Another point should be made about transitions in which questions lapse from the foreground of research because they seem to have been answered. There is obviously a close connection between answering a question and proving a theorem. Frequently, a question leaves the set of accepted questions, a new theorem is simultaneously added to the set of accepted statements and its proof joins the set of accepted reasonings. Yet this is not the only way in which problems are solved. Unrigorous reasonings can support answers, and the history of mathematics is full of cases in which mathematicians answered questions without being able to prove the associated theorems. The question "How many roots has a polynomial equation of degree n with rational coefficients?"

---

4. See N. Belnap and T. Steel, *The Logic of Questions and Answers.*

was answered long before the fundamental theorem of algebra was proved. And Euler solved the problem of finding the value of $\Sigma_{n=1}^{\infty} \frac{1}{n^2}$, without being able to prove that $\Sigma_{n=1}^{\infty} \frac{1}{n^2} = \frac{\pi^2}{6}$.

Questions can disappear from the set of accepted questions without being answered (or without appearing to be answered). Mathematicians may discover that a presupposition of a hitherto favored question is false. At some point in the late eighteenth century, the question "What is a method for solving the quintic in radicals?" lost its status as one of the important questions for mathematicians, because it became clear that no such method exists. A similar, more subtle, way for questions to be rejected is through modification of language so that some presupposition of the question seems at odds with the new scheme of classification. Conceiving of quaternions as numbers, Hamilton was encouraged to ask about them the kinds of questions standardly asked about other kinds of numbers—"What are the logarithms of quaternions?" and so forth. From our perspective, given the development of the modern approach to algebraic systems and the contraction of Hamilton's concept of number, many of his questions seem pointless. Similarly, in the *Cours d'Analyse,* Cauchy seems at least partly in the grip of traditional thinking about infinitesimals. For, after having given an explication of the crucial contexts in which 'infinitesimal' had traditionally occurred, he continues by taking very seriously questions about the ratios of infinitesimals. To a modern reader, the passages in which Cauchy discusses such issues offer a curious contrast to the other, more celebrated, parts of the book in which he offers definitions of 'limit,' 'convergence,' and 'continuity': Cauchy's questions about ratios of infinitesimals are questions which we have discarded thanks largely to his own penetrating insights about limits.

So far, I have concentrated on those types of change in the set of accepted questions which accord most nearly with the picture of mathematics as developing cumulatively. Someone who thinks of the mathematicians as piling up ever more true statements will conceive the stock of accepted questions to be progressively depleted, as answers are ticked off. Just as I contend that there are deletions as well as additions with respect to the set of accepted statements, I also claim that new questions enter the set of accepted questions. Let me mention two important ways in which this can occur. (I shall discuss the rationale for these transitions in the next chapter.) First, as the language of mathematics evolves, new questions arise by analogy with previous questions. After the introduction of expressions for complex numbers, the question of finding the logarithms of such numbers was an analogical extension of the traditional question of computing logarithms. Once the language of elliptic functions had been presented and the analogy between the elliptic and circular functions noted, mathematicians naturally asked "trigonometrical" questions about the elliptic functions. Examples of this type are legion. The major modifications of mathematical language usually prompt extension of old questions. Second, changes in the class of accepted statements may bring to the fore the question of characterizing entities which have previously been discussed. An especially striking example occurs in the nineteenth-century development of the theory of equations. After Abel's discovery that there is no method for solving the quintic in radicals and Gauss's discovery that some classes of equations of high degree can be solved in radicals, Galois posed the (immensely fruitful) question "Under what conditions will there be a method for solving an equation in radicals?''

These examples should suffice to show that we must consider the set of accepted questions as subject to expansion as well as contraction. There are also variations in the urgency of questions. Sometimes a question to which little interest had previously attached becomes endowed with a new significance. For Cauchy and Abel the question of why Euler's methods for series summation sometimes led to paradoxical results had an urgency which it had not possessed for earlier mathematicians, for whom the bizarre conclusions about infinite series were often dismissed as occasional peculiarities of a fruitful theory. Likewise, as Cantor's correspondence with Dedekind makes clear, Cantor originally saw little importance in questions about one-to-one correspondence. Only after it had been shown that the rationals are denumerable but the reals are not, did such questions come to appear crucial to the mathematical unfolding of the transfinite. Moreover, just as questions which were formerly viewed as legitimate but not particularly important may become endowed with new urgency, so too questions which were previously regarded as central may come to seem peripheral. We do not regard questions of transfinite arithmetic as wrongheaded (in the way that we conceive of questions about the logarithms of quaternions or the ratios of infinitesimals as misguided) but we do not attribute to them the importance that Cantor did. Nor are the questions which occupied the major analysts of the nineteenth century at the focus of contemporary analysis. Weierstrass's ardor for elliptic function theory is reflected in the fact that he taught courses in the subject virtually every year of his university career—far more frequently than he taught any other courses. The questions which dominated the attention of Weierstrass and his students concern an area of which even literate contemporary students of analysis are almost entirely ignorant.

IV

We can now turn our attention to the last component of a mathematical practice. I regard mathematical change as governed by very general considerations which are honored by all mathematicians at all times.[5] Mathematicians are

5. Similarly, Kuhn claims that there are very general considerations which are honored by all scientists at all times; a "set of commitments without which no man is a scientist" (*Structure of Scientific Revolutions*, p. 42). *If* Kuhn's work is to be understood as arguing that science inevitably

engaged in endeavors to advance true mathematical statements, to systematize the mathematical results which they have obtained, to offer proofs which serve the function of increasing mathematical understanding, and so forth. But these general aims are mediated by more specific metamathematical views, which can vary from community to community, and which represent the community's reflective understanding about how its ultimate goals are to be achieved. A parallel with the natural sciences may be useful here. We may conceive of natural scientists as striving to advance theories of high predictive and explanatory power, which are simple, fruitful, consilient with one another, and so forth. These desiderata are accepted by all scientists at all times. Yet they may be articulated in different ways by different groups of scientists. While agreeing to the basic demand that theories should be simple and explanatory, scientists of different eras may offer alternative construals of simplicity or of adequacy of explanation. To acknowledge this is not necessarily to endorse a relativistic view which would suggest that such interpretations are arbitrary. We can allow that there is a rational development of views about explanation and simplicity, and that such methodological views adjust themselves to the most general aims of science and to the substantive theories which are advanced in efforts to achieve those aims. In recognizing an analogous set of metamathematical views, I intend to present a similar picture of mathematical change. I do not suppose that these views are whimsical. They arise by reflecting on the ways in which past mathematical research has been most successful, endeavoring to isolate the secret of the success and to use it as a pattern for future mathematics.

The metamathematical views of a practice may not be explicitly formulated by the mathematicians engaged in the practice. Ideas about how one does mathematics may simply be inculcated in early training without any formal acknowledgment of their nature or defense of their merits. It is usually at times of great change that metamathematical views are focussed clearly, in response to critical questioning. The metamathematics of a practice is most evident when the practice is under siege.

What exactly does the set of metamathematical views include? I envisage it as containing proposals about *at least* the following issues: (i) standards for proof; (ii) the scope of mathematics; (iii) the order of mathematical disciplines; (iv) the relative value of particular types of inquiry. In all of these cases, the metamathematical views should be seen as proposals for articulating the overarching constraints on mathematical inquiry. Consider first the standards for proofs. As I have suggested above, the functions of proofs are to generate new knowledge and to advance mathematical understanding. The proof standards of a community are attempts to say how these functions are best fulfilled. So, for

---

involves subjective (irrational? arational?) decisions, then it cannot be viewed as advancing the simple subjectivism which asserts that there are no context-independent constraints. Rather, Kuhn has to be read as contending that the context-independent constraints are too weak to resolve revolutionary debates. In fact, I think that even this interpretation may represent Kuhn as more of a subjectivist than he is, but I shall not pursue such exegetical issues further here.

190

THE NATURE OF MATHEMATICAL KNOWLEDGE

example, it may be claimed that for a proof to generate knowledge or to provide complete insight into the connections among statements inferences of particular types must be employed. The claim may be advanced by describing the
kinds of inference which are held to be legitimate, or by indicating paradigms
of the type of reasoning which is preferred. In other words, to have a set of
standards of proof may be to have a logical theory, or it may simply be to
possess some approved examples of reasoning. Until relatively recently, the
latter approach to standards was more common in the history of mathematics:
proofs were sequences like those found in favorite texts (such as Euclid). Analogous considerations apply to the other issues which are resolved in the metamathematical views of a practice. Conceptions of the scope of mathematics, of
the ordering of mathematical disciplines and of the relative importance of questions, specify further the ideas which are common to those who participate in
the mathematical enterprise. Moreover, in each case, mathematicians may operate with a view which is descriptive (e.g. "mathematics is the science of
quantity") or by appeal to previous work ("mathematics consists of arithmetic,
geometry, and subjects like that").

Some examples will help to explain the metamathematical component of a
practice. Contemporary mathematicians appear to endorse something like the
following set of claims: (i) proofs should be, in principle, formalizable, and,
although formal proofs are not needed, the proofs which are presented should
allow for relatively mechanical formalization; (there may be some tacit disagreement among mathematicians about the type of formalization to be
achieved—whether, for example, second-order formalizations are permissible);
(ii) all of mathematics can be presented within a standard set theory (e.g. ZF);
(iii) set theory is the basic discipline of mathematics; in particular, set theory
is prior to arithmetic and analysis; analysis is prior to geometry; (iv) special
inquiries about particular cases are normally less useful than more general investigations; the theory of real functions should be developed as a special case
of function theory on metric spaces, the algebra of the complex numbers should
be investigated as part of the study of fields of characteristic zero, and so forth.
I suspect that, once these views are made explicit, most mathematicians will
want to qualify them in some way or another. However, I maintain that they
represent a consensus, which accords at least roughly with the ideas of contemporary mathematicians (who may divide into communities with slightly different sets of metamathematical views). This consensus contrasts with the metamathematical views of previous generations.[6] Newton and his successors of the
early eighteenth century held different positions on all four topics. Discounting
individual differences among them, such men as Newton, Cotes, MacLaurin,

6. Here I offer examples which I have articulated and defended in previous papers (but without
using the notion of a mathematical practice). For Newton's views, see "Fluxions, Limits and
Infinite Littlenesse"; for those of Bolzano, see "Bolzano's Ideal of Algebraic Analysis." The
Newtonian conceptions will be discussed briefly in Chapter 10.

and Robins held the following positions: (i) proofs are sequences of statements like those in Euclid (or *Principia!*); (ii) mathematics includes arithmetic, geometry, and kinematics; (iii) kinematics and geometry are fundamental mathematical disciplines; (iv) algebraic investigations are legitimate only insofar as they can be interpreted geometrically, arithmetically, or kinematically. Intermediate between this complex of views and the contemporary collection, is a conception of mathematics which became increasingly popular during the eighteenth century, and fully explicit in the writings of Bolzano. According to this conception: (i) mathematical proofs should not make appeal to intuition; (ii) mathematics is the science of quantity (where quantities are those things that can be added, subtracted, multiplied, and divided); (iii) algebra, the science of quantity in general, is the fundamental mathematical discipline; arithmetic and analysis are offshoots of it, and geometry is derivative from analysis; (iv) general algebraic investigations which lack geometrical interpretation are not only legitimate, but extremely important for the further progress of mathematics.

My purpose in juxtaposing these collections of assertions is to illustrate the character of the metamathematical component of a practice, rather than to explore any of them in detail. I suggest that a shared view of the nature and role of proof stands behind each of the particular theses advanced by Newton, Bolzano, and contemporary mathematicians. Thus I suppose that Newton, Bolzano, and their current successors would all agree that proofs are to generate new knowledge and to advance mathematical understanding.[7] Their differences concern the ways in which they understand these aims to be accomplished. For Newton, reasoning like that in Euclid shows the way to new knowledge and increased understanding. Bolzano offers elaborate arguments for the view that mathematical understanding is achieved by proceeding from the abstract and general to the specific and particular, not by beginning with intuitive special cases. Contemporary mathematicians develop Bolzano's point further, insisting that inferences must be presented in a fashion which allows for unproblematic decomposition into elementary steps. On Bolzano's approach, the Newtonian conception of proof is inadequate because it mistakenly ascribes priority to intuitive evidence; contemporary mathematicians would charge instead that Euclid's reasoning contains gaps which have to be filled by adding substantive mathematical premises. In retrospect we are able to judge that Newton's ideas about what proofs should be like if they are to enlarge our mathematical understanding were faulty. However, we are not tempted to dismiss those ideas as arbitrary. They are a perfectly reasonable, and comprehensible, response to the mathematics Newton knew.

Changes in metamathematical views are intertwined with large-scale changes in other components, and metamathematical change serves to indicate those

---

7. I do not attribute to these mathematicians the mistaken claim that proofs generate a priori knowledge. Bolzano is clearly innocent of the mistake, and I would like to believe that many other great mathematicians have also been guiltless in this regard.

episodes which are the closest analogs to scientific revolutions. When there are significant modifications of proof standards or of views about the scope of mathematics, we generally find substantial modification of the language of mathematics and of the set of accepted reasonings. Descartes's enrichment of mathematical language in his development of analytic geometry was accompanied by an extension of the set of accepted reasonings, and an explicit repudiation of some of the metamathematical views of his predecessors: Descartes campaigned against the dismissal of algebra as a ''barbarous art'' and against the banishment of many curves beyond the pale of geometry. In similar fashion, Leibniz's further broadening of the language and methods of mathematics was coupled with castigation of what he took to be the restrictive metamathematical ideas of his predecessors. A century later, Bolzano's reform of the language and proof-techniques of analysis was intertwined with the exposition of sophisticated metamathematical views. Finally, at the end of the nineteenth century, Cantor revolutionized mathematics, not only in changing its language, its questions, and its reasonings, but in modifying metamathematical views as well.

9
Patterns of Mathematical Change

I

My aim in this chapter is to expose some rational patterns of mathematical change in a way that will make it clear that they are rational. In most cases, the changes I shall consider will involve alteration of more than one component of the practice which is modified and the motivation for making them will reflect the character of several components of this prior practice (including components which are not themselves altered). Thus the enterprise undertaken here articulates the claim (made in Section III of Chapter 7) that the rationality of changes in one part of the practice must be understood by recognizing the relation of the modified component to other components of the prior practice.

Here is a general outline of an important way in which mathematical knowledge is transformed. Beginning from a particular practice, a mathematician (or group of mathematicians) proposes a method for answering some of the accepted questions of the practice. The proposal introduces some new language, which is initially not well understood in the sense that nobody is able to provide an adequate descriptive specification of the referents of some expressions; it advances some new statements, which may not only be unsupported by previously given reasonings but even threaten to contradict accepted statements; moreover, it may present some reasonings which cannot be integrated within the prevailing system of proofs, *unrigorous reasonings,* to use the terminology of the last chapter. Despite these deficiencies, the proposal wins acceptance because of its power to answer important questions, and the practice is extended to include it. After the extension, new questions are generated and mathematicians set about answering these questions. They also attempt to generalize the ideas put forward in the proposal. In the course of these activities, many new unrigorous reasonings may be added, new language may be developed, new statements added. Perhaps anomalous conclusions emerge. Eventually, the difficulties of understanding the language and the reasonings become sufficiently acute to threaten further attempts to answer the questions now hailed as important. At this point, mathematicians search for rigorous replacements for the unrigorous reasonings they have allowed themselves and try to achieve a systematic presentation of the results they have amassed. In the process, the language may again be changed, new statements (even new "axioms") may be added, and prior metamathematical views may be overthrown. The product may thus be a practice which appears completely different from that which initiated the process.

In Chapter 10, I shall provide a detailed account of major episodes in the history of the calculus from Newton and Leibniz to Weierstrass and Dedekind. It will become clear that the outline I have just given applies to the development of the calculus during this period. Moreover, I think that many other branches of mathematics have grown in a similar fashion, and that the constituent patterns of change which figure in my outline are even more widely applicable. Here I shall be concerned to isolate those constituent patterns of change and to illustrate them with brief examples. I shall attempt to explain how the activities of *question-answering, question-generation, generalization, rigorization,* and *systematization* yield rational interpractice transitions. When these activities occur in sequence, the mathematical practice may be dramatically changed through a series of rational steps.

I shall not discuss at any length two common types of interpractice transition, those in which mathematics is extended through adding new theorems and proofs, and those in which new statements are added through inductive generalization. These types of transition are epistemologically unproblematic. Provided we understand how axioms and standards of proof are justifiably adopted, we shall have no trouble with the extension of mathematical knowledge through the activity of proving. Similarly, just as we appreciate the rationality of inductive generalizations in some areas of scientific inquiry, so too we can recognize them as reasonable when they occur in mathematics. My neglect of these types of transition does not mean that they are excluded from my evolutionary epistemology for mathematics. Each of them will figure in part of the story. Since these parts of the story are relatively well understood, scepticism about my account will surely result from concern that *other* aspects of my theory cannot be coherently developed. To satisfy the sceptic, I need to exhibit methods of rational mathematical change beyond those of proof and inductive generalization.

II

I shall begin with *question-answering*. What occurs here is a transition from a practice $\langle L,M,Q,R,S \rangle$ to a practice $\langle L',M,Q',R',S' \rangle$ in which the metamathematical component of the practice is held constant, in which the set of accepted questions, reasonings, and statements change, and in which the language of the practice may also be changed. The modifications include: (i) deletion of some previously accepted questions, now taken to be answered; (ii) addition of some new statements, taken to answer the newly deleted questions; (iii) addition of some new reasonings, taken to support the new answers. The changes need not be exhausted by (i) through (iii), and, in the cases of principal interest to me, they will not be exhausted by (i) through (iii). Further new statements may be added because they occur in the reasonings used to support the new answers. Moreover the new reasonings may draw on language which has not previously been used and may employ expressions for which there is no adequate descriptive method of fixing the reference.

The rationale for making these modifications consists in the fact that they enable the mathematical community to answer questions previously recognized as important, and the value of providing answers may outweigh the difficulties involved in the ill-understood language, the new statements, and the unrigorous reasonings. I shall indicate in a moment some of the ways in which the costs and benefits may vary. Before I do so, I want to address an obvious concern about the possibility of extending mathematics through question-answering. It is tempting to argue as follows: "To answer a question is not simply to produce some statement but to produce a *true* statement bearing the appropriate relation (as determined by the form of the question) to the question posed. If the mathematical community is to accept new language, new reasonings, and new statements, by adopting some of the latter as answers to previously accepted questions, then there must be available to the mathematicians a method for showing that these alleged answers are correct. At the very least, they must have grounds for rational belief that these answers are correct. But if that is so, it is hard to see how an extension of mathematical language or the addition of new unrigorous reasonings can be justified in this way. For the extension would do no work beyond that already done by the available mathematical resources." To put the objection another way, suppose that I want to justify some new language, L^+, some new statements, S^+, and some new reasonings R^+, by claiming that L^+, S^+, R^+ together enable me to answer the antecedently accepted questions $q_1, \ldots, q_n$. Then I have to contend that some statements generated from my new resources, $p_1, \ldots, p_n$, provide answers to $q_1, \ldots, q_n$. How do I justify this claim? It appears that, if I am able to do so, I must have independent grounds for maintaining $p_1, \ldots, p_n$, so that the allegedly accepted questions have already been answered.

This objection fails for a number of reasons. First, it is possible that all that is available in the antecedent practice is a procedure for verifying solutions to problems and that the proposed extension gives a technique whereby solutions can be discovered. In the terminology of the last chapter, the antecedent practice contains no problem-solving reasonings for the questions although it does have the resources to check answers once they have been obtained. Second, the prior resources of the practice may not even provide complete verification of the newly achieved solutions. This can occur, for example, if there are established ways of achieving recognizably *approximate* answers to questions and the proposed extension generates answers which can be approximated by the familiar techniques. Or the extension may provide a general solution to a question many of whose *instances* can be checked by antecedently available techniques. A third possibility, which I shall examine in Section VI when I discuss the notion of *systematization,* is that, although the prior practice contains problem-solving techniques for all the questions to which the new procedures would be applied, those techniques are a motley, a bag of tricks whose workings can only be understood from the perspective provided by the proposed extension.

Let me illustrate my response with two examples.[1] The first of these involves a relatively small modification of mathematical practice: new statements and reasonings are added to resolve some antecedently accepted questions but there is no shift in mathematical language. Eighteenth-century mathematicians, most notably Euler, were concerned to compute the sums of infinite series. They tried to find general problem-solving techniques for this class of questions, verifying that the new method gave answers in accordance with the recognized approximations obtained by computing partial sums and that, in those cases in which exact solutions were already available, the new method agreed with the established results. One of the most celebrated methods is Euler's technique of representing $\sin x/x$ as an infinite polynomial and reasoning about its "roots."[2] Here is Euler's application of the technique to the previously unanswered question "What is the value of $\sum_1^\infty \frac{1}{n^2}$?"

If we expand $\sin x$ in a power series, we have $\sin x = x - \frac{x^3}{3!} + \frac{x^5}{5!} - \frac{x^7}{7!} + \dots$. Hence $\frac{\sin x}{x} = 1 - \frac{x^2}{3!} + \frac{x^4}{5!} - \frac{x^6}{7!} + \dots$. Consider this as an "infinite polynomial" whose roots are $\pm\pi$, $\pm2\pi$, $\pm3\pi$, $\dots$. This enables us to represent $\frac{\sin x}{x}$ as an infinite product $\left(1 - \frac{x^2}{\pi^2}\right)\left(1 - \frac{x^2}{4\pi^2}\right)\left(1 - \frac{x^2}{9\pi^2}\right) \dots$. When this product is expanded, the coefficient of x^2 will be $-\left(\frac{1}{\pi^2} + \frac{1}{4\pi^2} + \frac{1}{9\pi^2} + \dots\right)$. Identifying this with the coefficient of x^2 given by the power series,

1. It is easy to multiply the examples. An especially interesting case, which I have omitted here for reasons of space, is the acceptance of complex numbers.

2. For an excellent discussion of the example with similar epistemological emphasis to my own, see Steiner, *Mathematical Knowledge,* chapter 3. More mathematical details can be found in Polya, *Induction and Analogy in Mathematics.* For Euler's original arguments see his *Opera Omnia,* series 1, volume 14, pp. 75–85, 138–42.

we have $\frac{1}{\pi^2}(1 + \frac{1}{4} + \frac{1}{9} + \ldots) = \frac{1}{3!}$, whence we obtain $1 + \frac{1}{4} + \frac{1}{9} +$

$\ldots = \frac{\pi^2}{6}$.

Now one can achieve an approximate verification of Euler's result by comput-

ing (as Euler did) the sum of an initial segment of $\sum_1^\infty \frac{1}{n^2}$. But this is only one

source of evidence for the conclusion that $\sum_1^\infty \frac{1}{n^2} = \frac{\pi^2}{6}$. More evidence can be

gathered by using the same general idea—consideration of power series as infinite polynomials and identification of coefficients—in other cases. The results
obtained can themselves be checked either through computation of partial results, or, in some cases, by traditionally approved methods. Thus, for example,

Euler found that his technique gave the result that $1 - \frac{1}{3} + \frac{1}{5} - \frac{1}{7} + \ldots = \frac{\pi}{4}$,

a result which had been announced by Leibniz. This body of evidence amassed
by Euler made it reasonable for him to accept both the general method and the
particular statements it generated.

My second example involves a more major shift in mathematical practice.
Descartes (and, independently, Fermat) extended the geometrical language,
statements, and reasonings of the early seventeenth century by introducing analytic geometry.[3] In the *Géométrie,* the introduction is explicitly justified by
exhibiting the power of analytic geometry to solve traditional geometrical problems. Specifically, Descartes emphasizes the ability of his new methods to solve
problems of constructing points and lines and finding loci, problems which
figure prominently in Pappus' "collection," an important assembly of Greek
achievements. At the center of the *Géométrie* is a locus question which, according to Pappus, "neither Euclid nor Apollonius nor any one else had been
able to solve completely." Descartes indicates a complete solution which agrees
with the partial results of his predecessors.

Descartes presents the locus problem as follows: "Let AB, AD, EF, GH,
. . . be any number of straight lines given in position, and let it be required
to find a point C from which straight lines CB, CD, CF,CH, . . . can be
drawn, making given angles CBA,CDA,CFE,CHG, . . . respectively, with
the given lines, and such that the product of certain of them is equal to the
product of the rest, or at least such that these two products shall have a given
ratio, for this condition does not make the problem any more difficult."[4] What

---

3. In what follows I shall use translations from the D. E. Smith and M. Latham edition of Descartes's *Géométrie.* For an account of Fermat's contribution to the development of analytic geometry, see Michael S. Mahoney, *The Mathematical Career of Pierre de Fermat,* chapter 3. A more
extensive historical account of Descartes's work, compatible with the philosophical reconstruction
given here, is provided by Emily Grosholz in "Descartes' Unification of Algebra and Geometry."
4. *Géométrie,* pp. 26–29.

Descartes does, in effect, is to set up a system of coordinates (which will, in general, use oblique axes) taking $AB = x$, $BC = y$. In other words, the point C has coordinates (x,y) with respect to the axes AB and the line through A parallel to BC. (This line is determined since $\angle CBA$ is specified.) He then shows that the conditions of the problem determine the values of CB,CD,CF,CH, . . . , and so forth, as linear functions of x and y. Now, whatever the form of the condition which tells us that "the product of certain of them is equal to the product of the rest," substituting the values of CB,CD, . . . into the condition will give us an equation in x and y. The degree of the equation will depend, as Descartes remarks, on the number of terms occurring in the specified products. Whatever the form of the equation, it will represent the locus. Descartes expresses the point as follows: "Since this condition can be expressed by a single equation in two unknown quantities, we may give any value we please to either x or y and find the value of the other from this equation." [5] In the second book Descartes considers some special cases of the locus problem, showing that his solution agrees with that of the Ancients for the example of three or four given lines. In fact, Descartes can be more precise than his predecessors had been: he is not only able to declare that the locus, for the example of three or four lines, will be a conic section, but also able to specify the type of conic section obtained under various conditions. [6]

Both of my historical cases exemplify the same pattern. Extension of the prior practice to introduce new statements and new reasonings (and, in the Cartesian example, a modification of the old language) [7] is justified by pointing out that the extension enables one to answer questions antecedently identified as important. In both cases, reasonings which are already acceptable from the perspective of the prior practice can be used to check the new answers. However, these reasonings could not be used to *obtain* the answers (they are not

5. *Ibid.*, p. 34.

6. *Ibid.*, pp. 59–80.

7. It is easy to overlook the fact that the creation of analytic geometry by Descartes (and, independently, by Fermat) involved extensive changes in mathematical language. After Descartes and Fermat, expressions of form 'a^2,' 'ab,' 'a^3,' . . . . , where 'a' and 'b' designate line segments, acquire new meaning. No longer, as with Viète and earlier algebraists, are we compelled to interpret 'a^2' as the *area* of a square of side a, 'a^3' as the volume of a cube, and so forth. Descartes's book begins by giving a geometrical account of multiplication which interprets products as line lengths, and this linguistic change is a presupposition of his thesis that, after we have chosen lines (to serve as what we would call axes) curves are representable by algebraic equations. That thesis leads Descartes to a reclassification of curves. At the beginning of Book II he repudiates the ancient classification of curves, arguing that the old restriction to curves which are constructible by ruler and compass is groundless. Instead, Descartes proposes to classify curves according to the degree of the polynomial equation which represents them. (Curves which cannot be represented by polynomial equations are not included in Descartes's class of geometrical curves, on the grounds that there is no method for determining the value of one coordinate given an arbitrarily assigned value of the other.)

problem solutions). Moreover, in the Eulerian case they do not give *exact* answers, and in the Cartesian case they do not yield *general* answers.

There is little trouble in understanding the rationality of modifying one's practice so as to answer questions accepted as worth answering *under conditions in which the extension itself is unproblematic*. However, in the examples I have discussed, and, indeed, in the major examples from the history of mathematics, the extension is not unproblematic. Problem-solving benefits must be measured against the costs of the new methods. I claim that it is rational to accept the extended practice when the perceived benefits are greater than the perceived costs. But how is the accounting done? Here I shall simply draw attention to some obvious factors which determine the assessment of costs and benefits.

Plainly, the problem-solving dividends brought by a proposed extension of the practice will vary directly with the diversity and urgency of the questions which are answered. Sometimes the ability of the mathematical community to solve central mathematical problems is magnified enormously by an extension of the practice. Witness the examples of Cartesian geometry (briefly considered above) and the Newton-Leibniz calculus (to be examined in Chapter 10). Similarly, when problem-solving benefits initially appear to be small, in that few significant questions are answered, the proposed extension is not immediately accepted. Complex numbers were regarded with suspicion for a long time, partly because they appeared to bring only slight advances in question-answering: thus reference to them could be dismissed as "subtile and useless."[8] When the problem-solving power of complex number methods in analysis became apparent, the complaints about complex numbers became more muted.[9]

On the other side of the ledger stand the perceived costs of the extension. These will vary in ways which are not difficult to understand. Most dramatically, the new techniques of reasoning sometimes yield recognizably false results. This happened, for example, with eighteenth-century techniques for series summation. Less seriously, trouble may only arise in trying to reconstruct

8. The phrase is from Cardan's *Ars Magna* (published in 1545; available in English as *The Great Art*). See D. J. Struik, *A Source Book in Mathematics: 1200–1800*, pp. 67–69. Cardan takes a condescending attitude towards complex numbers because he sees their sole function as one of allowing for general formulation of equation-solving techniques.

9. The story of the introduction of complex number methods in analysis is fascinating, but, as I remarked in note 1 above, I have refrained from relating it on grounds of space. Here is a brief survey. Leibniz saw that complex numbers could be useful in evaluating indefinite integrals, and glimpsed the relation among complex numbers, logarithms, and trigonometric functions. (See *Mathematische Schriften*, p. 362.) The relation was also appreciated by Cotes, de Moivre, and especially Euler, who gave a full systematic presentation of it. For a short, but informative, account of Euler's achievement, see Morris Kline, *Mathematical Thought from Ancient to Modern Times*, pp. 406–11. Good discussions of the eventual acceptance of complex numbers can be found in chapter 1 of Michael Crowe, *A History of Vector Analysis*, and Ernest Nagel, "Impossible Numbers."

the new reasonings as gapless arguments from accepted premises: although the route to the conclusions is not well understood, the conclusions themselves are not recognizably incorrect. In such cases, the scope of the newly obtained solutions may make it reasonable to accept the extension and to maintain that the unrigorous reasonings will eventually be integrated into a system of proofs. As we shall see, the early adoption of the calculus was motivated by this assessment of costs and benefits. Even Berkeley, the most acute critic of the calculus, agreed that the widespread success of the calculus attested to the possibility of ultimately achieving a satisfactory explanation of its methods.[10] Nor should one suppose that it is unreasonable to adopt techniques of reasoning which generate recognizably false conclusions. If those techniques furnish sufficiently many problem solutions whose correctness can be checked, it may be reasonable to adopt them, subject to the proviso that their deliverances be verified, and to seek ways of restricting them which will no longer yield the anomalies.

Mathematicians, like natural scientists and like everyday people, are sometimes justified in using imperfect tools. Just as it is sometimes reasonable to adopt a solution to an everyday problem which is recognized as having certain deficiencies (or a scientific theory which faces certain anomalies) and to trust that, in time, one will be able to iron out the wrinkles, so it is reasonable for mathematicians to accept proposals for extending a practice which bring costs as well as problem-solving benefits. Qualitative though the foregoing discussion may be, I think it makes it clear that question-answering can be a rational way of extending mathematical practice even when the techniques introduced to answer the old questions bring problems of their own.

III

The next topic which I shall discuss concerns the generation of questions. As I indicated in the outline given in Section I, modifications of mathematical practice make it rational to pose new questions. We can envisage an indefinite sequence of rational changes taking a very simple form. At the first stage, some antecedently accepted questions are answered by the kind of extension of the practice discussed in the last section. The extension then generates new questions. A further extension of the practice is made in order to answer those questions. That extension, in turn, generates more new questions. To answer the latest questions we adopt another extension of the practice. And so it might go forever. In fact mathematics does not develop in so simple an oscillatory pattern. Other patterns of rational change interrupt the sequence I have envisaged. Nevertheless, precisely because question-answering is an important part

10. *The Analyst*, sections 19–22.

of mathematical change, the ways in which questions come to be accepted as important need to be understood.

There are many ways for a question to earn mathematical significance. Let me begin with the most obvious. Some mathematical questions have practical significance. It is helpful to know which solid with a given volume has minimal surface area if one is in the business of designing containers. Equally, the question of how to arrange a sequence of tasks so as to require the minimum expenditure of effort is relevant to the concerns of almost everyone. We can easily trace to a practical source the distinctive questions of various parts of mathematics. The theory of equations has its origin in problems about the division of legacies, the areas of fields, and other topics of everyday interest, questions which the Babylonians formulated and learned to answer. Simple questions about constructing figures provided a similar impetus to the development of geometry. Finally, questions of how to settle unfinished games of chance motivated Pascal's investigations in probability theory.

Slightly less obvious (but only *slightly* less) is the fact that importance may accrue to a mathematical question through the activity of natural scientists. As mathematical language is developed and mathematical principles are adopted, scientists are able to use that language and those principles in formulating hypotheses about the phenomena that interest them. These hypotheses can easily generate new questions for the mathematician. The fashioning of vector analysis in the nineteenth century, both in Hamilton's work and in the studies of Gibbs and Heaviside, was explicitly motivated by a desire to solve problems which arise for physicists and engineers.[11] Similarly, the rational mechanics of the eighteenth century excited interest in particular partial differential equations, and the question of finding the complete solutions of these equations played an extremely important role in the development of analysis.

However, it is evident that mathematicians do not only attach significance to questions when answers to those questions will have practical value or when their scientific colleagues are clamoring for solutions. Many mathematical questions are of interest only to mathematicians, and the principal part of our inquiry will consist in understanding what justifies attention to a problem which has no extra-mathematical significance. I regard such "impractical" questions as attempts to focus the general project of understanding the entities which mathematicians have so far discussed. Dropping the perspective of ontological neutrality and adopting the view of mathematical reality advanced in Chapter 6, we can articulate this suggestion as follows. At a given stage in the development of mathematics, mathematicians have achieved an account of the powers of an idealized subject, and, in their efforts to advance their understanding of the powers that have already been attributed, they may pose further questions concerning operations (or properties of operations) which are not found

11. Crowe, *A History of Vector Analysis*, especially chapter 6.

in the practical activities of actual people and which are not discussed in the current scientific story of the world. New questions continually arise as the account of the ideal subject is amended, as new language is added and new statements are accepted.

Let me now turn from this very general response to a more precise and detailed account. I shall divide transitions in which new questions are rationally generated into two classes. The first, *language-induced* question generation, covers those cases in which an extension of the language prompts the introduction of new questions. The second, *statement-induced* question generation, comprises cases in which modification of the set of accepted statements leads to new questions. Within these classes I shall describe the most prominent species of transitions which figure in the historical development of mathematics.

Language-induced question generation can be represented schematically as a transition from $\langle L, M, Q, R, S \rangle$ to $\langle L', M, Q', R, S \rangle$, where L' results from L through the addition of new expressions and Q' results from Q through the addition of questions involving these expressions. What makes such transitions rational? *One* important reason for extending the set of questions is to proceed by analogy with previously successful inquiries. Thus one species of rational transition consists in augmenting L with new expressions which are regarded as referring to members of a particular kind, other members of which have previously been studied, and to add to Q questions which ask about the newly distinguished entities of the kind the same things that have previously been asked about antecedently recognized entities of the kind. What motivates the transition is not hard to discern. Let us suppose that the new language is introduced on the basis of considerations like those discussed in Section II. The questions newly posed are those which it seems proper to ask about members of the kind, so that our understanding of the newly distinguished members will be advanced by finding answers to them. (Here, of course, I am not pretending to account for the acceptance of the traditional questions about the antecedently distinguished entities but to point out that that acceptance can serve as the basis for asking new questions.)

Straightforward examples of this type of transition can be found by examining episodes in which the number concept was extended. Those mathematicians of the early eighteenth century who took seriously the existence of complex numbers asked questions about the logarithms of such numbers. Note that the question of the logarithms of complex numbers, unlike that of the logarithms of real numbers, has no direct practical importance. Yet because the new entities were supposed to be *numbers,* it made sense to ask for the values of their logarithms. A century later, Hamilton's extension of mathematical language to embrace quaternions, which he construed as numbers, led him to pose questions about quaterions analogous to those previously asked about complex numbers. Work in other parts of mathematics exhibits the same pattern. When logicians introduce a new logical system they quickly ask themselves new versions of standard questions: Does the deduction theorem hold? Is the system complete? Is it sound? Does the compactness theorem apply? In all these cases, questions about members of a kind arise by analogy with traditional questions about familiar members of the kind.

However, analogy is not the only mode in which language change can induce new questions. Frequently, modification of language consists in introducing new expressions taken to refer to entities which have already been distinguished. The development of the explicit notation of polynomial equations, the introduction of the indefinite integral, the Cartesian interpretation of algebra in geometry, all offer new ways of talking about familiar entities. In their wake, new questions rationally arise: "For what values of x does $x^4 + 3x - 7 = 0$?"; "What is $\int \frac{dx}{1+x^2}$?"; "What curve has equation $x^3 + y^3 = 1$?" These questions are reasonable because, in each case, the modified language is viewed as providing a new way of talking about familiar entities and we want to find out *which* of the familiar entities the new expressions denote. (Perhaps we may even want a method for discovering the traditional representations of the referents of a set of newly introduced expressions: "How can we find, in general, those numbers x such that $ax^2 + bx - c = 0$?")

In general, this type of transition from $\langle$L, M, Q, R, S$\rangle$ to $\langle$L$'$, M, Q$'$, R, S$\rangle$ shows the following characteristics: (a) L$'$ is obtained from L by adding a set $\{\beta_i\}$ of new expressions; (b) the referents of the β_i are taken to be among the referents of a set $\{\alpha_i\}$ of expressions in L; (c) Q$'$ is obtained from Q by adding all questions of the form $\ulcorner$ What is (the value of) β_i?$\urcorner$, where these questions are to be answered by statements of form $\ulcorner\alpha_j = \beta_i.\urcorner$ The rationale for the transition is not difficult to find. Given that it is rational to extend one's language to introduce the β_i—perhaps because of considerations like those adduced in Section II—it is rational to attempt to compare the perspective which the new language provides with perspectives which were already available.

I have already offered three examples of this type of transition. Another, which will be discussed in my investigation of the development of analysis, is the genesis of questions about the value of infinite sums. These questions came to the forefront of mathematical attention because of the modification of mathematical language by Newton and Leibniz, both of whom had used infinite series representations in attempts to differentiate and integrate recalcitrant functions. Once the new language had won acceptance, it was perfectly rational to demand canonical representations of the entities (numbers, functions) to which infinite series expressions refer, and, generally, to ask for techniques for achieving such representations.

We have looked briefly at two important modes of language-induced question generation. Let us now consider statement-induced transitions, in which the emergence of new questions results from a change in the set of accepted statements. For someone who is concerned to fathom the nature of things of a certain kind, K, discoveries about members of that kind can rationally prompt various types of question. Suppose that it is found that all members of the kind hitherto investigated with respect to a property P have that property, but that it is also known that some entities (not belonging to the kind) lack P. Under these conditions it is rational to ask if all members of the kind have P. If one learns that some K's have P but some do not, then it is rational to inquire into what differentiates the subsets. Finally, if it is shown that a particular condition is necessary (sufficient) for membership in K, but it is not known whether or not that condition is sufficient (necessary) for membership in K, then it is reasonable to ask whether the condition is both necessary and sufficient. (I assume, in the case of each schema, that we can make sense of the notions of kind and property. No doubt there are parallel difficulties here to the familiar cluster of problems about kinds, properties, and projection in discussions of natural science. However, because such difficulties arise *überhaupt,* resolution of them is not a prerequisite for successful assimilation of the methodology of mathematics to scientific methodology in general.)

My point is that there are systematic ways in which specific questions emerge from the general project of understanding what it is to be a K. As we gain more information, the new statements we accept systematically yield new questions which make our general project more precise. To put it another way, the goal of fathoming the nature of K's finds expression in different particular inquiries as we modify our beliefs about K's. My three schematic examples expose straightforward ways in which local questions can result from our global endeavors.

Nothing I have said so far about statement-induced question generation is restricted to the emergence of questions in mathematics, and I think that my remarks present some general (if obvious) points about ways in which new questions come forth in science. My primary interest here is in the genesis of mathematical questions. Hence I want to show how the general suggestion of the last paragraphs and the three schemata invoked to illustrate it apply in the mathematical case. The goal of understanding mathematical reality issues in specific inquiries as mathematicians amend and extend their beliefs about various kinds of mathematical entities, and among the ways in which these inquiries are generated are the three I have explicitly noted.

One of the most striking examples of the evolution of questions in response to changing beliefs is the development of the central questions in the theory of equations. When mathematicians found that there are methods for solving the general quadratic equation, the general cubic equation, and the general biquadratic equation, in terms of the roots of the coefficients, it was rational for them to inquire whether the general polynomial equation of degree n admits of a similar method. Discovering that the general quintic is not solvable in radicals, but that some general classes of equations of high degree (for example, the *cyclotomic* equation, $x^p - 1 = 0$, where p is prime) do admit of a method of solution in radicals, it is rational to ask, as Galois did, under what conditions equations can be solved in radicals. (Galois's methods of attacking this question raise further questions in their turn.) In both cases, the new questions are a rational way to focus the general inquiry into the nature of equations, given what has been discovered. The first example poses the question of whether a property which has been found to apply to the members of a subset of a kind applies throughout the kind. Given Abel's discovery that the answer to that question is negative, and Gauss's demonstration that some members of the kind beyond those included in the original subset exemplify the property, it is reasonable to demand a criterion for separating the subkinds. In general, if we look at the history of the theory of equations from Babylon to contemporary Galois theory, we find an evolution of the questions accepted, from an initial set of practical demands to the refined, esoteric, and "impractical" questions posed today. I shall not attempt to tell the story of that evolution here, but will simply note that both statement-induced question generation and language-induced question generation, in the forms I have described, can be recognized as motivating the major changes.[12]

Similarly, in the development of analysis we find that the modification of beliefs rationally provokes new questions. Early nineteenth-century analysts saw that a necessary condition for a function to be differentiable on an interval is for it to be continuous on that interval. Under these circumstances, it was reasonable to inquire if the properties of continuity and differentiability are coextensive. Further investigations led to further questions. When it was found that a function continuous on an interval may not be differentiable at every point of the interval, it was reasonable to ask if the exceptional points are *rare* exceptions. So, quite rationally, analysts set themselves the question of whether continuous functions must be differentiable "almost everywhere"—and how to make sense of 'almost everywhere.'

At this point, it may be helpful to address an objection. On my account, new inquiries which lack practical significance result from the general project of understanding previously discussed mathematical entities. As language changes and as our beliefs are modified, this global project takes on new local forms. Perhaps it may be thought that this account is farfetched. After all, what mathematicians *say* is that certain questions are interesting, even that they have an aesthetic appeal. Would it not be simpler, and more in accordance with math-

---

12. For a discussion of the practical origins of the questions about equations which the Babylonians addressed, see O. Neugebauer, *The Exact Sciences in Antiquity* and *Vorgriechische Mathematik*. Eighteenth- and nineteenth-century work in the theory of equations is discussed in H. Wussing, *Die Genesis des Abstrakten Gruppenbegriffs*, and Mary Therese O'Malley's unpublished doctoral dissertation, "The Emergence of the Concept of an Abstract Group." An excellent survey of the history of Galois theory after Galois is available in B. M. Kiernan, "The Development of Galois Theory from Lagrange to Artin."

ematical practice, to claim that new questions are sometimes generated by the interests and aesthetic sensitivities of the mathematicians? The slogan *"L'art pour l'art"* would find its analog: *"Les mathématiques pour les mathématiques."*

Mathematicians do describe themselves as pursuing some inquiries because they are "interesting" or "beautiful." What is at issue is how we should understand these remarks. I suggest that we need not take the attribution of interest or aesthetic appeal to be groundless. The remarks of mathematicians about beauty and interest are not solutions to the problem of how mathematics develops, but material for philosophical explanation. We may view the mathematicians as responding to the fact that some questions are rationally generated, in the ways I have identified above, and characterizing their responses in terms of interest or beauty. To simplify, what makes a question interesting or gives it aesthetic appeal is its focussing of the project of advancing mathematical understanding, in light of the concepts and system of beliefs already achieved.

I do not claim to have given a complete response to the objection, and I shall return to consideration of it at the end of Section IV. At that point, I shall examine in more detail an example which appears to support the idea that brute considerations of interest and beauty play a role in the development of mathematics. If I can successfully show that, in this case, a rationale for change (and for attributions of interest and aesthetic appeal) can be given, then I shall have provided support for the general suggestions of the last paragraph.

Let us resume our main theme. So far in this section I have been concerned with the genesis of questions. Before I conclude my discussion of question generation it is important to recognize how assessment of the *urgency* of questions may rationally be modified. A major way for the significance attributed to a question to be increased is *problem reduction*. Suppose that, in modifying mathematical language or changing the set of accepted statements, we add a statement which shows how answers to certain questions q_i may be obtained from answers to other questions, q'_j. Then the questions q'_j will inherit at least some of the significance which had previously accrued to the q_i. Thus the urgency of a question may be dramatically increased, and a particular question (or form of question) may come to seem *the* question (form of question) of a particular field. If the q_i are numerous and diverse, and the q'_j are a small set (perhaps instantiating a common form), then the envisaged change will literally effect a reduction of problems, and it will be reasonable for mathematicians to attach great importance to the reducing questions (or question form), in the hope of reaping large dividends.

Examples of increase in urgency through problem reduction are legion. After Newton and Leibniz had reorganized the diverse set of geometrical problems which had occupied Descartes, Fermat, and their successors, the question "How can you differentiate an arbitrary function?" had special urgency, precisely because it was the key to solving so many of the old problems. Similarly, to extend a little my account of the evolution of investigations in the theory of equations, Galois's recasting of the theory of equations reduces the question of the solvability of a type of equation to questions about the possibility of finding subgroups of a particular order in the Galois group. Hence Galois's work lends urgency to the question "Under what conditions does a group have subgroups of a particular order?"

My central thesis in this section is that changes in mathematical language and in the set of accepted mathematical statements rationally produce changes in the set of accepted mathematical questions. More formally, there is a relation of *rational question modification* which can hold between the set of questions, Q', of a practice ⟨L', M, Q', R, S'⟩ and the set of questions, Q, of a practice ⟨L, M, Q, R, S⟩, and which depends on the relation between L' and L and the relation between S' and S. I have tried to identify some of the principal ways in which the relation of rational question modification obtains, and to show how the associated transitions develop the general project of mathematical inquiry. In doing so, I have focussed on those cases in which new questions are added (or in which the urgency of a question is increased) for two reasons. Because new questions provide the material for extension of the practice through question-answering, it is obviously important for my account of the growth of mathematical knowledge to show how the rational emergence of new questions is possible. Furthermore, my discussion of the ways in which questions may lapse from the forefront of research, given in the last chapter, should make it relatively easy to see how the deletion of questions can be rational. At first glance, the genesis of new questions appears epistemologically more problematic.

IV

One of the most readily discernible patterns of mathematical change, one which I have so far not explicitly discussed, is the extension of mathematical language by *generalization*. Thus, for example, Riemann undertakes the task of redefining the definite integral with the goal of retaining the previously achieved evaluations of definite integrals, relaxing unnecessary conditions on the class of integrable functions, and thereby embedding the prior work on integrals within a more general treatment. Hamilton's long search for hypercomplex numbers is motivated by a desire to generalize complex arithmetic. And Cantor is the most prominent—because the most successful—of those mathematicians who tried to generalize finite arithmetic. In this section, I want to try to understand the process of generalization which figures in these episodes and to see how the search for generalization may be rational.[13]

---

13. Accounts of the relevant work of Riemann, Hamilton, and Cantor are available in chapter 1 of T. W. Hawkins, *Lebesgue's Theory of Integration,* in chapter 2 of Crowe, *A History of Vector Analysis,* and in J. Dauben, *Georg Cantor,* respectively.

Generalization is usually difficult. That is, it is not readily apparent, in advance, how one can introduce into the language new expressions while conforming to the constraints imposed by previous usage. However, relaxing these constraints may make the "generalization" trivial. To focus our ideas, and to understand the nature and rationale of mathematical generalization, it will help to consider a simple example of *trivial* "generalization."

Imagine that a community of mathematicians has been working out the arithmetic of the natural numbers, and that they set themselves the task of generalizing arithmetic, "remedying its incompleteness,"[14] so that subtraction is always possible. There is an easy way to do this. They can introduce a new constant 'n' with the stipulation that, if $q > p$, then $p - q = n$. Notice that this does not lead to any inconsistency unless one assumes that general rules for operating with addition and subtraction carry over to the new language. It would be premature to reason as follows: $1 - 2 = n$; $1 - 3 = n$; so $1 - 2 = 1 - 3$; hence $1 + 3 = 1 + 4$, so that $3 = 4$. The fault in this reasoning is that we are not compelled to give a wider employment to '$+$,' '$-$,' in such a way that, for any p, q, r, s, if $p - q = r - s$ then $p + s = r + q$. But recognizing the fact that we are not *compelled* to accept this reasoning, and so be driven to unacceptable conclusions, only brings home to us the triviality of the extension. We wanted to extend the language so that *subtraction* is always possible. Thus we wanted to preserve the fundamental properties of subtraction in the wider employment of '$-$'; in particular, we wanted the usual rules of transposition of terms with change of sign to be forthcoming. Hence, while the extension of language does not land us in paradox, it fails to achieve the intended goal.

The general moral of this example is that, when we generalize by adding new expressions to mathematical language, we hope to preserve some features of the usage of old expressions, while relaxing other constraints. Sometimes, as in the case of Hamilton's quaternions, important aspects of the old usage are first articulated through the attempt to generalize it. What the generalization, if it is nontrivial, shows us, is that by abandoning certain constraints on prior usage we can obtain a theory yielding analogs of old results, and, in doing so, it brings to our attention fundamental properties of the entities which we have been discussing.

To appreciate the difference between the banal "generalization" of my example and the significant generalizations mentioned at the beginning of this section is to begin to see why generalization plays an important role in mathematics. Significant generalizations are explanatory. They explain by showing

14. The phrase comes from Dedekind, "Continuity and Irrational Numbers." In this stimulating paper, Dedekind not only offers his famous characterization of the continuum in terms of "cuts," but also suggests a general thesis about the extension of arithmetic. According to this thesis, new number systems arise as mathematicians successively "create" new numbers to "remedy the incompleteness" of the existing system. I hope that the present section indicates the possibility of a defensible interpretation of Dedekind's thesis.

us exactly how, by modifying certain rules which are constitutive of the use of some expressions of the language, we would obtain a language and a theory within which results analogous to those we have already accepted would be forthcoming. From the perspective of the new generalization, we see our old theory as a special case, one member of a family of related theories. It is helpful at this point to invoke the picture of mathematical reality which I have presented in Chapter 6. Our mathematical language embodies stipulations about the powers of an ideal subject. In attempting to generalize the language, we attribute to the subject a more inclusive set of powers and the attribution is defended by its ability to improve our understanding of the powers already attributed. Those "generalizing" stipulations which fail to illuminate those areas of mathematics which have already been developed—stipulations like that of my trivial example—are not rationally acceptable.

Let me illustrate the point by considering one of the most striking examples of generalization in the history of mathematics, a case in which the successful generalization was preceded by many failures. The question of how to generalize the language of arithmetic to include reference to transfinite numbers had baffled mathematicians before Cantor. Straightforward attempts to generalize quickly lead to trouble. Suppose that we introduce the symbol '∞,' stipulating that $\frac{1}{0} = \infty$. Then, applying the rules of arithmetic in the standard way, we can achieve the following: $1 = 0 \cdot \infty$; hence $a = a(0 \cdot \infty) = (a \cdot 0) \cdot \infty = 1$, for any number a. This kind of difficulty was recognized relatively early in the history of attempts at "infinite arithmetic." It can, of course, be avoided by restricting the arithmetical rules, but this yields a truncated arithmetic with no analogs of standard theorems. A major step forward is taken by Bolzano, who, instead of focussing on the employment of numbers in computations, saw the importance of using numbers to count the elements of a set: the analogy between finite and infinite numbers is to be grounded in the properties of finite and infinite sets, and Bolzano struggles mightily to show that there are infinite sets.[15] But, at this point, his program encounters a problem. Bolzano recognizes that, when two sets are infinite, they can satisfy both of the following conditions: (i) there is a one-to-one correspondence between them; (ii) one is a proper part of the other. This point (which goes back at least to Galileo) reveals that two conditions which coincide for finite sets can come apart in the case of infinite sets. Intuitively, it appears that the second condition is more important, so that Bolzano declares that two sets do not have the same number of members if one is a proper part of the other. Quite consistently, he goes on to claim that the existence of one-to-one correspondence between two sets is only a sufficient condition for the sets' having the same number of members when the sets are finite. Unfortunately, Bolzano's choice makes him unable to develop a theory

15. See *Paradoxes of the Infinite*.

of infinite numbers which will have analogs of standard theorems about numbers. His attempt to generalize casts no light on ordinary arithmetic, and, not surprisingly, no accepted theory of the transfinite results from his writing. Bolzano's stipulation of "sameness of size" for infinite sets fails to serve any explanatory ends, and so it is not rational to extend mathematical language by adding it.

The heart of Cantor's transfinite arithmetic is his rejection of the intuitive criterion for inclusion which Bolzano had adopted in favor of the condition on identity which Bolzano had repudiated. Cantor's first important stipulation in his presentation of set theory was his specification of the conditions under which two sets have the same *power:* M has the same power as N if and only if M and N can be set in one-to-one correspondence. Now what makes the concept of *having the same power* an interesting one? That is, what makes Cantor's stipulation reasonable? The answer comes from a pair of results which Cantor announced in his first paper on set theory (a paper which antedates the explicit introduction of the term 'power'). In that paper, Cantor showed that the set of algebraic numbers (the set of numbers which are roots of polynomial equations with rational coefficients—a set which obviously includes, but is not exhausted by, the set of rational numbers) has the same power as the set of natural numbers; but he also demonstrated that the set of real numbers does *not* have the same power as the set of natural numbers. Apparently, it was his achievement of these results which led Cantor to change his mind about the significance of questions concerning the possibility of one-to-one correspondence. It is easy to see why. The nondenumerability theorem suggests the possibility of ordering infinite sets in terms of their powers, thus pointing to the specification of a sequence of transfinite numbers. Cantor's subsequent work was to exploit this possibility in detail, showing that one could not only order infinite sets but also define analogs of the ordinary arithmetical operations. So, for example, in his final presentation of his set-theoretic work, Cantor defines the sum of two infinite cardinal numbers (or powers)[16] as the cardinal number (power) of the set which is the union of any two disjoint sets whose powers are the numbers to be summed.

Unlike Bolzano's attempt, Cantor's stipulation is rationally acceptable because it provides an explanatory generalization of finite arithmetic. Note first that the ordinary notions of order among numbers,[17] addition of numbers, mul-

16. For the sake of simplicity, I shall ignore Cantor's distinction of transfinite ordinals from transfinite cardinals and his development of transfinite ordinal arithmetic. However, it should be noted that the approach to ordinals is itself based on the notion of one-to-one correspondence. Two well-ordered sets have the same ordinal (order-type) if there is a one-to-one correspondence between them which preserves the order relations.

17. Strictly speaking, the claim that an order relation can be defined for the transfinite numbers presupposes a theorem which Cantor could not prove. For if we suppose with Cantor that the power of M is greater than the power of N ($\overline{\overline{M}} > \overline{\overline{N}}$) if N can be put in one-to-one correspondence

tiplication of numbers, and exponentiation of numbers are extended in ways which generate theorems analogous to those of finite arithmetic. (For example, if $\mathfrak{a}$, $\mathfrak{b}$ are infinite cardinals, we have the results: $(\exists x)x > \mathfrak{a}$, $\mathfrak{a} + \mathfrak{b} = \mathfrak{b} + \mathfrak{a}$, $\mathfrak{a} + \mathfrak{b} \nleq \mathfrak{a}$, $\mathfrak{a} \cdot \mathfrak{b} \nleq \mathfrak{a}$, $2^{\mathfrak{a}} > \mathfrak{a}$, and so forth.) By contrast, because he cleaves to the intuitive idea that a set must be bigger than any of its proper subsets, Bolzano is unable to define even an order relation on infinite sets. The root of the problem is that, since he is forced to give up the thesis that the existence of one-to-one correspondence suffices for identity of cardinality, Bolzano has no way to compare infinite sets with different members. Second, Cantor's work yields a new perspective on an old subject: we have recognized the importance of one-to-one correspondence to cardinality; we have appreciated the difference between cardinal and ordinal numbers; we have recognized the special features of the ordering of natural numbers. But we do not even need to go so far into transfinite arithmetic to receive explanatory dividends. Cantor's initial results on the denumerability of the rationals and algebraic numbers, and the nondenumerability of the reals, provide us with new understanding of the differences between the real numbers and the algebraic numbers. Instead of viewing transcendental real numbers (numbers which are not the roots of polynomial equations in rational coefficients) as odd curiosities, our comprehension of them is increased when we see why algebraic numbers are the exception rather than the rule.[18]

At this point, let me reconsider an objection raised in Section III. There I briefly discussed the hypothesis that irreducible attributions of interest and beauty play a part in directing the course of mathematical inquiry. If any case would support that hypothesis, it would seem to be that of the development of transfinite set theory and arithmetic. From Cantor on, defenders of the transfinite have waxed rhapsodical about the beauties of set theory. Hilbert's characterization of the realm of sets as a "paradise" is only one famous example. Transfinite arithmetic and transfinite set theory appear "impractical," and mathematicians testify to their aesthetic appeal—what more is needed to defend the idea of *"Les mathématiques pour les mathématiques"*?

I reply that the considerations I have adduced above show why Cantor's generalization of mathematics was rational. That generalization advanced our understanding in a number of ways, acquainting us with the distinction between

---

with a subset of M but not with M itself, then it is important to establish that there cannot be subsets M_1, N_1, of M, N respectively such that M can be put in one-to-one correspondence with N_1 and N can be put in one-to-one correspondence with M_1, unless M can be put in one-to-one correspondence with N. (This last claim is, of course, the Schröder-Bernstein Theorem.) If that were not so, then it would be possible to have $\overline{\overline{M}} < \overline{\overline{N}}$ and $\overline{\overline{N}} < \overline{\overline{M}}$, from which, by transitivity of the ordering, $\overline{\overline{M}} < \overline{\overline{M}}$, contradicting the antisymmetry of the ordering. In his last work on set theory, Cantor states the Schröder-Bernstein Theorem without proof. (See *Abhandlungen*, p. 285.)

18. The point is already made in the introductory section of Cantor's first contribution to set theory. See *Abhandlungen*, p. 116.

ordinality and cardinality, isolating one-to-one correspondence as fundamental to cardinality, displaying analogs of finite arithmetical operations, improving our grasp on the structure of the real numbers. So we can account for Cantor's extension without simply describing it as "beautiful" or "interesting." By recognizing the explanatory dividends, we expose what the aesthetic appeal and interest of transfinite set theory consist in: the theory is beautiful *because* of its forging of previously unsuspected connections among previously recognized entities and properties.

In a famous passage, Cantor defends his transfinite set theory against those who would restrict mathematical discourse to what can be translated into the language of the arithmetic of natural numbers.[19] He argues initially that mathematicians are free to fashion new concepts so long as they do not involve themselves in contradictions. But a second theme soon emerges. Unfruitful or pointless concepts will quickly disappear from mathematics. Cantor's celebrated concluding line, the assertion that the essence of mathematics lies in its freedom, should be understood in light of the two constraints he has imposed on that freedom. Mathematical stipulation is legitimate only if it is reasonable to think that the usage introduced is consistent and that it will prove fruitful. Cantor does not suggest that mathematicians should follow wherever whimsy leads them. Instead, he emphasizes the fact that mathematical concept formation has a built in corrective. If a new approach yields no new understanding of prior mathematics (or gives no new explanatory tools to the scientist), Cantor does not claim that appeals to "beauty" or "interest" can save it. Thus I regard his own defense of set theory as supporting my thesis that, insofar as we can honor claims about the aesthetic qualities or the interest of mathematical inquiries, we should do so by pointing to their explanatory power.

Generalization is a process in which the language of mathematics is extended. Prior constraints on the usage of expressions are relaxed, so as to advance our understanding of entities previously discussed. In the examples I have considered above, the constraints on prior usage are themselves first recognized through this process. Thus, by finding analogs of previously accepted theories, we achieve a more adequate descriptive specification of the entities to which we have been referring: *after* Cantor we have a clear way to state conditions on two sets' having the same size or on a set's being finite. We also extend our grasp on mathematical reality by recognizing how we would obtain results, analogous to those achieved in prior theorizing, if some conditions on usage were dropped. This is so because generalization reveals to us interconnections among properties of the entities under discussion. To put these points in terms of the ontological perspective I favor, the attempt to generalize the powers already attributed to the ideal subject enables us to achieve a more adequate specification of the powers already attributed and to understand those powers from the perspective of a more inclusive attribution.

19. *Abhandlungen*, pp. 182–83.

V

The next pattern of rational transition which I shall consider is *rigorization*. Sometimes mathematical practice is modified with the avowed aim of making the practice rigorous. More exactly, the set of accepted reasonings is changed, and new language and new statements may be introduced, on the grounds that the changes will replace unrigorous parts of the practice with rigorous substitutes. I shall try to understand this type of transition and its rationale.

Before offering my own account, I think it will be helpful to note explicitly that there is a traditional philosophical view of the search for rigor. Inspired by mathematical apriorism, many philosophers suggest that the goal of achieving rigor is to turn mathematics into a body of genuine a priori knowledge. But, despite the "philosophical" remarks mathematicians sometimes insert into their prefaces,[20] I see no reason to assume that those mathematicians have any exalted epistemological interests and that they become concerned when the reasonings in some branch of mathematics are incapable of furnishing a priori knowledge. We can provide an explanation of demands for rigor which does not assume that the mathematicians who make the demands are motivated by (misguided) commitment to apriorism. Moreover, my discussion of the calculus will show that this explanation makes better sense of the history of research in the foundations of analysis than its apriorist rival.

To demand rigor in mathematics is to ask for a set of reasonings which stands in a particular relation to the set of reasonings which are currently accepted. For want of a better word, I shall call the relation in question the *rigorization* relation. A rigorization of a set of mathematical reasonings is a set of reasonings which contains a *rigorous replacement* for each of the reasonings in the original set. A rigorous replacement for a reasoning is an argument meeting the following conditions:

(i) the reasoning consists of a sequence of elementarily valid steps;
(ii) all of the premises are true;
(iii) the conclusion is the conclusion of the reasoning which is replaced;
(iv) on the basis of the replacement, we can explain how the original reasoning led to its conclusion and thus see why it led to a true conclusion (or, more briefly, we can use the replacement to explain the success of the original reasoning).

There is a natural way to construe clause (iv) if the reasoning replaced was an argument with clearly stated premises and an explicit commentary (that is, a sequence of remarks telling us how each of the steps was obtained). A rigorous replacement should preserve the structure of the original argument but it should

20. See, for example, the passage quoted from Cauchy on pp. 247–48. Also Lagrange, *Théorie des Fonctions Analytiques*, p. 3.

"fill in the gaps." I have not adopted this as a general requirement because I think it is inapplicable to the interesting cases. Unrigorous reasonings often deploy language which is ill understood, and mathematicians are rarely in a position to present the structure of their unrigorous reasonings until they are ready to rigorize them. (For example, the reasonings employed in the method of infinitesimals were not presented as arguments with clearly stated premises and explicit commentary. Nor were the proposals advanced by Newton, Berkeley, and MacLaurin wrong *in principle*, even though these proposals amended the premises and the language of the original reasonings.)[21]

On my account, a necessary condition for rational demands for rigor is that the current set of accepted reasonings should contain reasonings which cannot be reconstructed as arguments involving elementary steps from true premises, *given prevalent ideas about what steps are elementary and what statements are acceptable as true*. The goal of rigorization is to replace the unrigorous reasonings with reasonings meeting (i) through (iv), where clauses (i) and (ii) are construed in accordance with the background metamathematical views of the practice. I anticipate the complaint that demands for rigor are frequently formulated as demands for clear and precise definitions of mathematical concepts, so that my linkage of rigorization to reconstruction of the set of accepted reasonings is incorrect. However, this complaint overlooks the source of the requests for definition. Mathematicians want clear and precise definitions because the present unclarity of their language has been identified as the source of their failure to understand certain reasonings that they accept. (Thus, for example, the demand for a rigorous definition of 'group' emerges from the search for rigorous replacements of group-theoretic reasonings.)

There are two distinct, but related, questions which I hope to answer: What makes a demand for rigor rational? How is it possible rationally to extend one's practice through rigorization? Let me begin with the first question. I have suggested that perceived mismatch between the set of accepted reasonings and the constraints imposed by background mathematical views is a necessary condition for rational demands for rigor to arise. I want to deny, however, that this condition is sufficient. Perhaps we are inclined to think of the failure to explain the success of a reasoning or to understand an expression—failure which is revealed in inability to reconstruct the reasoning or to give an identifying description of the referent of the expression—as an important blemish which must be removed. Yet this is to overlook the simple point that mathematicians often manage quite well, even though they do not entirely understand what they are doing. Expending part of the energy of the mathematical community on a program of rigorization is not always a rational activity, because, at times, the project of removing defects of understanding is less important than attending to outstanding questions of other kinds. To acknowledge this point is to see

21. For these proposals, see Newton, "Treatise on Quadrature"; Berkeley, *The Analyst*, sections 24–29; MacLaurin, *Treatise on Fluxions*, especially pp. 52–90, 412–16. I shall consider them briefly in Chapter 10.

that the explanation of the rationality of demands for rigor is not finished until we have shown why increased understanding is sometimes needed.

I do not want to deny that increased understanding is a good thing. My point is simply that sometimes it is not a sufficiently good thing to divert attention from specific, urgent mathematical questions. Indeed, there are occasions on which it is eminently rational for the mathematicians to believe that they will not be able to attain complete understanding of their unrigorous reasonings until they have applied those reasonings to a broad range of particular problems— "charting the limits of those reasonings," as it were. The attitude to the calculus of the early Leibnizians exemplifies this type of rational appraisal. Yet, by the same token, there are times at which the question of rigor becomes urgent. When mathematicians recognize that important problems cannot be solved without some clarification of language and techniques of reasoning, they rationally attend to foundational issues. Similarly, we do not catch mathematicians saying to themselves, "This concept is really unclear; I must define it properly," unless they believe that their current grasp of the concept is sufficiently tenuous to interfere with their reasonings about issues which interest them. Thus Cauchy and Abel were not suddenly sensitive to unclarities in the notion of infinite sum which had been ignored by their predecessors; their demands for rigorous definition sprang from a desire to reconstruct reasonings involving infinite series expressions, a desire which had its source in their assessment of the centrality of such reasonings to research in analysis.

Of course, once a foundational program has been set in motion, it acquires its own momentum. When the reconstruction of certain reasonings has been identified as an important task—on the grounds that mathematical research cannot proceed without such reconstruction—mathematicians quite rationally attend to the questions which are generated by attempts to complete the task. The set-theoretic paradoxes had a large impact on the mathematical community at the beginning of the twentieth century because the *apparently successful* systematization of the set-theoretic references of classical analysis, achieved by Dedekind and Cantor, was suddenly seen to be beset by serious internal difficulties. In their definitions of 'set' (or 'system'), Cantor and Dedekind had brought into the open the principle which seemed to underlie the practice of nineteenth-century analysts. Mathematicians from Bolzano to the students of Weierstrass had taken it for granted that they had the right to talk of the collection of things meeting any arbitrary condition. Dedekind and Cantor *defined* 'set' in a way that would ground this right. The paradoxes not only showed that the new set-theoretic concepts forged by Cantor, which initially seemed so fruitful, must be handled with care; they also revealed the inadequacy of what had seemed to be a successful—and needed—rigorization. No wonder, then, that the paradoxes were viewed as an important challenge.[22]

22. At least they were viewed in this way by some members of the mathematical community— especially those most interested in Cantor's work. Other mathematicians, who felt that they could carry on their work without the new set-theoretic tools, were unmoved by the paradoxes. It is all

Let me now take up the second question raised above, the question of how rigorization can yield a rational extension of mathematical practice. Rigorization involves a shift from a practice ⟨L,M,Q,R,S⟩ to a practice ⟨L',M,Q,R',S'⟩ where, in the interesting cases, both L' and S' will differ from L and S, and where R' differs from R through the provision of rigorous replacements for some of the reasonings belonging to R. The new language (new expressions or proposals for redefinition) and the new statements (including, perhaps, new "first principles") win acceptance because, by adopting them, we can find rigorous replacements for reasonings, antecedently recognized as successful, whose success was not understood.

In the ideal case, we begin with a set of successful reasonings, which we are initially unable to reconstruct as sequences of elementary inferences from premises we can accept, and we accept new statements and/or new language because the new statements (new language) enable us to provide rigorous replacements for the successful, but unrigorous reasonings. Moreover, there is no evidence against the new statements, no difficulty in understanding the new language. However, life is usually less than ideal. The introduction of a new principle or a new definition may force us to reject some of the successful reasonings of the old practice (that is, it may only allow for *corrective* rigorization), and it may not enable us to solve the problems at which those reasonings were directed. Or there may be alternative incompatible principles which give different corrective rigorizations (that is, which enable us to reconstruct different proper subsets of the set of accepted reasonings). Worse still, the proposed principle (or new language) may have difficulties of its own. Foundational disputes arise when the significance of these "interfering factors" is weighed differently by different mathematicians.

---

too easy for historians and philosophers to suppose that panic swept through the entire mathematical community. Thus Joseph Dauben writes as follows:

> To a mathematician like J. Thomae, this was a shattering discovery. As a result of the paradoxes he felt that nothing had become so uncertain as mathematics [*Georg Cantor*, p. 242].

Ironically, Thomae was quite unperturbed. His line "Mathematics is the most obscure of all the sciences" occurs in a reply to Frege's attack on formal theories of arithmetic, a reply written with tongue-in-cheek. It is worth quoting in full the closing passage of Thomae's paper:

> And after Frege has "once and for all done away with formal number theory," and after he has recognized that even his own attempt to give a logical foundation to numbers has failed (p. 253, Epilogue), it results that we are left with no numbers at all and according to him must come to the sad conclusion that
>
> Mathematics is the most obscure of all the sciences.
>
> Written in the dog-days of the year 1906 [*On the Foundations of Geometry*, pp. 119–20].

The fact that a sensitive historian like Dauben can so misconstrue Thomae's remark (which occurs in a paper characterized as "a holiday chat"!) testifies to the influence of the idea that demonstrations of defects in rigor must always provoke crisis. As I have been emphasizing (and will continue to emphasize in Chapter 10), mathematicians need motivation before they will bother with foundations.

As in the case of question-answering, discussed in Section II, rational rigorization involves reckoning costs and benefits. Here the benefits are to be measured in terms of the understanding that the proposed modification of the practice would bring, both in its revelation of why previously ill understood reasonings were successful and in its removal of obstacles from the path of research into problems currently identified as significant. The costs are counted in terms of previously accepted problem solutions which must now be jettisoned without replacement, and in terms of the unclarities (threats of error) which the proposal itself may bring. Our examination of the introduction of the concept of convergence by Cauchy and Abel will provide a clear example of this accounting. Cauchy and Abel were forced to abandon some Eulerian series summations and Cauchy's revised analysis yielded some anomalous results. Nevertheless, the production of clear techniques for reasoning about infinite series, conceived as the key to central issues in analysis, outweighed these defects and made the rigorization rational.

I shall close this section by giving a systematic series of questions and answers on the subject of unrigorous reasonings. "How do unrigorous reasonings become rationally accepted?" By proving their worth in solving problems which have previously been identified as important. (See Section II above.) "When does it become rational to worry about defects in rigor?" When mathematicians come to appreciate that their current understanding of parts of their language and of the workings of their problem solutions is so inadequate that it prevents them from tackling the urgent research problems that face them. "Under what conditions is it rational to accept a proposal for replacing unrigorous reasonings?" A proposal for rigorization is rationally acceptable when the benefits it brings in terms of enhanced understanding outweigh the costs involved in sacrifice of problem-solving ability, difficulties of newly introduced language, and so forth. Ideally, a proposed rigorization wins rational support by providing rigorous replacements for all the old unrigorous reasonings, thereby paving the way for investigation of the research problems whose solution was blocked by prior defects of rigor. Under ideal conditions, the proposal will also only introduce new language and new principles against which there is no evidence. However, in cases where the previous demand for rigor was urgent—when the interesting mathematical questions cannot be tackled because of deficiencies in mathematical understanding—mathematicians can rationally accept a proposed rigorization which does not account for the success of all the problem-solving techniques previously adopted or which introduces new puzzles of its own.

Vi

The last type of rational interpractice transition to be considered is *systematization*. Here one begins from a practice in which some component(s) is (are) disordered. So, for example, the results amassed in a particular field may be derived by methods which seem to have little in common; the field divides into a myriad of special techniques, each of which generates a few conclusions. In the transition, one introduces new language or new principles, amending the set of reasonings and possibly even the set of questions and metamathematical views. The outcome is a practice in which one has achieved a unifying perspective on the results, questions, or reasonings previously regarded as disparate. The new practice is justified in virtue of its providing this unified perspective.

There are two main types of systematization. In the first, *systematization by axiomatization,* systematization is achieved by introducing new "axioms," new "definitions," and new "proofs" in order to derive the scattered statements of a field from a small number of "first principles." (My use of quotes is intended to indicate that the statements or reasonings in question *earn* this status through the exhibition of the system of derivations.) The second type of systematization, *systematization by conceptualization,* consists in modifying mathematical language so as to reveal the similarities among results previously viewed as diverse or to show the common character of certain methods of reasoning.

I shall begin with the more prominent type, systematization by axiomatization. Let us recall my characterization of proofs. To be a proof is to be a member of a system of reasonings serving two functions: (a) providing optimal generation of new knowledge from old; (b) providing increased understanding of statements previously accepted. When we are thinking about the introduction of new axioms or new definitions, we cannot defend them by pointing out that acceptance of them will fulfil the former function. In such cases, (b) is crucial. We accept new axioms (either by giving already accepted principles a new status or by adopting new principles which simultaneously acquire this status), new definitions, and new proofs employing these axioms and definitions, because, by doing so, we can derive a large number of antecedently accepted statements from a small number of statements. The ability to do this is important to us because the unification of a field enhances our understanding of it.

The rationality of systematization by axiomatization is precisely the rationality of accepting scientific theories on the basis of their power (or their promise) to unify the phenomena. Elsewhere I have attempted to give an analysis of the notion of explanatory unification.[23] Independently of the merits of that analysis, there are two points which I take to be uncontroversial. First, much of the evidence for some scientific theories consists in demonstrating their ability to unify—or at least their promise for unifying—statements already accepted. The Newtonian programs of eighteenth-century chemistry, Darwin's evolutionary theory, and the modern theory of plate tectonics all owed much of their justification to the explanatory unification they provided (or suggested). Second,

23. In "Explanatory Unification."

the gross features of appeals to unification in science are obvious: to unify a field is to derive a lot from a little, to generate a large number of consequences from a few statements accepted as basic. To adapt an important insight of Michael Friedman's, unification advances our understanding because it reduces the number of types of fact that we have to view as basic (or brute).[24] In a more venerable terminology, we might say that unification exposes to us the order of nature by showing how diverse phenomena result from a small number of elements.[25]

On the view I recommend, the same applies in mathematics. Our understanding of mathematical reality is enhanced by achieving the same unification that we value in the natural sciences. I shall support this view by considering very briefly two of the most important cases of systematization by axiomatization in the development of mathematics.

The most famous example is that of Euclid. Unfortunately, Euclid's *Elements* is not explicit about the rationale for adopting the axioms chosen (or, indeed, any axiomatic presentation at all). When later writers, such as Proclus, descant on Euclid's accomplishment, their praises usually mix together two ideals which Euclid is supposed to have achieved. On the one hand, Euclid is credited with having achieved a systematization of earlier results which enables them to be grasped more easily: he has provided "a means of perfecting the learner's understanding with reference to the whole of geometry."[26] But at other times, the axioms are attributed the property of self-evidence.[27] Now if the axioms of geometry, or other parts of mathematics, were indeed knowable on the basis of some process which generated knowledge independently of prior mathematical beliefs, then there would be no need to argue that they are rationally acceptable because they bring order to a diverse collection of previously accepted statements. However, we have seen that the traditional accounts which take our knowledge of mathematical axioms to be the result of processes which warrant us independently of background experiences and beliefs are misguided. My present claim is that the systematization which they bring provides a *sufficient* reason for rationally accepting mathematical axioms. I regard the first compliment that Proclus pays to Euclid as the significant epistemological point. What Euclid accomplished was an advancement of our understanding of geometry. He did it by showing how five postulates, a few "common notions," and some definitions would suffice for the derivation of a host of accepted statements. Perhaps we may even suggest that the attribution of self-evidence

24. See "Explanation and Scientific Understanding," pp. 23–24.

25. This traditional view of the explanatory order is prominent in Bolzano, who applies it to mathematics. For a discussion of these ideas of Bolzano's see my "Bolzano's Ideal of Algebraic Analysis."

26. Quoted in Heath's introduction to his edition of *Euclid's Elements*, p. 116. The quote is from Proclus.

27. *Ibid.*, p. 121. Again Heath quotes Proclus.

to Euclid's first principles is the *result* of the enhanced understanding they provide in unifying geometry.

The power of systematic considerations to justify the adoption of axioms and definitions can be seen more clearly in a more recent example, where the acceptance of the new axioms and definitions was postponed until its unifying power became clear. In the late 1840's and early 1850's, Cayley offered a characterization of the fundamental properties of groups, taking a group to be a set closed under an associative operation (multiplication) and containing an identity. (At this stage, Cayley did not recognize explicitly the need for inverses.) This proposed characterization was defended by appealing to its power to bring together researches in three different parts of algebra, the study of permutations, the theory of matrices, and the theory of quaternions. Indeed, Cayley's explicit recognition that multiplication need not be commutative seems to be prompted by his desire to accommodate the latter cases. At the time, Cayley's proposal fell on deaf ears. Why? The answer, I think, is that few of Cayley's readers were familiar with matrices or quaternions, so that his attempt to unify the results of disparate areas was not appreciated. By the 1870's and 1880's, with the incorporation of group-theoretic ideas into geometry by Klein and Lie, there was a wealth of diverse material to be systematized by the formulation of the concept of an abstract group. During this period, the mathematical community came to endorse a characterization of the group, and they were justified in so doing because of the ability of the group concept to subsume results in a variety of areas.[28]

My thesis is very simple. The examples of Euclid and Cayley (as well as other important axiomatizations such as those of Zermelo[29] and Kolmogorov) should be understood as analogous to the uncontroversial cases in which scientific theories are adopted because of their power or promise to unify. The mathematical examples share with the scientific examples the essential idea of deriving a large number of diverse conclusions from a small set of first principles. Moreover, in both types of case we can see why unification advances our understanding, and thus why it is rational to search for a unified theory.

Let us now consider systematization by conceptualization. Since this is less familiar than axiomatization, some examples may help to fix ideas. Viète's introduction of an explicit method for representing equations made it possible to unify questions which algebraists had previously posed. No longer, as in Cardan's *Ars Magna*, were the questions which we would formulate as "What is the solution of $x^3 + ax = b$?," "What is the solution of $x^3 = ax + b$?," "What is the solution of $x^3 = ax^2 + b$?," and so forth, regarded as distinct problems.

28. For a clear account of the axiomatization of group theory and Cayley's contribution to it, see H. Wussing, *Die Genesis des Abstrakten Gruppenbegriffs*, pp. 171ff., especially p. 173.

29. The case of Zermelo has been discussed from a similar point of view by Hilary Putnam in "What Is Mathematical Truth?"

Instead, they were grouped together under the rubric of finding the solution to the general cubic, and the set of accepted questions was presented in a simpler and more systematic way. Similarly, Lagrange made decisive changes in the language used to study the solution of equations. These changes were justified by their provision of a unified perspective on techniques for problem solutions already achieved. Consider the situation which Lagrange inherited. His predecessors had discovered apparently haphazard techniques for solving cubic and biquadratic equations. So, for example, to solve $x^3 + ax + b = 0$, one sets $x = y - \dfrac{a}{3y}$; we obtain the equation $y^6 + by^3 - \dfrac{a^3}{27} = 0$, and, since this is quadratic in y^3, it can be solved by using the standard formula for quadratic equations. Now this looks like a lucky trick. Why should there be any such substitution which leads to an equation reducible to an equation of lesser degree than the one with which we began? Lagrange answered this question by modifying the language of the theory of equations—introducing the concepts of *resolvent equation,* and of equations' *admitting permutations*—showing how, within the extended language, the bag of tricks assembled by his predecessors could be seen as techniques instantiating a common pattern. The unification he had offered by introducing new concepts (concepts which were to play an extremely important role in the development of group theory) convinced him, and his contemporaries, that he had isolated the form that equation-solving techniques should take, that he had discovered "the true principles of the solution of equations."[30]

In both these cases, one adopts new language which allows for the replacement of a disparate set of questions and accepted solutions with a single form of question and a single pattern of reasoning, which subsume the prior questions and solutions. Generally, systematization by conceptualization consists in modifying the language to enable statements, questions, and reasonings which were formerly treated separately to be brought together under a common formulation. The new language enables us to perceive the common thread which runs through our old problem solutions, thereby increasing our insight into why those solutions worked. This is especially apparent in the case of Lagrange, where, antecedently, there seems to be neither rhyme nor reason to the choice of substitutions and thus a genuine explanatory problem.

I want to conclude my discussion of systematization by considering two topics, both of which concern the interplay between systematization and the metamathematical component of the practice. (Changes in this component have received little attention in earlier sections.) The first issue concerns the way in which systematization can rationally provoke revision of metamathematical views. The second examines the way in which changes in metamathematical

30. An excellent treatment of Lagrange's enterprise and his proposals can be found in O'Malley, "The Emergence of the Concept of an Abstract Group."

views can lead to dissatisfaction with prior attempts to systematize a part of mathematics, thereby prompting us to engage in new systematization.

The metamathematical views of a practice are claims about how the ultimate goals of the mathematician are to be achieved. These views are typically justified by reflection on the ways in which previous mathematics has been successful. Certain disciplines have always made up the set of mathematical subjects; previous work has counted some of them as fundamental, some as independent of one another; proofs have always been presented in a particular way. Implicitly generalizing on prior work, a mathematical practice may adopt certain metamathematical views which then conflict with a proposal for systematization. At this point, the conflict may be resolved either way: previous mathematical practice, and the general morals culled from it, may block acceptance of the new systematization; alternatively, the new systematization may be viewed as exposing prior metamathematical views as shortsighted. What concerns me here is to argue that the latter reaction is sometimes rational, so that we can understand the rational evolution of the metamathematical views which mathematicians have held.

In many ways it is easier to understand why it is reasonable to change one's metamathematical views in response to proposals for systematization than to appreciate the rationality of some of the other transitions I have discussed. Consider the following scenario. The mathematical community of a particular time, reflecting on the mathematics they know, justifiably claim that proofs should exemplify a particular feature. In solving important problems, they introduce some reasonings that cannot be reconstructed as reasonings which accord with the accepted constraints on proofs. Attempts to systematize the resultant practice which honor the commitment to the idea that proofs should exemplify the feature may continually fail. Eventually, a proposal for a unified treatment of the practice which abandons this commitment may emerge. Its adoption can be justified on the grounds that it unifies, and thus accords with the overarching aim to construct an explanatory system of proofs, and that the old criterion of proof is based only on part of mathematics and is unable to accommodate the rest. Intuitively, the old metamathematical view can be seen as an inductive generalization ("All proofs are to have feature . . ."), which is *disconfirmed* by the successive failure of attempts to turn successful reasonings into proofs, so construed, and *overthrown* by showing how arguments lacking the feature can serve the explanatory function of proofs. (Indeed, as we shall see, the metamathematical view, espoused by Newton's successors, that proofs in analysis should be geometrical, was disconfirmed well in advance of the provision of a system of proofs in analysis, as it became evident that the successful reasonings of the subject would not submit to geometrical reconstruction.)

Again, the kinship between mathematics and natural science should be evident. It is a commonplace that reflection on how simplicity, explanatory power, and other scientific desiderata are to be achieved plays an important role in scientific decision making, but that views about these notions are revised as new theories, which satisfy the desiderata but violate methodological generalizations, become accepted. (Think of the influence that claims about teleological explanation, deterministic explanation, and micro-reductive explanation have had in the course of the history of science, and how those claims have been revised and amended.) Methodological generalizations are justified so long as it appears that successful science achieves its goals by satisfying them. They are overthrown when new theories are recognized as achieving those goals in ways which the generalizations do not permit. I am contending that the same goes for mathematics.

One example, which I have already mentioned, is the eighteenth-century abandonment of the thesis that proofs in analysis should be given in geometrical terms. Another is the late nineteenth-century endorsement of set theory as a framework for mathematics. Throughout the nineteenth century, the official metamathematical doctrine was that proofs in analysis should ultimately be traceable to arithmetical premises. Now the slogan "Arithmetize analysis!" is actually less clear than it may initially appear, since the notion of arithmetic which it employs was never clearly defined before the time at which the slogan itself became an issue. Only in the second half of the century, when vague references to "quantities" were replaced with precise references to numbers, did the slogan receive a clear formulation as the requirement that all results of analysis should be presented in language which makes reference to natural numbers alone. (In Kronecker's words: ". . . and all the results of significant mathematical research must ultimately be expressible in the simple forms of properties of whole numbers.")[31] Mathematicians quickly recognized that, in this specific form, the enterprise of arithmetization must fail. What Dedekind and Cantor showed was that both arithmetic and analysis can be systematized within set theory. In light of their work, the mathematical community had three options: (i) to repudiate classical analysis and to insist that all significant results of analysis be reformulable as statements about natural numbers; (ii) to regard arithmetic and set theory as equally basic and to develop analysis by combining them; (iii) to regard set theory as the basic mathematical discipline and to develop both arithmetic and analysis from it. Kronecker's option, (i), is vulnerable to the charge that it truncates mathematics. The pursuit of (ii), which, in retrospect, one can see as the approach of much nineteenth-century mathematics, yields a less unified theory than (iii). But we do not have to have either a unified, but truncated, analysis based on arithmetic alone, or a heterogeneous theory in which set theoretic concepts and principles are tacked on to arithmetic as needed. Instead we can opt for an approach in which analysis and arithmetic are treated from the unifying perspective of set theory. This approach *does*

31. Kronecker, *Werke*, vol. 3, p. 274.

require that we abandon prior metamathematical views: proofs in analysis are no longer required to be purely arithmetical nor is arithmetic considered as a fundamental discipline. I think it is obvious both that the abandonment of these views is analogous to the revision of methodological beliefs at some stages in the history of science, and that it is an eminently rational response to the unified mathematical treatment of arithmetic and analysis which Cantor and Dedekind proposed.

Let us now briefly consider the issue of how changes in our metamathematical views may force us to amend what we had previously accepted as systematic presentations of parts of mathematics. These changes themselves are frequently prompted by recognition that inferences previously considered as elementary are not fully elementary. How can that occur? Suppose that mathematicians have been accustomed to view the inferences in certain reasonings as paradigms of elementarily valid steps. (For centuries, of course, Euclid's inferences were viewed as paradigms.) One day they discover that reasoning which seems to accord with the paradigms generates recognizably false conclusions from true premises. At this point, it is rational for them to raise the requirements on inferential steps in proofs. Given the new requirements, old systematizations may have to be revised, and further axioms or definitions accepted, in order to bridge gaps of which the mathematical community was previously unaware.

Attention to the presentation of inferences may be forced by the discovery that previously accepted methods of reasoning can lead one astray. In consequence, one may seek new systematization of areas of mathematics in which adequate axiomatizations had apparently already been achieved. The rationality of this process is so easy to understand that I shall not dwell on it at any great length. It will suffice to mention three examples of it. Saccheri's faulty attempt to vindicate Euclid by *reductio* brought home to his successors the importance of exposing more clearly the assumptions and rules on which they depended. Those who developed non-Euclidean geometry—Gauss, Bolyai, and Lobatschevsky—were forced to take greater pains with their inferential steps than had previously seemed necessary, and the felt need for a finer decomposition of inferential transitions led to overhauling of previously accepted proofs. Similarly, as we shall see, Weierstrass was led to insist on stricter standards for inferences in proofs in analysis by his diagnosis of mistakes in Cauchy's reasonings, and his requirements led to a reform of analysis. Finally, the paradoxes of naïve set theory gave impetus to the demand for formal proofs. Ironically, it was the collapse of Frege's own foundation for arithmetic which made his contemporaries at last take seriously the strictures on inference which he had long been advocating, and which led to the formal systematization of the theory of sets.

VII

It is time to draw the threads together. In Sections II through VI, I have attempted to describe certain general patterns of mathematical change—forms of interpractice transition—which will be recognizably rational. To claim that they are recognizably rational is to suggest that they should be accommodated by a full theory of rationality. I have tried to support that claim by presenting them in ways which reveal their kinship to familiar cases of rationality, both in everyday decision-making and in the activities of the natural scientist, and by showing how they would accord with the general goals of asserting true mathematical statements and explaining what has already been accepted.

I set out intending to answer the question "How do people have the mathematical knowledge they do?" The present chapter gives an important part of my answer. In general, knowledge is appropriately grounded (*warranted*) true belief. In Chapter 6, I addressed the question of mathematical truth. Since then, my enterprise has been to examine what makes for "appropriate grounding." In particular my identification of rational interpractice transitions is an attempt to tell part of the story (the hardest part!) about how mathematical beliefs are warranted.

Most cases of warranted mathematical belief are cases in which an individual's belief is warranted in virtue of its being explicitly taught to him by a community authority or in virtue of his deriving it from explicitly taught beliefs using types of inference which were explicitly taught. This reduces the task of explaining most cases of warranted mathematical belief to that of accounting for the warranted belief of community authorities. The account which I offer will also be used to explain the exceptional cases of knowledge, the knowledge of the pioneers whose work typically changes the authoritative corpus of beliefs.

The beliefs of community authorities are warranted in virtue of their relation to the beliefs of previous community authorities. We should envisage a chain of communities, beginning with a community whose beliefs were perceptually warranted, such that the links in the chain transfer warrant from a community to its successor. To understand the relations which can transfer warrant, we must consider more than the set of beliefs associated with each community. Thus I introduce the notion of a *mathematical practice*. My account of warranted mathematical belief takes the following form.

      (A) A set of (mathematical) beliefs is *directly warranted* if its members are all perceptually warranted.

      (B) A (mathematical) practice is *grounded* if it results from a directly warranted set of (mathematical) beliefs by means of a sequence of rational interpractice transitions.

(C) A set of (mathematical) beliefs is *warranted* if it is directly warranted or if it is the set of accepted statements of a grounded (mathematical) practice.

Thus I see the growth of mathematical knowledge as a process in which an *ur*-
practice, a scattered set of beliefs about manipulations of physical objects, gives
rise to a succession of multi-faceted practices through rational transitions, leading ultimately to the mathematics of today. Once we have this account, it is
easy to explain the knowledge of the great innovators. Even in advance of
community endorsement of their proposals, they obtain warranted belief by
adopting a practice which results from the traditional (grounded) practice by
means of a rational interpractice transition.

Hence, given an account of mathematical truth which would allow for [A],
and assuming that warranting *via* perception and education are epistemologically unproblematic, the task of answering my initial question reduces to that
of describing what kinds of interpractice transitions are rational interpractice
transitions. Given a specification of the rational interpractice transitions, we
can articulate [B] and thus complete our explanation of how mathematical
knowledge is obtained. In this chapter, I have attempted to describe the major
rational interpractice transitions which have played a role in the actual development of mathematics.

My description of various types of interpractice transition has been admittedly qualitative and imprecise in certain respects. This admission should not
be damning. Philosophers of science have often yearned for a complete formal
model of scientific inference, something which, when applied to actual scientific situations, past and present, would yield appraisals of rationality which
accord with our intuitions (or at least with our "educated" intuitions). Despite
the philosophical attention which has been lavished upon the growth of scientific knowledge—*vastly* more than has been paid to the growth of mathematical
knowledge—our best understanding of the most interesting cases of scientific
decision is thoroughly qualitative (and imprecise). Where we can attain precision, as in various systems of inductive logic, our models seem to be inapplicable to the principal achievements of major scientific figures. I hasten to
add that I am not discounting the ultimate possibility of a full formal model of
scientific reasoning. My point is simply that philosophers have had to learn that
scientific reasoning is more complex than had originally been supposed. Hence,
if a formal model of scientific reasoning (or mathematical reasoning) is attain-
able, and if we are to attain it, we do well to begin by investigating *qualita-
tively* the kinds of specific (mathematical) reasoning involved in decisions we
take to be rational, identifying those desiderata which rationally motivate the
community in modifying its beliefs (or, more generally, its practice).

Perhaps we shall one day achieve a more precise model of scientific inference. If so, so much the better. My aim here has been to show how inference

in mathematics would be accommodated within our model. By uncovering various types of mathematical inference, exposing their kinship to patterns of scientific inference, I also hope to have advanced the project of eventually reaching a model of rational inference.[32] My examples should serve as previously neglected data for our projected model, and I hope that the concepts and distinctions I have used to discuss them may be refined against further mathematical and scientific examples. Yet I should emphasize that attaining a formal account is not necessary for my project here. What I have been attempting to show is that there are recognizably rational patterns of interpractice transition— *data* for a theory of rational inference, if you like—in virtue of which our current mathematical practice may be grounded. Thus the principal project of this book is not to construct a global theory of knowledge or a complete account of rational inference—that would be far too large a task!—but *to show how mathematics is to be integrated within a general theory of knowledge and a complete account of rational inference.*

I began this chapter by indicating one way in which the transitions I have hailed as rational could be combined. It is not hard to conceive of others, and I believe that the patterns of change described above, occurring in various sequences, can be discerned in the evolution of major fields of mathematics. My goal in the final chapter will be to show how the sequence described at the beginning of this chapter is instantiated in the historical development of the calculus. This is needed to complete my evolutionary epistemology for mathematics because that epistemology is committed to two theses: (a) certain patterns of interpractice transition (question-answering, and so forth) are recognizably rational; (b) we can account for the mathematical knowledge we have by tracing a sequence of these interpractice transitions from rudimentary mathematics to current mathematics. In this chapter I have tried to establish (a). Even if I have been successful, someone might still legitimately harbor doubts about (b). To assuage those doubts completely, I should have to reconstruct the entire history of each mathematical field, and that is clearly out of the question. However, I think it will be sufficient to reconstruct a major period (two centuries)

---

32. Also, perhaps, for a model of explanation. The account I have offered distinguishes three types of explanation in mathematics. The discussion of generalization in Section IV indicates how we can sometimes explain mathematical theorems by recognizing ways in which analogous results would be generated if we modified our language. (This type of explanation has some affinity with that discussed by Steiner in "Mathematical Explanation.") My notion of rigorization introduces a second type of explanation: explanation by removal of previous inability to recognize the fine structure of connections. Finally, Section VI is explicitly concerned with explanation by unification. Thus, at first sight, mathematical explanations, like scientific explanations, appear heterogeneous. Whether we shall some day achieve a single model which covers all cases of scientific explanation—or even of mathematical explanation—I do not know. However, I suggest that any adequate account of explanation in general should apply to the mathematical cases ("data") presented here.

in the history of a major field (the calculus), showing how a sequence of the transitions, claimed here to be rational, transformed mathematical practice.

Hence the case study of the last chapter is a response to the worry that my brand of empiricism cannot account for "higher" mathematics. Yet I think it will also reinforce the conclusions of this and earlier chapters. For my appraisal of certain types of interpractice transition as rational will be underscored as we examine concrete examples of those types of transition. Moreover, I hope to show how comprehensible the history of mathematics can be, once we take off our apriorist spectacles and adopt the perspective which I have been developing.

10
The Development of Analysis:
A Case Study

I

I shall examine some of the major episodes in the history of the calculus from the middle of the seventeenth century to the end of the nineteenth century. The development of the calculus during this period is closely connected with a transformation of the whole of mathematics. It is no exaggeration to suggest that the vast differences between the language of contemporary mathematics, the statements mathematicians now accept, the questions they take to be important, the reasonings they employ, and the metamathematical views they hold, and the corresponding features of the mathematical practice of the early seventeenth century, are, to a very large extent, the result of the mathematical activity inaugurated by Newton and Leibniz. Hence the study of this one example should provide the best single illustration of my account of rational mathematical change.

At the beginning of the investigation, it will be useful to fix some terminology. I shall use the term 'calculus' to refer to the mathematical theory (or theories) proposed by Newton, Leibniz, Jean and Jacques Bernoulli, and the Marquis de l'Hôpital. By the beginning of the nineteenth century, these early researches had been transformed into a new discipline, and it seems appropriate to call that discipline 'analysis.' Characterizing the eighteenth-century endeavors is more tricky, and I shall sometimes use one term, sometimes the other, depending on whether the work I am discussing seems closer in spirit to the early work or to the modern analysis which began to emerge in the nineteenth century.

The first task which confronts us is to account for the original acceptance of the calculus. Because of the fact that the calculus was developed independently by Newton and Leibniz, and because of the differences in their presentation of it, our account will offer two variations on one theme. There are certainly common features of the methods proposed by Newton and Leibniz—enough to group them together as the originators of a distinctive part of mathematics and to contrast them with their predecessors—and there is also a common structure to the acceptance of their proposals. Both Newton and Leibniz introduced new language, new reasonings, new statements, and new questions into mathematics. Some of the new expressions were not well understood and the workings of some of the new reasonings were highly obscure. Despite these defects, the changes they proposed were accepted quite quickly by the mathematical community, and the acceptance was eminently reasonable. Newton and Leibniz were able to show that they could obtain, systematically, the solutions to problems previously recognized as important, solutions which had been achieved in bits and pieces by their predecessors; that they could answer, in the same fashion, questions which others had unsuccessfully attempted to answer; and that they could solve problems which "nobody previously had dared to attempt."

To understand how the power of the methods introduced by Newton and Leibniz outweighed the unclarities which attended them, we must begin with the problems which interested the mathematicians of the early seventeenth century. Greek geometry bequeathed to the mathematicians of the renaissance a number of general classes of questions, which the Ancients had been able to answer in particular cases, but for which no general method had survived. Among these are the problems of constructing tangents, constructing normals, computing the length of arcs (rectification), computing areas (quadrature), and finding maxima and minima. The mathematicians who preceded Newton and Leibniz were interested in solving these problems both in special cases of particular interst and, if possible, in general. Their endeavors were motivated by a number of factors of the kind with which we are already familiar. First, there is generalization of questions posed and answered by the Ancients. Second, growing interest in certain physical problems makes some curves (notably the conchoid, cissoid, and cycloid) seem especially worthy of treatment. Third, Descartes's *Géométrie* appeared to provide a key for solving the problems: algebraic representation could recast the old geometrical questions in forms which held out the promise of solution. Indeed, Descartes himself had claimed that the problem of finding the value of the subnormal to a curve (the distance between the foot of the ordinate and the point where the normal meets the axis) is the central problem of the entire complex of Ancient difficulties, and he proposed a method for solving it. Vigorous research led to other techniques for tackling instances of the geometrical questions—Fermat had a method for finding maxima and minima, Hudde proposed a rule for applying Descartes's method of subnormals, Roberval achieved some success wih the problem of tangents, Cavalieri devised a technique for "quadratures." Two points about this collection of techniques need to be recognized. Before Newton and Leibniz, there was no general method for solving any of the important problems, nor was there any appreciation of the relationships among the various problem classes. Moreover, some of the techniques employed reasoning which was not well understood or computations which proved difficult to effect in the most interesting cases. So, for example, Cavalieri found the areas under curves by using infinitesimals: he divided the area into "very thin" segments. In other cases, such as the methods of Descartes for subnormals, dubious references to "very small increments" could be avoided, in favor of the more comprehensible idea of finding equal roots of an equation, but the cost of this technique lay in the difficulty of applying it. To summarize: the mathematicians of the early and middle seventeenth century inherited a class of significant questions, and they made some progress with those questions, but they lacked the general technique which they desired.

The success of the Newtonian and Leibnizian calculus consisted precisely in its satisfaction of this prior demand. Newton and Leibniz provided algorithms which could be applied to generate answers to two fundamental questions—inversely related—and subsidiary algorithms for using these answers to resolve all of the geometrical problems. The staggering complexity of the techniques devised by earlier mathematicians (and by Newton in some of his first papers) thus gave way to a simple and unified method. To see how this was achieved let us examine Newton's approach.

Basic to Newton's algorithmic advance is the concept of a *fluxion*. Newton thinks of a *fluent* as any quantity which is in the process of changing, and he takes the fluxion of the fluent to be the rate at which it is changing. So, for example, if someone draws a straight line with a pencil, we can take the length of the line as a fluent whose fluxion, at any time, is the instantaneous velocity of the pencil at that time. Newton's simplification of the problem complex with which earlier mathematicians had struggled consists in making the central questions that of computing fluxions given fluents, and that of computing fluents given fluxions. As we shall see, these problems are, respectively, those of differentiating and integrating. In his "Method of Fluxions," Newton provides a clear statement of the effectiveness of the fluxion concept in systematizing the cluster of traditional geometrical questions.

> Now in order to this, I shall observe that all the difficulties hereof may be reduced to these two problems only, which I shall propose, concerning a Space describ'd by local Motion, any how accelerated or retarded.
>
> I. The length of the space describ'd being continually (that is, at all times) given; to find the velocity of the motion at any time propos'd.
>
> II. The velocity of the motion being continually given; to find the length of the Space describ'd at any time propos'd.[1]

1. *Mathematical Works of Newton*, pp. 48–49.

Newton goes on to show how problems of tangents, normals, curvature, maxima, and minima reduce to the first of these problems. Quadrature problems reduce to the second. Rectification problems involve both.

The reduction is effected by Newton's kinematic approach to geometry. Like several of his predecessors (most notably Roberval and Barrow), Newton favors the idea that curves are generated by the motion of an ordinate segment as the foot of that segment moves along a base line. Here is Newton's account of how the solution to the first fundamental problem answers questions about tangents:

> In ye description of any Mechanicall line what ever, there may bee found two such motions wch compound or make up ye motion of ye point describeing it, whose motion being by them found by ye Lemma, its determinacon shall bee in a tangent to ye mechanicall line.[2]

The reduction works as follows. Suppose that we wish to find the tangent to the curve $f(x,y) = 0$ at the point (X,Y). (It is assumed that ordinary Cartesian coordinates are used.) Consider the curve as swept out by the motion of a point, and resolve the motion along the axes into components $p(t)$, $q(t)$ respectively. Let the components of velocity when the point is at (X,Y) be P,Q. If we can solve the first fundamental problem, we can derive from the relation $f(x,y) = 0$ a general relation $g(x,y,p,q) = 0$. In particular, we have $g(X,Y,P,Q) = 0$, giving us a relation between P and Q. Let α be the angle which the tangent at (X,Y) makes with the x-axis. Since the instantaneous velocity of the moving point at (X, Y) is in the direction of the tangent, we know that $\tan \alpha = \dfrac{Q}{P}$. We calculate $\dfrac{Q}{P}$ from the equation $g(X,Y,P,Q) = 0$, thus solving our problem.[3]

This example is typical of the way in which Newton was able to break down the traditional geometrical problems into two stages. He showed how the geometrical questions can be reduced to the fundamental questions by posing the geometrical questions in kinematic terms. His task therefore reduced to that of giving algorithms for the fundamental problems. Newton's papers are full of rules for tackling these problems. The methods of infinitesimals and infinite series play an important role in the discovery and justification of Newton's rules.

Let us concentrate on the first fundamental problem, asking ourselves how from an explicit equation $y = f(x)$ we may derive a relation between the fluxion

2. *Mathematical Papers of Newton*, vol. 1 (1664–1666), p. 377. The lemma referred to is the parallelogram of velocities lemma.

3. Newton's method for solving the first fundamental problem ensures that the ratio Q/P will be obtainable from the derived equation $g(X,Y,P,Q) = 0$. It is a general feature of Newton's approach that only ratios of fluxions need to be considered, and Newton sometimes explicitly assigns an arbitrary value to the fluxion of x.

of y and the fluxion of x. We think of the situation kinematically, imagining a particle moving along the curve $y = f(x)$. Whatever values of x and y we consider, we can suppose that, through a small time interval of length o, the velocity of the particle remains constant. In this time interval, y increases to $y + \dot{y}o$, x increases to $x + \dot{x}o$.[4] Since the particle remains in contact with the curve, $y + \dot{y}o = f(x + \dot{x}o)$. But now, subtracting $y(= f(x))$ from both sides, we have: $\dot{y}o = f(x + \dot{x}o) - f(x)$. This gives the derived equation $\dot{y} = \dfrac{f(x + \dot{x}o) - f(x)}{o}$. When we work out examples, we are entitled to omit (or, to use Newton's term, "reject") expressions which contain powers of o after the division, since "o is supposed to be indefinitely little." Thus, from $y = x^3$, we obtain:

$$\dot{y} = \frac{(x + \dot{x}o)^3 - x^3}{o} = \frac{3x^2\dot{x}o + 3x\dot{x}^2o^2 + o^3}{o} = 3x^2\dot{x} + 3x\dot{x}^2o + o^2 = 3x^2\dot{x}.$$

Hence $\dfrac{\dot{y}}{\dot{x}} = 3x^2$.

The technique works. When we apply it to the first fundamental problem we get answers which generate recognizably correct solutions to the traditional geometrical questions. Yet there is an obvious puzzle about this way of proceeding. Why are we entitled to make the assumption that the fluxions remain constant through small intervals of time? Why are we allowed to neglect some terms? These questions challenge us to reconstruct the method of infinitesimals so that it will proceed by steps which are recognizably valid and so that it will employ no premises which are known to be false. For we do know that, in general, if a particle moves along a curve its velocity will not remain constant, even through small intervals of time, and we understand also that dropping terms from equations is not usually valid.

We shall return later to Newton's attempts to answer the challenge. For the moment I want to point out that Newton took a further step which makes matters even worse. The problem-solving technique which represents $\dot{y}$ as $f(x + \dot{x}o) - f(x)/o$ will not enable us to compute the desired ratio $\dot{y}/\dot{x}$ in all cases. It works beautifully when $f(x)$ is a polynomial in x. However, consider a more tricky example. If $y = \sqrt{(1 + x)}$, then the technique yields $\dot{y} = \dfrac{\sqrt{1 + (x + \dot{x}o)} - \sqrt{1 + x}}{o}$, and there is no obvious way to obtain the ratio $\dot{y}/\dot{x}$. Newton's approach to such cases is to use infinite series representations of recalcitrant functions. He convinced himself of the truth of the general binomial theorem $(1 + z)^\alpha = 1 + \alpha z + \dfrac{\alpha(\alpha - 1)}{2!}z^2 + \ldots$ (where, to use anach-

4. Here I follow Newton's notation. $\dot{y}$, $\dot{x}$ are the fluxions of y, x respectively.

234 THE NATURE OF MATHEMATICAL KNOWLEDGE

ronistic terminology, α can be any real number).[5] Once he had accepted such representations, he was able to extend the method of infinitesimals to handle examples which do not succumb to direct attack. Series representations make it possible to apply the rules for polynomials.

There are many passages in Newton's writings which show his appreciation of the difficulty of understanding the method of infinitesimals. As we shall see below, he distinguished between his solutions to the fundamental problems and genuine proofs, and took seriously the problem of rigorizing his technique. By contrast, he seems to have seen no difficulty in the use of expressions for "infinite sums." What is important for our purposes is the fact that even though embarrassing questions could be asked about the new language and the new reasonings, the power of the calculus to provide a unified approach to the complex of geometrical questions and to generate answers to a far greater number of those questions made its acceptance by the mathematical community entirely rational.

A similar story can be told about the endorsement of Leibniz's version of the calculus. Unlike Newton, Leibniz does not adopt the kinematic conception of curves, and his early papers make few concessions to the reader. (Those who studied Newton's presentation could at least use the concept of an instantaneous velocity to obtain some understanding of the symbols figuring in the computations.) Leibniz proposes a calculus of differences and a calculus of sums, both of which give directions for moving from a given algebraic equation to a new equation. The relationships found in the new equation then lead to the solution of the geometrical problems. Consider, once again, the problem of tangents. Given the equation of a curve, $f(x,y) = 0$, we construct the equation of differences, or differential equation, by applying rules which Leibniz gives us. This leads to an equation $g(x,y,dx,dy) = 0$, from which we can compute $\frac{dy}{dx}$, and this ratio (for Leibniz thinks of it as a ratio) gives the slope of the tangent. (Leibniz's own example will reveal his way of presenting the point: "When $dv = dx$, then the tangent makes half a right angle with the axis.")[6] There is only the barest hint of the pedigree of the rules which Leibniz sets down. At one point, in his first published article on the differential calculus, Leibniz indicates that the problem-solving technique which underlies his rules is the method of infinitesimals:

> We have only to keep in mind that to find a *tangent* means to draw a line that connects two points of a curve at an infinitely small distance, or the continued

5. Newton's reasons for thinking that the theorem is true seem to have been analogical extrapolation from the case where α is a positive integer, and verification of the result for special cases. Thus Newton considered the coefficients of powers of x in the product of $(1 + \frac{1}{2}x + \frac{\frac{1}{2}(\frac{1}{2} - 1)}{2!}x^2 + \ldots)$ with itself, and satisfied himself that the product yields $1 + x$.

6. D. J. Struik, *A Source Book in Mathematics: 1200–1800*, p. 274.

side of a polygon with an infinite number of angles, which for us takes the place of the *curve*. This infinitely small distance can always be expressed by a known differential like dv. . . .[7]

This is the closest Leibniz comes, at this stage in his career, to the claim that the differences with which his calculus is concerned are infinitesimals. His calculus is truly a *calculus,* and he seems to shun the idea of endorsing an interpretation of it. Later work, both in his own papers and those of the Bernoullis, and, above all, in the textbook authored by l'Hôpital, makes it clear that the symbols 'dv,' 'dx,' and so forth are supposed to designate infinitesimals, and that the preferred explanation for the success of the Leibnizian techniques articulates the account hinted at in the passage just quoted.

Leibniz's employment of the method of infinitesimals is different from that which we have discerned in Newton. The Leibnizian technique for constructing the important ratio $\frac{dy}{dx}$, can be formulated by the following sequence of instructions:

(1) Compute the difference $f(x + dx) - f(x)$, where dx is "small," representing it as $dx(A(x) + B(x, dx))$, where $B(x, dx)$ contains all and only those terms with dx as a factor.

(2) Divide the difference by dx, obtaining $A(x) + B(x, dx)$.

(3) Ignore terms containing dx as a factor, thus obtaining $A(x)$. If $y = f(x)$ then $dy/dx = A(x)$ [or $dy = A(x)dx$].

Obviously, this shares the two important features of Newton's use of the method. (Indeed, we frequently find Newton abandoning his official kinematic conception and giving instructions for computing the ratio $\frac{\dot{y}}{\dot{x}}$ which are just notational variants of (1) through (3).) The same troubles arise in understanding exactly what is being computed. If $dx \neq 0$, then it appears that the ratio $\frac{dy}{dx}$, will be the slope of a chord joining neighboring points on the curve, instead of the slope of the tangent as intended. Moreover, if $dx \neq 0$, it is hard to see why we are entitled to ignore terms which contain dx as a factor. On the other hand, if $dx = 0$, then we are, presumably, allowed to write, at step (1), $f(x + dx) - f(x) = f(x + 0) - f(x) = f(x) - f(x) = 0$, irrespective of the function f. And, of course, if $dx = 0$ then step (2) involves division by zero, which is illegitimate. Hence, there are acute problems in understanding the new language. Symbols have been introduced without any specification of their referent, and reasonings involving those symbols are proposed which, given the available methods of interpretation, appear to rest on false principles.

Leibniz's defense of his calculus consists of forthright claims about its power

7. *Ibid.*, p. 276.

to solve important mathematical problems. Advertisements of this kind begin with the first paper on the differential calculus. Leibniz points out that his method gives a systematic way of solving problems, like those of maxima and minima and tangents, which had hitherto been treated separately. He emphasizes that it is not subject to the restrictions of previous techniques, and that it can be applied to curves specified by transcendental equations (i.e. to curves, like the cycloid, which are not given by polynomial equations). He concludes by promising more:

> And this is only the beginning of much more sublime Geometry, pertaining to even the most difficult and most beautiful problems of applied mathematics, which without our differential calculus or something similar no one could attack with any such ease.[8]

The same attempt at justifying his procedures pervades Leibniz's subsequent papers. Throughout his published articles and his correspondence we find the same themes. The calculus delivers a systematic and fully general way of answering the traditional questions of geometry, and the difficulty of understanding how the technique works should not be allowed to stand in the way of our using it. Indeed, Leibniz constantly exhorts his followers to *extend* the calculus: he sees this as the most important task facing the scientific world, "because of the application one can make of [the calculus] to the operations of nature, which uses the infinite in everything it does."[9]

II

The Leibnizian and the Newtonian calculus developed in very different ways. When we read the papers of Leibniz's followers—especially the Bernoullis—and examine the letters which were exchanged among members of the Leibnizian circle, we find that the mathematicians involved heeded Leibniz's exhortations. There is a constant effort (and competition) to use the new techniques to tackle problems, especially those problems which have physical significance. In addition, new questions are generated by analogy with problems already posed. Leibniz finds that the calculus enables him to compute the sums of certain infinite series, and his research shows increasing interest in series summation. In correspondence with Jean Bernoulli, another analogy is suggested: can "fractional differentiation" be introduced into the calculus? Thus Leibniz and Bernoulli consider the possibility of introducing the expressions '$d^{1/2}x$,' '$\sqrt[3]{(d^6y)}$,' and so forth, and they explore such statements as '$d^3y/dx^2 =$

8. *Ibid.*, p. 279.

9. From a letter to L'Hôpital. Leibniz, *Mathematische Schriften*, volume 2, p. 219. For Leibniz's defenses of the calculus, see also volume 1, p. 155; volume 2, pp. 43–165 *passim*, pp. 216–18, 228–29; volume 4, p. 54; volume 5, pp. 229, 259, 290, 307, 322, 386.

$d^3y \int {}^2x$,' in the hope of using this generalized calculus to help with the solution of differential equations.[10] This spate of activity is interspersed with occasional reflections on the question of rigor. Leibniz's own writings rotate among three different suggestions for reconstructing the method of infinitesimals: at times, he is happy to admit the existence of infinitesimal quantities; at other times, he suggests that the method introduces errors but that these errors can be made as small as one pleases; and, at a few conservative moments, he proposes that his method can be vindicated by a classical geometrical proof, using the Ancient method of exhaustion.[11] None of these approaches is pursued in any detail, and one can only conclude that Leibniz did not regard the problem of rigorization as particularly important. Two of Leibniz's followers—l'Hôpital and Jean Bernoulli—did produce an attempted rigorization of the calculus (to which Leibniz sometimes refers with approval, even though it is at odds with some of his remarks about infinitesimals as "useful fictions"!). Their strategy, announced in the textbook which l'Hôpital published in 1691, is to suppose that there are infinitesimal quantities which satisfy the following conditions:

(a) The slope of the tangent to the curve $y = f(x)$ at $(x, f(x))$ is
$$\frac{[f(x + \alpha) - f(x)]}{\alpha},$$ where α is infinitesimal.

(b) If $f(x + \alpha) - f(x) = \alpha(A + B)$, where B contains all and only terms which occur with α as a factor, and if α is infinitesimal, then
$$\frac{[f(x + \alpha) - f(x)]}{\alpha} = A.$$

However, the l'Hôpital-Bernoulli program faces the problem of avoiding obvious inconsistencies generated by applying standard algebraic laws to infinitesimals. Moreover, the Leibnizians were all aware of these difficulties (as their correspondence shows), and, unable to specify principles on infinitesimals which would enable them to deduce just the consequences they wanted, and none of the ones they needed to avoid, they proceeded in practice by selectively applying standard algebraic laws. Hence the Leibnizian calculus was developed unrigorously, and there is little reason to think that anybody regarded the situation otherwise.

Newton took the problem of rigor much more seriously. He believed that it was possible to give a fully geometrical explanation, couched in the terms of his kinematic geometry, for the success of the method of infinitesimals. The treatment of the method of fluxions in *Principia* hides the dubious manoeuvres which Newton usually allows himself by deploying a geometrical conception of limit. Lemma I of Principia announces an important new concept: "Quan-

10. See *ibid.*, volume 3.1, pp. 180ff.

11. For these three different attitudes see *ibid.*, volume 5, pp. 322–23, 327, 350, 385, 389, 392, 407; volume 2, pp. 287–88, 294; volume 3.2, pp. 536–39; volume 4, pp. 96, 105–6, 218.

tities and the ratios of quantities, which in any finite time converge continually to equality, and, before the end of that time approach nearer to one another by any given difference become ultimately equal.'' Here Newton comes close to advancing the modern definition of a limit, and it is tempting to read him as *defining* two quantities $X(t)$ and $Y(t)$ to be equal in the limit as t goes to 0 just in case the following conditions hold:

(i) $(t)(t')(t < t' \rightarrow |X(t) - Y(t)| < |X(t') - Y(t')|)$

and

(ii) $(z)(\exists t')(t)((z > 0 \ \& \ t < t') \rightarrow |X(t) - Y(t)| < z)$.

On this reading, Newton has introduced the notion of monotonic passage to a common limit. The importance of the new concept is that it paves the way for what Newton takes to be a satisfactory explanation of the success of the method of infinitesimals. In his 1693 "Treatise on Quadrature," he first gives a qualitative statement of the connection between fluxions and the procedure for computing them which is supplied by the method of infinitesimals.

> Fluxions are very nearly as the Augments of the Fluents, generated in equal, but infinitely small parts of Time; and to speak exactly, are in the *Prime Ratio* of the nascent Augments:
> Tis the same thing if the Fluxions be taken in the *ultimate Ratio* of the Evanescent Parts.[12]

Newton follows this with a more extensive examination of how the method of infinitesimals works:

> Let the Quantity of x flow uniformly, and let the Fluxion of x^n to be found. In the same time that the Quantity x by flowing becomes $x + o$, the Quantity of x^n will become $(x + o)^n$, that is, by the method of Infinite Series's $x + nox^{n-1} + \frac{nn - n}{2} oox^{n-2} + \&c.$ and the Augments o and $nox^{n-1} + \frac{nn - n}{2} oox^{n-2} + \&c.$ are to one another as 1 and $nx^{n-1} + \frac{nn - n}{2} ox^{n-2} + \&c.$ Now let those Augments vanish and their ultimate Ratio will be the Ratio of 1 to nx^{n-1}; therefore the Fluxion of the Quantity x is to the Fluxion of the Quantity x^n as 1 to nx^{n-1}.[13]

How should we read this argument? Uncharitably construed, it is simply a reiteration of the method of infinitesimals: one divides by o and then sets $o = 0$. But, in the light of Newton's definition of 'ultimate ratio,' we can give a more favorable interpretation. Newton can be taken to be claiming that the usual method works because the ratio of the fluxions of x^n and x is the ultimate ratio of the "Augments," and that this ultimate ratio is nx^{n-1} because the dif-

12. *Mathematical Works of Newton*, p. 141.
13. *Ibid.*, p. 142.

ference between nx^{n-1} and the ratio of the "Augments," to wit, $nx^{n-1} +$ $\dfrac{nn - n}{2} ox^{n-2} + $ &c., can be made as small as one pleases by choosing o to be sufficiently small. Even if we adopt this reading, Newton's argument would still be defective on two grounds. First, he would not have shown explicitly that $\dfrac{n(n - 1)}{2} ox^{n-2} + \dfrac{n(n - 1)(n - 2)}{3!} o^2 x^{n-3} + \ldots$ can be made as small as one pleases by taking o sufficiently small. When n is a natural number, there is no great difficulty in constructing an argument to this end. Two mathematicians who drew inspiration from Newton's argument, Benjamin Robins and Simon L'Huilier, provided the requisite reasoning. However, there is no indication that n is to be a natural number, and Newton's reference to his method of infinite series indicates that he is prepared to advance his explanation even when the series of terms is infinite. The deep problem with his approach is that, without any clear specification of the meaning of infinite series expressions, Newton has no way of showing that the sum of the terms in o can be made as small as one pleases in cases where there is an infinite number of such terms. So Newton's treatment will not solve the problem of rigorizing the calculus because he cannot provide rigorous replacements for all the infinitesimalist reasonings he uses. However, his failure stems from difficulties with the concept of *convergence,* not with the notion of infinitesimal.

Newton was plainly convinced that he had shown why the method of infinitesimals worked, and he considered himself entitled to use it to avoid "the tediousness of deducing involved demonstrations *ad absurdum,* according to the manner of the ancient geometers."[14] The importance Newton ascribed to the problem of rigor and his confidence that he had solved this problem had indirect effects on the course of eighteenth-century British mathematics. After the acrimonious wrangling about priority between Newton and Leibniz, the mathematicians who succeeded Newton were inclined to contrast their own "rigorous" approach to the calculus with the algebraic approach of the Leibnizians. Newton was perceived as following "the true methods of the Ancients," while Leibniz's followers were criticized for their free use of algebraic manipulation of "empty symbols." Hence, when Berkeley challenged the rigor of the Newtonian calculus in 1734, his critique was taken very seriously. Berkeley's criticisms were presented clearly, they displayed a competent reading of Newton's mathematics,[15] and Berkeley claimed that, while Newton's methods were successful, no explanation had been given for their success. The

14. *Principia,* p. 38.

15. Competent but uncharitable. Berkeley's reading presupposes the unfavorable interpretation of the argument of the *Treatise on Quadrature,* which we considered in the last paragraph. Berkeley also proposes his own explanation of the success of Newton's techniques, in terms of compensation of errors.

publication of *The Analyst* precipitated a flurry of writings on the calculus, as Newton's successors rallied to his defense. Their concern contrasts sharply with the attitude of the Leibnizians when faced with similar challenges, and indicates the significance which the eighteenth-century British mathematical community attached to the problem of rigor. Berkeley's critique had unfortunate consequences. Colin Maclaurin, the most talented of Newton's successors, presented his major work on the calculus, the *Treatise on Fluxions,* in cumbersome geometrical style, explicitly responding to Berkeley's objections. In the course of his attempt at defense, Maclaurin was drawn into philosophical issues which are largely irrelevant to mathematical research, and, in an effort to make his mathematics conform to his philosophical presuppositions, he developed a style for the Newtonian calculus which widened the gap between British and Continental mathematics. For Newton, the problem of rigor was to provide a geometrical interpretation of the algebraic techniques used in the calculus, so that those techniques could be freely employed in the confident expectation that they could always be replaced if necessary. Maclaurin's attitude towards algebra is far more sceptical: he is haunted by Berkeley's complaint that algebraic procedures may deploy "empty" symbols to gloss over defects in reasoning. His efforts at reconstruction cover only a fragment of the analysis developed by his Leibnizian contemporaries. Priding itself on its rigor and its maintenance of a proper geometrical approach to mathematics, the British mathematical community fell further and further behind.

It was not irrational for Newton to give serious thought to issues of rigor in the calculus. Given the set of background metamathematical views and the corpus of questions identified as important, it was reasonable for him to wonder if efforts to continue research by tackling the unanswered questions might break down unless the mismatch between the techniques of the calculus and the standards of proof were not remedied. As we have seen above, it was also reasonable for Newton to think that a geometrical rigorization of his calculus could be given. Nor was Leibniz unreasonable in his comparative neglect of questions of rigor and in his emphasis on stretching the new methods without rigorizing them. Leibniz's successors produced so many apparently successful reasonings, which could not readily be reinterpreted using the methods favored by eighteenth-century Newtonians, that, by 1750, the Newtonian claim that all the unrigorous reasonings of the calculus could be reconstructed, in the terms that Newton had preferred, was no longer defensible. The achievements of the Bernoullis, Euler, and other Continental mathematicians also undercut the thesis that the question of rigorizing the calculus was an important one. That question only became urgent for the Newtonians because of an accidental sequence of events. Newton's original attention to the problem of rigor combined with the squabble about priority to produce a situation in which Newton's successors proclaimed that the Newtonian calculus, unlike its Leibnizian counterpart, constituted rigorous mathematics. Berkeley's charge stung because it undermined this vaunted superiority.[16]

Leibniz's successors took his directives to heart, and the Leibnizian calculus engendered results in profusion. The techniques introduced by Leibniz were quickly extended, as Continental analysts generated questions by analogy with problems which had previously been identified. How can one find explicit representations of indefinite integrals? How can infinite series be summed? What methods are available for solving differential equations? How do you differentiate and integrate functions of more than one variable? These questions (among others) are central to the research of Continental mathematicians in the early decades of the eighteenth century. In some cases, they are important not only because they generalize traditional questions, but also because they have scientific significance. Much of the important work of the Bernoullis is directed at solving problems in physics. But there are some questions, to which much attention is given, whose significance is purely mathematical. Euler's work, the full flowering of Leibnizian analysis, ranges from the most abstract topics to mathematical physics. The next step in our account will be to see how two questions which occupied Euler set the stage for later developments.[17]

III

Some of Euler's most imaginative research consists in the summation of infinite series of numbers. Interest in the questions of computing the values of infinite sums, or, more generally, in finding techniques for computing sums, goes back at least to Leibniz. As we saw above, both Leibniz and Newton introduced infinite series representations to help with the differentiation of recalcitrant functions. Leibniz was able to use such representations to find an answer to a long-established question, the question of finding expressions for π. In one of his earliest "quadratures," Leibniz shows that $\int_0^1 \frac{1}{1+x^2}\,dx = \frac{\pi}{4}$. The technique of infinite series enables him to expand the integrand as a power series:

$$\frac{1}{1+x^2} = 1 - x^2 + x^4 - \ldots$$

16. I do not intend to deny that Berkeley's critique was more penetrating than that which Nieuwentijdt directed at Leibniz. My point is that the Newtonians were extremely sensitive on this score, whereas the Leibnizians were not. I conjecture that a far less competent criticism could have set the cat among the Newtonian pigeons, while a more acute attack on the Leibnizian calculus would have extracted comparatively little response.

17. I shall not attempt any systematic account of Euler's achievement. It is instructive to compare his works with the laborious researches of his (rough) contemporary Maclaurin. The density of results in Euler's pages is astounding, and his writings are a gold mine for anyone interested in uses of nondeductive argument in mathematics—as readers of Polya are surely aware!

Leibniz then proceeds to integrate the series term by term, claiming that

$$\frac{\pi}{4} = \int_0^1 \frac{1}{1+x^2}\,dx = \int_0^1 (1-x^2+x^4-\ldots)dx = [x - \frac{1}{3}x^3 + \frac{1}{5}x^5 - \ldots]_0^1$$
$$= 1 - \frac{1}{3} + \frac{1}{5} - \frac{1}{7} + \ldots$$

This example is historically significant, because it introduces an important technique for series summation. The methods of power series expansion provide representations of functions, and, given clever substitutions, combined with appropriate differentiation or integration, one can obtain values for many sums. Euler turned this technique into a fine art.

However, as Euler and his contemporaries knew perfectly well, using the Leibnizian approach blindly can lead to paradoxical conclusions. One of the earliest signs of trouble came with a series which Leibniz had investigated.

Expanding $\frac{1}{1+x}$ as a power series, Leibniz claimed that $\frac{1}{1+x} = 1 - x + x^2 - x^3 + \ldots$ Setting $x = 1$, he obtained the result that $1 - 1 + 1 - 1 + \ldots = \frac{1}{2}$. In a paper of 1713,[18] he discusses this result and the explanation which one of the Leibnizian circle, Grandi, had given of it. For a contemporary reader, the discussion is entertaining. Grandi had proposed that the result represents the method used by God at the creation, and Leibniz rejects this on the grounds that creation cannot be a simple repetition of zeros. After comparing the case at hand with the valuable example that $1 - \frac{1}{3} + \frac{1}{5} - \frac{1}{7} + \ldots = \frac{\pi}{4}$, an example which Leibniz acknowledges as using a similar technique, he feels obliged to vindicate the method by explaining the correctness of the apparently anomalous statement that $1 - 1 + 1 - 1 + \ldots = \frac{1}{2}$. He suggests that we can view this infinite series as arising from the superposition of two series: $1 - (1-1) - (1-1) - \ldots$ and $(1-1) + (1-1) + (1-1) + \ldots$. The series represents the "average" of these, and thus has value $\frac{1}{2}$. As Leibniz puts it:

> . . . hence since $1 - 1 + 1 - 1 + 1 - 1 + \ldots$ in the case of an even finite number of terms is equal to 0 and in the case of an odd finite number of terms is equal to 1, it follows that when we proceed to the case of an infinite number of terms, where the even and odd cases are mixed, and there is equal reason for it to go either way, we obtain $\frac{0+1}{2} = \frac{1}{2}$.[19]

18. "Epistola ad V. Cl. Christianum Wolfium, Professorem Matheseos Halensem, circa scientiam infiniti," *Mathematische Schriften* volume 5, pp. 382ff.

19. *Ibid.*, volume 5, p. 387. This explanation was accepted by at least one of Leibniz' disciples, namely Varignon. See *ibid.*, volume 4, p. 388.

Thus Leibniz is prepared to accept the consequences generated by his infinite series techniques and to explain away their apparent peculiarities.

Euler's attitude is different. Writing some forty years after Leibniz, he is prepared to admit that, on occasion, the techniques of substituting in power series representations, of term by term differentiation and integration, and of series rearrangement, can lead to trouble. Considering the Leibniz-Grandi result, he suggests that the expansion $\frac{1}{1+x} = 1 - x + x^2 - \ldots$ holds only when $x < 1$, noting that the terms do not tend to zero unless $x < 1$.[20] Euler is not convinced by Leibniz's attempt at explanation. He dismisses the idea that infinite numbers can be seen as an "average" between odd and even. Moreover, he points out that Leibnizian methods can produce even more peculiar results: if in the expansion $\frac{1}{1-x} = 1 + x + x^2 + \ldots$, one sets $x = 2$, one obtains $-1 = 1 + 2 + 4 + \ldots$. Euler remarks, somewhat scornfully, that some people are prepared to explain away even this by taking infinity to be a boundary between positive and negative numbers!

Euler's awareness of anomalies which can arise from Leibnizian techniques is accompanied by an acute diagnosis of how the troubles are caused. He is accustomed to verify the results which he obtains by his summation methods, and his way of verification is to compute the value of an initial segment of the series, showing that it approximates the value he has obtained. Thus, if Euler has claimed that $V = \sum_1^\infty a_n$, his practice is to compute $\sum_1^N a_n$ for sufficiently large N, and verify that it is close to V. Of course, this will only be a satisfactory procedure if one supposes that, as one takes N larger and larger, one achieves successively better approximations. Euler is clear on the point. Apropos of a discussion of the series $1^m - 2^m + 3^m - 4^m + \ldots$ (where m is allowed to be greater than 1), Euler begins by apologizing for talking about the sums of these series at all:

> . . . it is quite true that one cannot form an exact idea of their sum so long as one understands by *sum* a value such that one approaches it more closely the more terms of the series one takes.[21]

However, he goes on at once to defend the use of series which do not meet his condition:

> But I have already remarked . . . that one must give to the word *sum* a more extended significance and understand by it a fraction or other analytic expres-

---

20. Euler, *Opera Omnia*, series 1, volume 14, pp. 589–90. I think it is fair to assume that Euler is only considering positive values of x and that he realizes perfectly well that analogous complaints about the series can be made for $x < -1$.

21. *Ibid.*, series 1, volume 15, pp. 70–71.

sion which, being developed, according to the principles of analysis produces the series whose sum is sought.[22]

In our terms, Euler is suggesting that the referent of '$\Sigma_1^\infty a_n$' can be fixed in two ways. On one way of fixing the referent, '$\Sigma_1^\infty a_n$' only refers if the sequence of partial sums, $\Sigma_1^N a_n$, tends to a limit as N goes to infinity, (i.e. the series is convergent), and the referent is this limit. On the other way of fixing the referent, '$\Sigma_1^\infty a_n$' refers if '$\Sigma_1^\infty a_n$' can be obtained by substitution of some term 'b' for 'x' in a power series expansion '$\Sigma_1^\infty g_n(x)$,' where $f(x) = \Sigma_1^\infty g_n(x)$, and in this case the referent is $f(b)$. Now Euler seems to argue that it would be unwise to restrict ourselves to the former usage, because, by doing so, we would eliminate the use of divergent series (series for which $\Sigma_1^\infty a_n$ does not tend to a limit) in analysis. These series are useful for achieving results about convergent series. Some of Euler's most striking results are obtained by using them. Thus, by expressing a convergent series as the difference of two divergent series, he is able to show that $\frac{1}{2} = \frac{1}{3} + \frac{1}{4} + \frac{1}{5} - \frac{3}{6} + \frac{1}{7} + \frac{1}{8} + \frac{1}{9} - \frac{3}{10} + \ldots$, and he also proposes a "beautiful relationship" between the sums of the series $1^m - 2^m + 3^m - 4^m + \ldots$ and $\frac{1}{1^m} - \frac{1}{2^m} + \frac{1}{3^m} - \frac{1}{4^m} + \ldots$.[23] Euler believes that we should recognize that the sums of divergent series are not approximated by partial sums—and, in consequence, abandon spurious attempts to explain the results we obtain by using divergent series, attempts like those of Grandi and Leibniz—but he insists that divergent series are a proper part of the problem-solving apparatus of analysis.

Euler accepted the general problem of summing infinite series as an important problem, and, indeed, gave greater emphasis to it than Leibniz had done. We can recognize this problem as emerging rationally from the late seventeenth-century situation: the Leibnizian calculus introduced new language, notably the infinite series expressions, which made it possible to generate canonical representation questions of computing the sums of infinite series. Interested in these questions and ingenious at answering them, Euler refused to allow the restriction of the methods by which he found his answers. Hence, although he rejected the idea that some questions about divergent series which had worried his predecessors made sense, he defended the use of divergent series, thereby bequeathing to the analysts of the early nineteenth century a wealth of papers containing statements (such as "$\frac{1}{4} = 1 - 2 + 3 - 4 + \ldots$"), to which they could react in pious horror.

22. *Ibid.*
23. See *ibid.*, series 1, volume 14, pp. 89ff., 592–95; volume 15, pp. 70–84.

The second issue which Euler investigated and which plays an equally important role in the history of analysis was the problem of the vibrating string. In 1747, d'Alembert had pointed out that the equation of motion of the vibrating string is $\dfrac{\partial^2 y}{\partial x^2} = \left(\dfrac{1}{c^2}\right)\left(\dfrac{\partial^2 y}{\partial t^2}\right)$, and had arrived at a "general solution" of it, namely $y = f(x + ct) + g(x - ct)$. (D'Alembert's formulation of the problem itself owes a debt to Euler, who had investigated partial differentiation more thoroughly than anyone else.) Interest in differential equations in general, and this equation in particular, is readily comprehensible in the light of the scientific significance of the problem. The rational mechanics of the Bernoullis and Euler had already provided substantial support for Leibniz's claim that the calculus is a key for unlocking the secrets of nature. What occupied Euler and d'Alembert is the question of how to find the complete solutions to partial differential equations, and their attention to this question led them into an important dispute.

To summarize a complicated debate,[24] the central point of disagreement between Euler and d'Alembert concerned the conditions which must be met by the functions f and g. In d'Alembert's treatment, only differentiable functions were admitted, whereas Euler—who seems to have been moved by the physical possibility of setting the string in motion by plucking—allowed for functions, which he called "discontinuous," made up of segments which are everywhere differentiable but not differentiable at the points of junction. Euler insisted that his solution was more general than d'Alembert's, and d'Alembert responded by questioning Euler's interpretation (how can one make sense of the wave equation itself at the "corner points"?). For our purposes, what is important about this controversy is Euler's reaction to a proposed solution offered by Daniel Bernoulli (the son of Jean Bernoulli). In 1753, Daniel Bernoulli suggested that the full solution of the wave equation is given by $y = a_1 \sin \dfrac{\pi x}{l}$ $\cos \dfrac{\pi c t}{l} + a_2 \sin \dfrac{2\pi x}{l} \cos \dfrac{2\pi c t}{l} + \ldots$ Euler himself had previously given cursory attention to the possibility of using a trigonometric series to solve the equation, and had rejected it. He replied to Bernoulli by claiming that the character of the trigonometric functions imposes certain restrictions on the form of the solution so that it cannot be fully general. As we shall see, Bernoulli's proposal and Euler's reaction to it generate an important question for nineteenth-century analysis, a question which will play a crucial role in the development of new definitions and new principles. That question concerns whether an arbitrary function can be given a particular type of series representation.

24. For an excellent survey of the debate, see the first chapter of I. Grattan-Guinness, *The Development of the Foundations of Analysis from Euler to Riemann.*

IV

We are now ready to consider the next episode in the history of the calculus for which I hope to provide a rational reconstruction. In the 1820's, Cauchy introduced several important definitions which were accepted almost at once by the mathematical community and which produced a dramatic change in analysis. Specifically, Cauchy proposed that the notions of continuity, derivative, and series sum should be approached in terms of an algebraic concept of limit. Let me hasten to note that this proposal represents only a fragment of Cauchy's important work in analysis. Like Euler, Cauchy was prolific, and his contributions far outrun the particular suggestion on which I shall concentrate. My aim in emphasizing the "foundational" parts of Cauchy's writings in real analysis is twofold: first, I hope to show how the ideas elaborated in earlier chapters can help us to understand why Cauchy proposed, and others accepted, his new definitions and principles; second, I think that examination of this episode will underscore my thesis that foundational work is not usually undertaken by mathematicians because of apriorist epistemological ideas, but because of mathematical needs.

For ease in future reference, let us begin by setting forth the new definitions which Cauchy proposed.

(1) f is *continuous* on $[a,b]$ if and only if $|f(x+h)-f(x)|$ tends to 0 as h tends to 0 (i.e., $\lim_{h\to 0}|f(x+h)-f(x)|=0$);

(2) the series $\Sigma_{n=0}^{\infty} s_n$ is *convergent* if and only if the sequence of partial sums $\Sigma_{n=0}^{N} s_n$ tends to a limit as N tends to infinity; and, if the sequence of partial sums $\Sigma_{n=0}^{N} s_n$ does tend to a limit as N tends to infinity then the *sum* of the series is $\lim_{N\to\infty}\Sigma_{n=0}^{N} s_n$;

(3) the function f has a *derivative* at each point of the interval $[a,b]$ if and only if, for any point x in $[a,b]$, $\dfrac{f(x+h)-f(x)}{h}$ tends to a limit as h tends to zero; if $\dfrac{f(x+h)-f(x)}{h}$ tends to a limit as h tends to zero, then $\lim_{h\to 0}\dfrac{f(x+h)-f(x)}{h}$ is the *value of the derivative* at the point x; the *derivative*, $f'(x)$, is the function which takes at each point of $[a,b]$ the value $\lim_{h\to 0}\dfrac{f(x+h)-f(x)}{h}$.

Each of these definitions uses the notion of limit, so that it is important for us to recognize the significance which Cauchy gives to this term.

(4) "When the values successively attributed to the same variable approach indefinitely a fixed value, eventually differing from it by as little as one could wish, that fixed value is called the *limit* of all the others." [25]

Cauchy's definition of 'limit' is immediately followed by a definition of 'infinitesimal.'

(5) "When the successive absolute values of a variable decrease indefinitely in such a way as to become less than any given quantity, that variable becomes what is called an *infinitesimal*. Such a variable has zero for its limit." [26]

Armed with (5), it is easy for Cauchy to provide "streamlined" definitions of some of the most important concepts of his analysis. Thus after offering (1) as the definition of continuity, Cauchy offers a reformulation: "In other words, the function $f(x)$ will remain continuous relative to x in a given interval if [in this interval] an infinitesimal increment in the variable always produces an infinitesimal increment in the function itself." [27] Should such references to infinitesimals worry us? Well, (5) clearly recognizes that infinitesimals are "variable quantities," and if we set aside Fregean worries about the 'variable quantity' locution, we might think that Cauchy has *legitimized* the old language of infinitesimals (just as Newton tried to do) and that references to infinitesimals can always be unpacked when needed. Unfortunately, as Cauchy proceeds he is sometimes seduced into departing from his official doctrine, and he sometimes treats infinitesimals as constants. Moreover, as we shall see, the use of the language of infinitesimals allows for dangerous ambiguity, so that the Newtonian strategy of providing an initial vindication of the language and then using it will prove ultimately unsatisfactory.

We can now present our main epistemological questions. Why did Cauchy think it important to advance (1) through (5)? Why did his contemporaries quickly accept these proposals? Now it may appear that there is an easy answer to these questions, that we can identify a current of feeling of dissatisfaction with the central concepts of analysis which begins in the early eighteenth century and finds its expression in Cauchy's prefatory remarks.

> As for my methods, I have sought to give them all the rigor which is demanded in geometry, in such a way as never to run back to reasons drawn from what is usually given in algebra. Reasons of this latter type, however commonly they

25. *Cours d'Analyse*, p. 19. The translation is from G. Birkhoff, *A Source Book in Classical Analysis*, p. 2.

26. *Ibid*.

27. *Cours d'Analyse*, p. 43; Birkhoff, *A Source Book in Classical Analysis*, p. 2.

are accepted, *above all in passing from convergent to divergent series and from real to imaginary quantities,* can only be considered, it seems to me, as inductions, apt enough sometimes to set forth the truth, but ill according with the exactitude of which the mathematical sciences boast. We must even note that they suggest that algebraic formulas have an unlimited generality, whereas in fact the majority of these formulas are valid only under certain conditions and for certain values of the quantities they contain.[28]

Historians and philosophers have often fastened on such passages—which occur in the letters and the prefaces of great mathematicians—interpreting them as indicative of a sense of scandal and uncertainty which demanded the reform of analysis. Yet we must be cautious. Eighteenth-century analysis, developed with great speed and scope by Euler, Lagrange, and others, shows no signs of insecurity.[29] When serious attention was given to the problem of rigorizing the unrigorous reasonings of the calculus, as it was in L'Huilier's prize-winning essay of 1786, the study was ignored. It is a gross caricature to suppose that Cauchy's work was motivated by a long-standing perception that mathematics had lapsed from high epistemological ideals and that it was accepted because it brought relief to a troubled mathematical community. Instead, we must ask why, in the early decades of the nineteenth century, *some* problems of rigor were suddenly urgent and how Cauchy's proposals solved those problems.

The first point to appreciate is that Cauchy mentions particular difficulties: in the passage I have quoted, the italicized sentence identifies two problematic areas. Only one of these, the question of how to handle infinite series, will concern us here. (I shall not discuss Cauchy's important work in complex analysis.) Notice that Cauchy does *not* mention the method of infinitesimals: he does not declare it to be a scandal that mathematicians use infinitesimalist reasoning, and, as noted already, he will himself develop the language of infinitesimals and put it to use. Why, then, is the issue of the rigorous treatment of infinite series singled out as an important problem to be solved by workers in real analysis? I suggest that the answer is that Cauchy recognizes the importance of infinite series representations of functions to the questions in real analysis which he hopes to pursue, and that he appreciates that the available algebraic techniques for manipulating infinite series expressions sometimes lead to false conclusions. I shall support this answer in several different ways. First, I shall show how problems about the representation of functions by infinite series were at the forefront of research in the early nineteenth century. Second, I shall

28. *Cours d'Analyse,* pp. ii–iii. My translation. My emphasis.

29. The brief discussion of Euler above should have made this clear. Lagrange, like Cauchy, fulminates against the failure of rigor in contemporary analysis in his preface—and quickly allows himself virtually any algebraic technique that seems useful! For a succinct summation of the character of eighteenth-century analysis see S. Bochner, *The Role of Mathematics in the Rise of Science,* p. 142. The usual, misguided, assessment of eighteenth-century attitudes is expressed in the title of the relevant chapter of Carl Boyer's *The Concepts of the Calculus,* where the eighteenth century is labelled as ''the period of indecision.''

explain how the available series techniques were perceived as unreliable tools for tackling these problems. Third, I shall examine the ways in which Cauchy and Abel devoted themselves to fashioning new techniques.

Daniel Bernoulli's suggestion for the solution of the vibrating string problem was not accepted by his contemporaries, and none of the major eighteenth-century analysts developed his fruitful idea. However, in the hands of Joseph Fourier, at the very beginning of the nineteenth century, trigonometric series expansion of functions became a major research tool in solving important partial differential equations generated in mathematical physics. Fourier provided a method for calculating coefficients for trigonometric series expansion of arbitrary functions, and he showed how his expansions could be employed in mathematical physics. His work immediately generated an important question. Can any function be given a trigonometric series representation? An informal argument, whose pedigree goes back to Euler's response to Bernoulli, suggests that the answer is "No." Trigonometric functions have special properties—they are periodic and continuous. How can a *sum* of trigonometric functions—even an "infinite sum"—lack these properties? But, lacking any clear notion of continuity, and in the context of a theory of infinite series which was known to generate bizarre consequences, the issue could not be definitely resolved. The "Fourier question" (as I shall henceforth call it) thus invited analysts to rigorize certain parts of their subject. Cauchy accepted the invitation.

The Fourier question was only the tip of the iceberg. Late eighteenth-century analysis had generated numerous interesting functions whose properties seemed to require infinite series techniques for their full disclosure. Abel expresses the perspective of the 1820's in a letter to his friend Holmboe, a letter written after his visit to Paris and his acquaintance with Cauchy's work on convergence: ". . . in the whole of mathematics there is scarcely one infinite series whose sum has been determined in a rigorous way, that is to say that *the most essential part of mathematics* is without foundation."[30] This attitude will help us to see why, in 1821, Cauchy accepted a definition of the sum of an infinite series—namely (2)—which had been considered, and rejected, by Euler over half a century earlier. Euler had needed divergent series to achieve some of his summations of *convergent* series. Cauchy and Abel were prepared to declare that "a divergent series has no sum," banishing some Eulerian results from analysis.[31] They viewed this sacrifice as necessary if they were to have tools for tackling the problems which concerned *them,* problems about the representability of functions by particular kinds of series (of which the Fourier question is an example). Abel, who writes with the zeal of a new convert,[32] is especially

30. Abel, *Oeuvres* volume 2, p. 257. My translation and emphasis.

31. These were partially rehabilitated in the theory of summation of divergent series developed at the beginning of this century.

32. In his youthful writings Abel frequently used Eulerian methods for the summation of infinite series, interchanging operations without any worries about the validity of doing so.

forthright both on the need for a reliable technique for infinite series and on the
faultiness of the old methods:

> Divergent series are generally quite fatal, and it is shameful to base any dem-
> onstration on them. You can use them to prove anything you like and they are
> responsible for making so many mistakes and engendering so many paradoxes.
> Can one imagine anything more horrible than to set down $0 = 1 - 2^n + 3^n - 4^n
> + . . .$ where n is a positive number?[33]

The statement which Abel repudiates with disgust is one which Euler had viewed
as bizarre—but which he had used to obtain uncontroversial conclusions. Concerned to find sums of infinite series of numbers, Euler could always check to
see if his conclusions made sense. For Abel and Cauchy, occupied with a
different employment of infinite series in probing the possibilities of representation for arbitrary functions, such comforting verification was no longer available. Needing reliable instruments, they trimmed the Eulerian series techniques
to fit.

Thus, if we want to understand the acceptance of (2) in the 1820's we must
begin by seeing it as part of a rigorization which *eliminated* some successful
problem solutions but which was geared to the questions perceived as important
by nineteenth-century analysts. Euler had rationally rejected (2), because (2)
curtailed his ability to solve the problems which occupied him. In the next half-
century, changes in the set of interesting questions tipped the balance towards
(2). The questions about sums of numbers which Euler loved were relegated to
a subsidiary place, and questions about functional representation, many of which,
like the Fourier question, grew out of mathematical physics, became ever more
prominent.

Given that Cauchy and Abel were moved to rigorize the theory of infinite
series because they needed tools for tackling certain kinds of problems, we
should expect to find them investigating ways in which (2) can be made applicable to research in analysis. This expectation is confirmed. The pragmatic
streak in Cauchy's interests appears in his important chapter on the convergence of infinite series (Chapter 6 of Part I of the *Cours d'Analyse*). What
occupies Cauchy in the chapter is the issue of providing tests for determining
if a series is convergent, and this is exactly the topic which must be treated if
the fundamental definition, (2), is to be applicable in research. He gives a
variety of tests for convergence, usually justifying them by using the result that
the geometric series $1 + x + x^2 + . . . + x^n + . . .$ converges for $|x| < 1$ and
diverges for $|x| \geq 1$.[34] Most of the chapter is taken up not with questions of
proof but with examples and problems. As we shall see shortly, the fundamental theoretical result, the criterion of convergence which bears Cauchy's name,
is treated quite casually.

---

33. *Ibid.*, volume 2, pp. 256–57. My translation.

34. See, for example, *Cours d'Analyse*, p. 125, where the "nth root test" is justified.

V

So far I have been exploring one theme: Cauchy's rigorization of the theory of infinite series was motivated by the problem-solving requirements of the research in analysis of the 1820's. Now I want to turn to the other definitions which Cauchy proposed. What led him to put forward (1), (3), and (5), and what prompted other mathematicians to accept them? In one case, that of the definition of 'continuous function,' (1), partial support for the new definition accrues from the Fourier question. To elaborate the issue of whether the sum of an infinite series of continuous functions is inevitably continuous, one requires both an explication of 'infinite sum' and an explication of 'continuous function.' The appeal of (1) lay in the fact that it answered this need, providing a reconstruction in algebraic terms of reasonings which had previously been carried out under the guidance of geometrical representation. The new criterion for continuity enabled mathematicians to replace geometrical arguments by algebraic arguments to the same conclusions, in areas where the geometrical approach had been applicable, and it allowed them to extend the notion of continuity into regions where geometrical thinking failed. Such questions as the Fourier question made the issue of finding rigorous replacements for reasonings about continuity urgent, and Cauchy's thesis (1) proved acceptable because it allowed these rigorous replacements to be given.

With respect to (3) and (5) we must tell a slightly different story. The language of infinitesimals did not give rise to any urgent problem of rigorization. Despite nearly 150 years of occasional complaint about the difficulties of understanding what is going on when the method of infinitesimals is used to differentiate a function, there was no reason for the analysts of the 1820's to believe that use of the method would pose problems for their research. One of the great ironies of the history of analysis is that the first substantial mathematical difficulty of the method emerges from Cauchy's reconstruction of it. The fact that Cauchy, Abel, and their contemporaries freely employed the language of infinitesimals should not surprise us. What requires explanation is the reconstruction of the language which Cauchy gave. I suggest that Cauchy proposed (3) and (5), not in response to an urgent problem of rigorization, but simply because his approach to *other* analytic problems in terms of the concept of limit permitted him to incorporate reconstructions of reasonings about infinitesimals which had previously been offered. The definition of the derivative is part of a unified treatment of analytic concepts, in terms of the fundamental notion of limit, and it allows Cauchy to extend a search for rigorous replacements for infinitesimalist arguments begun by some of his predecessors.

Let us assemble some striking facts about the pedigree of Cauchy's definition of the derivative. First, on a charitable interpretation, that definition is already at work in Newton's work *De Quadratura*. We saw above that Newton can be construed as identifying the notion of monotonic passage to the limit, as specifying the operation of differentiation (or of finding fluxional ratios) in these terms, and as using his specification to reinterpret the infinitesimalist derivation of the rule for differentiating powers of a variable. But Newton's argument is easily misconstrued. It seems to employ a geometrical rather than an algebraic conception of limit, and it is open to the Berkeleian reading that obtains the conclusion not by showing that certain terms can be made arbitrarily small but by setting those terms equal to zero. In a reply to Berkeley, one of Newton's successors, Benjamin Robins, improved the Newtonian argument by adding an explicit demonstration that the terms rejected by the inifinitesimalist argument can be made "as small as one pleases."[35] However, although some Continental mathematicians were sympathetic to the Newtonian approach, it never became a major part of the eighteenth-century calculus. D'Alembert remarks in the *Encyclopédie* that the concept of limit is the proper basis for the calculus but he does not explore this idea in any detail.[36] The most thorough attempt to give a mathematical articulation of this suggestion was given by L'Huilier, whose prize essay of 1786 expounds in detail the rigorous replacement for infinitesimalist arguments which had been anticipated by Newton and Robins. The striking feature of L'Huilier's treatment is that it begins from almost exactly the definition of derivative which Cauchy offers in 1823.[37] Like New-on—and unlike Cauchy—L'Huilier defines 'limit' so as to allow only for monotonic passage to the limit. Having done so, he specifies the derivative in terms of limits just as Cauchy does: $\frac{dy}{dx} = \underset{\Delta x \to o}{\text{limit}} \frac{\Delta y}{\Delta x}$. Armed with his specification, he completes the Newton-Robins argument. Having shown earlier that, if $f(x) = A + Bx^b + \ldots + Nx^n$, where $0 < b < \ldots < n$, then $\underset{x \to o}{\text{limit}} f(x) = A$ (L'Huilier argues this by showing that $f(x) - A$ can be made arbitrarily small by taking x as small), he uses the general binomial theorem to expand $(x + \Delta x)^n$. Thus we have:

If $y = x^n$, then

$$\frac{\Delta y}{\Delta x} = \frac{(x + \Delta x)^n - x^n}{\Delta x} = \frac{nx^{n-1}\Delta x + \frac{n(n-1)}{2!}x^{n-2}\Delta x^2 + \ldots}{\Delta x}$$

$$= nx^{n-1} + \frac{n(n-1)}{2!}x^{n-2}\Delta x + \ldots.$$

So, by the previous result,

$$\frac{dy}{dx} = \underset{\Delta x \to o}{\text{limit}} \frac{\Delta y}{\Delta x} = nx^{n-1}.$$

35. See *The Mathematical Tracts of Benjamin Robins*, Vol. 2, p. 15.

36. See the Article "Différentiel."

37. S. L'Huilier, *Exposition Élémentaire des Principes des Calculs Supérieurs*. For the reasoning presented below see pp. 7, 21, 25–26, 31–32.

Now, even though he is sensitive to his dependence on the general binomial theorem, L'Huilier fails to appreciate that his demonstration works only for those values of n for which the series expansion is finite. Like Newton before him, L'Huilier has offered a "proof" which is flawed because he has no way to argue for the values of certain limits for "infinite sums." There are two views we can take of his reconstruction: either it provides rigorous replacements for a small subset of the infinitesimalist reasonings of the calculus or it is a faulty reconstruction of the entire corpus of infinitesimalist arguments.

L'Huilier's memoir was largely forgotten. Cauchy's *Résumé de Leçons Données à l'Ecole Royale Polytechnique sur le Calcul Infinitésimal* continued the transformation of analysis begun two years earlier in the *Cours d'Analyse*. Cauchy improves on L'Huilier's definition of 'limit,' but his account of the derivative in terms of limits is parallel to that of his predecessor. Why, then, was this old account suddenly popular? The answer lies in the power of Cauchy to integrate the limit approach to the derivative into an account of other problematic notions, and thus to provide a sweeping reconstruction of infinitesimalist reasonings in a fashion that had been previously unavailable. Using his notion of convergence, and drawing on Euler's systematic presentation of the exponential function, Cauchy quickly establishes some basic results about limits— notably the result that $\lim_{\alpha \to 0} (1 + \alpha)^{1/\alpha} = e$—from which the standard rules for differentiation quickly and painlessly emerge as special cases.[38] Cauchy has cashed the promise of d'Alembert's suggestion, not simply by offering the definition of the derivative, (3), but by using the concept of limit as a unifying concept in analysis. The definitions (1) through (4) work together to allow for a massive reconstruction of traditional reasonings as well as the development of arguments for settling those issues of the greatest concern to Cauchy's contemporaries.

I hope that this will suffice to show how Cauchy's proposals for rigorization became accepted. My next task is to explain how the new approach led to difficulties and generated problems of rigorization for Cauchy's successors. I shall focus on two topics. First, Cauchy's vindication of infinitesimalist reasoning led him to employ infinitesimalist language. Such language not only seemed harmless, but it also appeared to effect convenient simplifications in discussions which *we* see as involving multiple limits. Second, the use of an algebraic concept of limit and the central role of this concept in establishing theses about convergence, continuity, and derivatives introduced into analysis an important set of new questions, questions about how to show that limits exist in particular cases. These questions, and Cauchy's sketchy attempts to answer them, played an extremely important role in late nineteenth-century analysis.

VI

Let us begin with the more obvious of the two "loose ends" left dangling in Cauchy's treatment. The first big result of the chapter of the *Cours d'Analyse* devoted to convergence is an attempt to resolve the Fourier question. Cauchy aims to reconstruct—and to endorse—an intuitive argument similar to that which Euler had directed against Bernoulli: no discontinuous function can have a Fourier expansion because the sum of an infinite series of continuous functions is continuous. Here is Cauchy's "proof":

> When the terms of the series u_o, u_1, u_2, . . . , u_n , . . . contain a variable x and its different terms are continuous functions of x in the neighborhood of a particular value given to the variable, s_n , r_n , and s are three functions of the variable x, of which the first is obviously continuous with respect to x in the neighborhood of the chosen value. [In our notation, $s_n(x) = \Sigma_{i=o}^{n} u_i(x)$, $r_n(x) = \Sigma_{n+1}^{\infty} u_i(x)$, $S = \Sigma_{o}^{\infty} u_i(x)$; Cauchy is supposing that, for all values of x in question, $\lim_{N \to \infty} \Sigma_{o}^{N} u_i(x)$ exists.] Given this, consider the increases which these quantities have when x is increased by an infinitely small quantity. The increase in s_n will, for all values of n, be an infinitely small quantity, and that in r_n will become insignificant just as r_n does if one gives n a large enough value. As a result the increase in s must be an infinitely small quantity.[39]

The reasoning flows easily in the language of infinitesimals. Since the u_i are continuous, infinitesimal increments in x will produce infinitesimal increments in each $u_i(x)$ and hence in $s_n(x)$. Since $\Sigma u_i(x)$ is convergent, $r_n(x)$ is infinitesimal for large n. The sum of two infinitesimals is infinitesimal. Hence, infinitesimal increments in x produce an infinitesimal increment in $s_n(x) + r_n(x)$ (i.e. in $s(x)$). Therefore $s(x)$ is continuous. However, the conclusion is false. As Abel pointed out in 1826, the series $\sin x - \frac{1}{2} \sin 2x + \frac{1}{3} \sin 3x - \ . \ . \ .$ converges to a function discontinuous at each value of $x = (2m+1)\pi$. Yet it took considerable talent to say where Cauchy had erred.

From a contemporary perspective (or from a Weierstrassian perspective), we can see that Cauchy's argument trades on concealing a logical mistake—a fallacious interchange of quantifiers—through the use of the *derived* language of infinitesimals. Suppose that we were to expound Cauchy's reasoning in terms of his primitive notions, using the precise formulations we owe to Weierstrass. Then we would obtain the following:

$\Sigma_{o}^{\infty} u_i(x)$ converges for every value of x. This means that
   (1) $(\epsilon > 0) \ (x) \ (\exists N) \mid \Sigma_{N+1}^{\infty} u_i(x) \mid < \epsilon$.
Each of the $u_i(x)$ is continuous for all values of x. This means that
   (2) for each i, $0 < i$, $(x) (\epsilon > 0) (\exists \delta) (y) (\mid x - y \mid < \delta \to \mid u_i(x) - u_i(y) \mid < \epsilon)$.

We want to show that $s(x) = \Sigma_0^\infty u_i(x)$ is continuous. In other words,

(3) $(x)(\epsilon > 0)(\exists\delta)(y)(|x - y| < \delta \rightarrow |s(x) - s(y)| < \epsilon)$.

Consider any particular value of x, i.e. x_0. Take any value of ϵ. By (1) there will be a value of N, N_0, dependent not only on ϵ *but also on x_0*, such that

(4) $\left|\Sigma_{N_0+1}^\infty u_i(x_0)\right| < \dfrac{\epsilon}{4}$.

Since each of the u_i are continuous, we know from (2) that we can find δ_0 such that, for $0 < i < N_0$

(5) $(y)(|x - y| < \delta_0 \rightarrow |u_i(x) - u_i(y)| < \dfrac{\epsilon}{2(N_0 + 1)})$.

Suppose now that we choose y such that $|x_0 - y| < \delta_0$. By elementary results on inequalities we have:

(6) $|s(x_0) - s(y)| < |u_0(x_0) - u_0(y)| + |u_1(x_0) - u_1(y)| + \ldots + |u_{N_0}(x_0) - u_{N_0}(y)| + |\Sigma_{N_0+1}^\infty u_i(x_0)| + |\Sigma_{N_0+1}^\infty u_i(y)|$.

From (4), (5), (6), we can get:

(7) $|s(x_0) - s(y)| < \dfrac{\epsilon}{2(N_0 + 1)} + \dfrac{\epsilon}{2(N_0 + 1)} + \ldots + \dfrac{\epsilon}{2(N_0 + 1)} + \dfrac{\epsilon}{4}$

$+ |\Sigma_{N_0+1}^\infty u_i(y)|$,

i.e.

(8) $|s(x_0) - s(y)| < \dfrac{3\epsilon}{4} + |\Sigma_{N_0+1}^\infty u_i(y)|$.

We could use (8) to establish (3), if we could only obtain

(9) $|\Sigma_{N_0+1}^\infty u_i(y)| < \dfrac{\epsilon}{4}$.

But we could achieve (9) very easily *if the choice of N_0 were independent of the value of x_0*. Then, just as we inferred (4) we could infer (9). What this means is that if, *instead of (1)*, our starting point was:

(10) $(\epsilon > 0)(\exists N)(x)|\Sigma_{N+1}^\infty u_i(x)| < \epsilon$,

we could obtain both (9) and (4) and the reasoning would go through. Of course (10) does not follow from (1) (as students of quantification theory quickly learn). However, Cauchy's language of infinitesimals enables him to formulate both (1) and (10) using the same sentence, so that, in his presentation of the reasoning, the difference is obscured and he manages to reach (9) and so obtains (3).

It was extremely hard for Cauchy's successors to sort this out. The use of the language of infinitesimals buries Cauchy's unfortunate interchange of quantifiers, providing no clue that the mistake will emerge once the argument is recast in primitive terms. We owe to Dirichlet, Seidel, and especially Weierstrass the demonstration that Cauchy's reasoning is incorrect and the formulation of an idiom (the idiom used above) in which the error can be clearly presented. Weierstrass began a new style of analysis, insisting on the "ϵ-δ" language which is so familiar to contemporary students of the theory of functions. The change in language was itself motivated by the need to resolve the puzzles which Cauchy's treatment had brought forth.

Despite the quick recognition of the fact that Cauchy's resolution of the Fourier question was faulty, his mathematical contemporaries did not abandon the definitions he had introduced. The treatment of the *Cours d'Analyse* and the *Résumé* was so powerful, offering a systematic reconstruction of traditional arguments in terms of the limit concept and providing methods for tackling the interesting questions of analytic research (techniques which were recognizably successful—except in isolated cases like that of the Fourier question), that mathematicians were convinced that the proper course to adopt was to follow Cauchy's lead, and, ultimately, to iron out the wrinkles. Just as the mathematical community had appreciated the power of Newtonian and Leibnizian techniques, and had shelved worries about the explanation of their success, so now Cauchy's new analysis won adherents in spite of its occasional anomalies. The new analysis did not settle all the issues, and it was unclear in places, but it indicated the way in which the subject was to be pursued. Abel expresses this attitude very clearly in a letter of 1826:

> Cauchy is crazy (*fou*), and it's impossible to make oneself heard by him, though at the moment he's the man who knows how mathematics should be done. What he does is excellent, but very confused (*très brouillé*).[40]

We are so accustomed to the legend of Cauchy as the savior of clarity in analysis that Abel's characterization jars. Yet this contemporary assessment fits neatly into the account of the development of analysis which I have given. Cauchy fashioned the research tools for the analysis of thirty years (1820–50), and in his zest to put those tools to work he was not inclined to concern himself with criticisms of his results. His own vast output, and the achievements obtained by others following his lead, paid ample tribute to Abel's compliment. There were indeed confusions and anomalies, but his problem-solving success established him as "the man who knows how mathematics should be done."

Cauchy's style in analysis gave way to that of Weierstrass. To simplify a complex story I shall ignore the changes—some of them significant—introduced into analysis by Abel, Jacobi, Dirichlet, and others. The traditional account of the history of nineteenth-century mathematics, from the obituaries of Mittag-Leffler and Poincaré, through Hilbert's laudatory remarks, to contemporary studies, credits Weierstrass with the founding of modern analysis, and the tradition is, in this respect, correct.[41] However, there is an important dif-

---

40. *Oeuvres*, volume 2, p. 259. I conjecture that Abel's reasons for reproaching Cauchy stem from his recognition that Cauchy's "anti-Fourier theorem" is false.

41. Mittag-Leffler, "Une page de la vie de Weierstrass" (Proceedings of the 2nd International Congress of Mathematicians), Poincaré in *Acta Mathematica* 22:1–18; Hilbert "On the Infinite"; Boyer, *Concepts of the Calculus;* Grattan-Guinness, *Development of the Foundations of Analysis from Euler to Riemann;* Dugac, *Éléments d'analyse de Karl Weierstrass.*

ference between the change inaugurated by Cauchy and that authored by Weierstrass. Consultation of the seven volumes of Weierstrass's works can easily lead to perplexity, for it is not easy to locate the papers which "founded modern analysis." Weierstrass gave us no systematic presentation of the principles of analysis which we can compare with Cauchy's *Cours d'Analyse* and *Résumé*. There is no paper with the explicit aim of expounding the foundations of analysis in a systematic way. The correct answer to the question of where Weierstrass presents the new analysis is "Everywhere." Instead of looking for an expository treatise, we should see that Weierstrassian rigor is exemplified in Weierstrass's technical writings, and that the point of Weierstrassian rigor is established by showing how the new approach is capable of answering interesting technical questions.

There are large differences between the techniques used in Weierstrass's papers and those employed even by Dirichlet. All talk of infinitesimals has vanished, to be replaced with precise "ϵ-δ" formulations. The discussions are austere and abstract, and, because the abbreviatory idiom of infinitesimals has been forsworn, Cauchy's condensed arguments give way to lengthy and intricate chains of reasoning. What motivated Weierstrass to adopt the new language, and why did it gain acceptance so quickly? The answer can partly be gleaned from an episode we have already examined. Cauchy's quick "resolution" of the Fourier question was clearly faulty, and indicated that something was amiss with his treatment of series expansions. Weierstrass's development of the concept of uniform convergence, and his use of "ϵ-δ" notation to distinguish uniform convergence from convergence *simpliciter* (that is, to formulate and thus to separate (1) and (10)), responded directly to the major known anomaly. In cleansing the theory of series, Weierstrass was fashioning tools for tackling the abstract and technical questions in function theory which were his main concern. He was spurred into foundational activity by his desire to extend the elliptic function theory of Abel and Jacobi, and, to his students and colleagues, Weierstrass's treatment of elliptic functions came to be seen as impressive testimony to the problem-solving power of his approach.

Weierstrass provides us with an explicit account of his reasons in the inaugural address which he delivered on the occasion of his appointment at the University of Berlin. After years of work as a village schoolmaster, Weierstrass's important and thorough studies of analytic problems earned him a professorship. Weierstrass used the speech to explain to his audience the nature of his mathematical development. He begins by referring to his early enthusiasm for Abel's theory of elliptic functions, and he recalls how he felt the importance of Jacobi's extension of the theory.

> These quantities of a completely new kind for which analysis had no example were something I wished to represent properly, and from then on it became my chief task to determine their properties more precisely, a task I was prevented from pursuing until I had become clear about their nature and meaning.[42]

42. *Werke*, volume 1, p. 224.

Weierstrass had learned the lessons of eighteenth- and nineteenth-century analysis. Correct procedures in fields like elliptic function theory, which could only be tackled formally, had to be established. The process of establishment required the investigation of methods of the more general theory of functions, to ensure that the techniques which were to be employed reliably generated true conclusions.

> Clearly it would have been foolish, if I had only wanted to think about the solution to a problem of this kind without having investigated thoroughly the available tools (*Hülfsmittel*) and without practice with less difficult examples.[43]

Although analytic techniques had to be sharpened on "less difficult" examples, cases which are susceptible to independent methods of verifying results, the general studies were, initially, only to be means to the true end (elliptic function theory). At this stage, Weierstrass may even have seen his apprenticeship as tedious.

> In this way years flew past before I could devote myself to the proper work (*die eigentliche Arbeit*), which I have only been able to begin slowly because of the difficulty of my circumstances.[44]

Thus, in 1857, the year of his inauguration, Weierstrass's attitude seems to have been that the problem of rigorizing Cauchy's analysis obtained its urgency because a set of reliable algebraic procedures was needed for "the proper work" (elliptic function theory).

In the next decades, Weierstrass's view of the matter changed. Although technical areas of analysis continued to occupy him—between 1857 and 1887, Weierstrass gave thirty-six sets of lectures on elliptic function theory—he became more inclined to stress the value of his work in providing a systematic treatment of the entire theory of functions. The change of heart is evident in a letter, written to Schwarz in 1875, in which Weierstrass comments on a proof which his former student had sent him:

> The more I reflect on the principles of the theory of functions—and I do this unceasingly—the firmer becomes my conviction that it must be built on the basis of algebraic truths, and that, for this reason, it is not right to claim the "transcendent"—to express myself briefly—as a foundation for the simpler and more basic algebraic theorems—however attractive may seem at first glance those ways of thinking through which Riemann has demonstrated so many of the most important properties of algebraic functions. (That, to the discoverer, who searches so long, any way may be allowed as permissible, is naturally understood; this concerns only the systematic foundation.)[45]

43. *Ibid*.
44. *Ibid*.
45. *Werke*, volume 2, p. 235.

This passage makes it clear that, by the 1870's, Weierstrass regarded the methods and language which he had brought to analysis as indispensable to the subject. They could be used as more prosaic ways of confirming those results discovered by the imaginative flights of genius. The reasons for Weierstrass's increasing firmness of conviction that his stark algebraic arguments represent the proper way to do analysis are not hard to find. The 1860's and 1870's saw the proliferation of analytic theorems by Weierstrass and his students (the profusion of results continued into the early decades of the twentieth century). Gone are the old controversies about representability of functions which had occupied Cauchy's successors. The results of Weierstrassian analysis are unquestioned.

Weierstrass effectively eliminated the infinitesimalist language of his predecessors. In so doing, he finally answered the question of rigor which had arisen for the early calculus of Newton and Leibniz. Yet the context of Weierstrass's reform is leagues away from Berkeley's objections or Newton's sensitivities. Infinitesimalist references were discarded when it was clear that they were no longer useful and that their use could hide mistakes in reasoning about multiple limits. The problem of the infinitesimal first became a mathematical difficulty as the result of the work of Cauchy—work which appeared initially to make infinitesimalist discussions innocuous.

VII

We have now examined one of the two topics which Cauchy's work failed to resolve, and which prompted his successors to further development of analysis. The second episode I shall investigate is the emergence of the theory of real numbers, and, here again, we must begin with Cauchy's approach to convergence. Although the primary emphasis of that chapter of the *Cours d'Analyse* which deals with convergence is on the provision of techniques which will enable mathematicians to decide when individual series are convergent, Cauchy also states a fundamental general condition on convergence, the convergence criterion which bears his name. This criterion is introduced quite casually:

> . . . in other words it is necessary and sufficient (for convergence) that, for infinitely large values of the number n, the sums s_n , s_{n+1} . . . differ from the limit S, and consequently among themselves by an infinitely small quantity.[46]

If we unpack this statement using the equivalences which Cauchy has laid down, making the criterion fully explicit, we obtain:

(6) $\Sigma_1^\infty u_i$ is convergent if and only if for all $\epsilon > 0$ there is an N such that, for all $r > 0 \left| \Sigma_N^{N+r} u_i \right| < \epsilon$

46. *Cours d'Analyse*, p. 114. Recall that s_n, in Cauchy's usage, is $\Sigma_0^n u_i$ (or $\Sigma_1^n u_i$), i.e. s_n is a partial sum of the series.

As every student of real analysis knows, it is easy to prove one half of the equivalence. If $\Sigma_1^\infty u_i$ is convergent, then there is an S such that for any $\epsilon > 0$ we can find N such that $\left| S - \Sigma_1^n u_i \right| < \dfrac{\epsilon}{2}$ if $n > N$. But now, $\left| \Sigma_N^{N+r} u_i \right| = \left| (S - \Sigma_1^N u_i) - (S - \Sigma_1^{N+r} u_i) \right| < \left| S - \Sigma_1^N u_i \right| + \left| S - \Sigma_1^{N+r} u_i \right| = \dfrac{\epsilon}{2} + \dfrac{\epsilon}{2} = \epsilon$. Hence, if a series is convergent, the Cauchy criterion holds. What is difficult to show is that the Cauchy condition is *sufficient* for convergence—and, as we shall see, this difficulty exposes deep shortcomings in Cauchy's untroubled references to "quantities."

Cauchy does not append to his introduction of the criterion anything that could be construed as an attempt at proof. He is content to illustrate it by example, showing that three series which are known to converge (or diverge) are classified correctly by the condition he proposes. It is almost as though Cauchy were claiming that his criterion works for these cases and therefore should be accepted—although that would be one of the "inductions" of which he complains in the Preface to the *Cours d'Analyse!* [47]

The Cauchy condition for convergence makes an important appearance in an Appendix to the *Cours d'Analyse*, and, by studying the role it plays in this passage, we can begin to understand both the importance of the condition for real analysis and the complex of reasons which led Cauchy to adopt it. The aim of the appendix is to provide a purely analytic proof for an important special case of a fundamental general theorem. The general theorem, the Intermediate Value Theorem, states that if $f(x)$, $g(x)$ are functions continuous on an interval $[a,b]$, such that $f(a) < g(a)$ $g(b) < f(b)$, then there is at least one c such that $a < c < b$ and $f(c) = g(c)$. The special case, the Intermediate Zero Theorem, restricts the result to the special case in which $g(x)$ is identically zero on $[a,b]$: that is, if $f(x)$ is continuous on $[a,b]$ with $f(a) < 0$, $f(b) > 0$, there is at least one c such that $a < c < b$ and $f(c) = 0$. This pair of theorems had been of great concern to one of Cauchy's contemporaries, Bolzano, who in 1817 published an article (*Rein analytischer Beweis*) inveighing against traditional endeavors at proving the theorems, and presenting a new algebraic proof (or attempted proof) of them. [48] The text of the *Cours d'Analyse* presents just the kind of reasoning which Bolzano hoped to replace. Having introduced his *algebraic* notion of the continuity of functions, Cauchy immediately links it to the intuitive geometrical concepts.

47. See the passage cited above, pp. 247–48.

48. Ivor Grattan-Guinness has offered the intriguing suggestion that Cauchy knew of Bolzano's work, and quietly took over some of Bolzano's ideas. (See *The Development of the Foundations of Analysis from Euler to Riemann*, pp. 51–52.) It is hard to evaluate this suggestion. Although Cauchy's treatment of the Intermediate Value Theorem resembles Bolzano's in certain central ideas, it differs in important respects. In the face of inconclusive evidence, charity inclines me to suppose that Cauchy did not plagiarize!

A remarkable property of continuous functions of a single variable is to be able to be represented geometrically by means of straight lines or continuous curves. From this observation we can easily deduce the following theorem.[49]

The theorem is the Intermediate Zero Theorem, and the proof is indeed easy. Suppose we assume that $f(x)$ is continuous according to Cauchy's definition of continuity (see (1) above, p. 246). By the "remarkable property," f can be represented by a continuous curve, and if $f(a) < 0 < f(b)$, then, geometrical considerations tell us that there is a c such that $a < c < b$ and $f(c) = 0$: a continuous curve must cross the axis between a and b. Let me hasten to point out that the alleged "remarkable property" is a mistake. As we know from Weierstrass's later investigations into continuity, functions which satisfy the Cauchy criterion for continuity may be radically unrepresentable by continuous curves (in the intuitive sense of 'continuous curve' on which Cauchy's reasoning illicitly trades).[50] Yet what may seem even more puzzling than this particular mistake is the general strategy applied here. Why does Cauchy inject into a volume of *algebraic* analysis a *geometrical* argument of the kind popular with his eighteenth-century precursors? The oddity is resolved by recalling the pragmatic streak in Cauchy's thought which I have been emphasizing. Cauchy wanted to fashion algebraic tools for developing analysis in domains to which the traditional geometrical reasonings are inapplicable. To attain this end he did not need to abandon those reasonings in the areas where they had previously seemed to be successful.

However, Cauchy did want to confirm his new techniques by showing that they would be consonant with the old geometrical arguments. At the end of his "proof" he advertises that Note III to the *Cours d'Analyse* provides a "direct and purely analytic" proof which even has the advantage of giving the appropriate value of c. When we turn to this note we discover that the attempted proof encounters the same difficulty in establishing the existence of limits that besets any attempt to prove the condition on convergence, (6). Cauchy proceeds as follows. Let $b - a = h$, and let m be an integer ≥ 2. Consider the sequence $f(a), f(a + \frac{h}{m}), \ldots, f(b - \frac{h}{m}), f(b)$. There must either be a zero in this sequence (in which case we have established the existence of an intermediate zero), or there are two consecutive terms with different signs. In the latter case, suppose the (first such) terms to be $f(x_1), f(y_1)$. We know that $y_1 - x_1 = \frac{h}{m}$.[51] We now construct the sequence $f(x_1), f(x_1 + \frac{h}{m^2}), \ldots,$

49. *Cours d'Analyse*, p. 50.

50. See Weierstrass's celebrated paper which shows the existence of everywhere continuous nowhere differentiable functions (*Werke*, volume 2, pp. 71–74). I conjecture that Weierstrass's discovery of such functions was a further product of his tying together of the loose ends of Cauchy's analysis. However, I shall not pursue this conjecture here.

51. At this point in the proof, Cauchy interpolates the false assertion that $a < x_1 < y_1 < b$. I sus-

$f(y_1 - \dfrac{h}{m^2}), f(y_1)$, either finding a zero or else obtaining consecutive terms

$f(x_2), f(y_2)$ which differ in sign. Repeating the process, we have a pair of sequences $\{x_n\}$, $\{y_n\}$; the former is monotonic increasing, the latter monotonic decreasing. The difference between y_n and x_n can be made as small as one pleases by choosing n sufficiently large. More exactly, the conditions of the construction make it clear that, for any $\epsilon > 0$, we can find N such that, for all $n > N$, $y_n - x_n < \epsilon$. Cauchy proceeds from this to the essential point.

> We must conclude from this that the terms of the sequences converge to a common limit.[52]

Setting the limit to be c, it is relatively easy to show that $f(c) = 0$.

What sanctions Cauchy's inference? Why must we conclude that the terms of the sequences converge to a common limit? One way of answering the question would be to invoke the Cauchy criterion for convergence, for the inference can be recast so that it is authorized by that criterion. (Bolzano's version of the proof considers only the monotonically increasing sequence, shows that it satisfies (6), and concludes that it converges to a limit. Bolzano also attempts—unsuccessfully—to provide a direct proof of (6).)[53] Alternatively, we can view the situation geometrically. Imagine a line on which we mark the points corresponding to the two sequences. The conditions of the construction of the sequences make it seem inevitable that the sequences of points approach a common limit. I think that Cauchy did view the situation in this fashion, supplementing his algebraic maneuvers with geometrical representations when it became useful. These appeals to geometry established for him the existence of limits, which he could not guarantee algebraically, and, by doing so, they confirmed some of his algebraic principles. In *formulating* those principles algebraically but *justifying* them geometrically, Cauchy bequeathed to his successors a new problem. How can one show algebraically that the requisite limits exist?

For Cauchy and his contemporaries the question would have smacked of pedantry and excessive purism. It needed further investigation of the algebraic concepts Cauchy introduced to show that the simple isomorphisms which he had envisaged—such as the "remarkable property" of continuous functions—were not as straightforward as they had seemed. As the nineteenth century grew older, mathematicians became more wary of the traditional geometrical reasonings to which Cauchy had helped himself when the algebra became difficult.

---

pect that this reflects the influence of a geometrical representation of the situation. He is presumably thinking of a curve which crosses the axis in the middle of the interval!

52. *Cours d'Analyse*, p. 379.

53. For a reconstruction of Bolzano's proof, see my paper, "Bolzano's Ideal of Algebraic Analysis."

Thus there arose the slogan of arithmetizing analysis, a slogan prompted in large measure by a desire to free analysis from geometrical analogies which had once seemed helpful but had proved misleading all too frequently. The new attitude is well expressed in Weierstrass's letter to Schwarz (which I have quoted above) and in Dedekind's introduction to his memoir on continuity. Dedekind deplores his need to rely on "geometric evidences" or "to depend on theorems which are never established in a purely arithmetic manner."[54] So Weierstrass, Dedekind, and others set themselves the task of showing that one could avoid geometrical analogies, which might prove faulty, and replace Cauchy's mixture of algebra and geometry with a systematic analysis, developed arithmetically. The principal focus of their efforts was that corpus of results which Cauchy's new language had formulated, but for which he had had to rely on geometrical evidence, to wit, those theorems asserting the existence of limits. It is no accident that the two theorems of "infinitesimal analysis," with which Dedekind ends his memoir, and for which he claims to have given genuine arithmetical proofs, are intimately related to the theorems examined above. Dedekind first demonstrates that "If a magnitude x grows continually but not beyond all limits it approaches a limiting value" (a result which could have completed Cauchy's proof of the Intermediate Zero Theorem), and he concludes by showing that "If in the variation of a magnitude x we can for every positive magnitude δ assign a corresponding position from and after which x changes by less than δ then x approaches a limiting value" (which yields Cauchy's criterion for convergence).[55] Exactly those limit existence theorems which Cauchy brought to the fore are theorems which play a major role in the development of real analysis, so that any fully algebraic treatment of the subject must come to terms with them.

Dedekind and Weierstrass were not the first mathematicians to exercise themselves with the tricky task of giving algebraic demonstrations of the existence of limits. As I have already noted, Bolzano, writing in 1817, insisted that appeals to geometry do not belong in a proper treatment of analysis.[56] Yet Bolzano's reasons for repudiating geometrical arguments for general results about functions are somewhat different from those that moved his successors. Adopting a general philosophical picture of the proper structure of the sciences, Bolzano argues that deriving theorems of general sciences (function theory) from principles of special sciences (geometry) is an "intolerable offence against correct method." What exactly is the nature of the offence? On Bolzano's view, although the appeal to the special science may generate conviction *that* the theorem of the general science is true, it will not show *why* it is true. To put

---

54. *Continuity and Irrational Numbers*, pp. 1 and 2.

55. *Ibid.*, pp. 25–27.

56. For a much more detailed presentation of Bolzano's accomplishments see "Bolzano's Ideal of Algebraic Analysis."

the point in a slightly different way, Bolzano's complaint is that, by continuing the eighteenth-century practice of using geometry in analysis, we shall fail to understand properly the results we announce. In this respect, Bolzano may be seen as prophetic. Nineteenth-century analysts indeed found that the mixture of algebraic formulation and geometrical justification which Cauchy favored did not provide full comprehension of the major results of function theory, and that the failures of understanding showed up in analytic mistakes. So, by a different route, Dedekind and Weierstrass came to espouse the goal which Bolzano had already adopted.

Bolzano campaigned (unsuccessfully) for a reform in standards of proof. In philosophical and mathematical writings, he argued that genuine proofs should conform to the proper ordering of the mathematical disciplines, an ordering he had obtained by reflecting on the nature of eighteenth-century analysis. Bolzano saw more clearly than any of his contemporaries or immediate successors that the calculus had outgrown its geometrical beginnings and that, if it was to be systematized on a unified basis, then the systematization must be achieved in algebraic terms. Unfortunately, Bolzano's "argument from above" for the revision of standards of proof was ignored by most of his contemporaries. Only when supplemented by an "argument from below," which showed the need to avoid appeals to geometry if research in analysis was to be clearly conducted on the basis of firm understanding of elementary theorems, did mathematicians come to deny that geometrical arguments amounted to "scientific proof" and to regard Bolzano's critique as anything more than misguided fussiness. Moreover, writing half a century later than Bolzano, Dedekind and Weierstrass could arrive at a more precise formulation of the central issue. Where Bolzano had argued for an *algebraic* version of analysis, viewing algebra as a general science of quantities, Weierstrass and Dedekind saw in the notion of quantity the stumbling block which had made it necessary for Cauchy to take his detour into geometry. Claiming that the theorems of analysis are theorems about *numbers,* they proposed that analysis must be done *arithmetically,* and that the central task for the analyst who wishes to prove limit existence theorems is to give a characterization of the real numbers which will enable one to show, by drawing on principles of arithmetic, that there are real numbers satisfying the conditions laid down in Cauchy's theorems.

I shall concentrate on Dedekind's response to the problem of characterizing the real numbers in a way which would allow for the proof of limit existence theorems. (I choose Dedekind because he gives the most explicit account of his motives and reasoning, and because his solution to the problem has been the most popular in subsequent analysis.) What Dedekind intends to accomplish is the reconstruction of the intuitive reasoning on which Cauchy had relied: he will analyze the concept of the continuity of the real numbers so that this is arithmetically expressed, and, in consequence, he will be able to avoid Cauchy's detour into geometry. The problem, then, is to formulate a "principle of continuity" which will state in arithmetical language that "gaplessness" of the reals which can so easily be represented geometrically and can thus induce conviction about the existence of limits. Dedekind solves this problem by suggesting that the "essence of continuity" of the line resides in the principle that "If all points of the straight line fall into two classes such that every point of the first class lies to the left of every point of the second class, then there exists one and only one point which produces this division of all points into two classes, this severing of the straight line into two portions."[57] This principle provides a transition from the intuitive geometrical conception of continuity to a formulation of the continuity of the real numbers from which geometrical language has been banished. (Indeed it could pave the way for defining a general notion of the continuity of a set with respect to an ordering relation R, although Dedekind does not pursue so abstract a treatment.) Dedekind forthrightly notes that he cannot *prove* his principle, and I interpret his memoir as an argument for its acceptance. For, on the basis of the account of the continuity of the real numbers which he generates from it, Dedekind demonstrates those limit existence theorems which had foiled previous attempts to arithmetize analysis. To use the terminology of previous chapters, a new way of fixing the reference of 'continuous domain' is justified by showing that it squares with prior views and enables the reconstruction of previous reasonings.

Inspired by his principle of continuity, Dedekind introduces the notion of a *cut* on the set of rationals R. A cut is a separation of R into two classes, A_1, A_2, such that for any x, y, if x belongs to A_1 and y belongs to A_2 then $x < y$. (A_1 and A_2 are implicitly assumed to be nonempty. Dedekind uses the notation '(A_1, A_2)' for cuts.) The heart of Dedekind's proposal is that we should think of the real numbers as specified in the following way:

(7) For any cut (A_1, A_2) there is a unique real number which "corresponds to" (or "produces") the cut.[58]

Now, as it stands, (7) is too vague to enable us to reconstruct the arguments for limit existence in arithmetical terms. What Dedekind has to do is to provide rules for the usage of real number expressions, which will enable him to defend principles about addition, multiplication, and so forth, analogous to the standard laws of arithmetic. Thus he shows how to define the ordering of the reals, addition of reals, multiplication of reals, in ways which will allow for the preservation of the familiar arithmetical laws. For example, ordering is introduced as follows. If (A_1, A_2), (B_1, B_2) are two cuts such that A_1 contains at

57. *Continuity and Irrational Numbers*, p. 11.

58. *Ibid.*, pp. 12–15. Dedekind does not identify real numbers as, say, the first members of cuts, but engages, throughout his essay, in talk of the "creation" of numbers. This talk led to Russell's famous complaint. I believe that Dedekind's ideas are more reasonable than they may appear at first glance, but to pursue this issue would be to lose sight of my main concern, to wit, Dedekind's development of language which could help him to reformulate and prove limit existence theorems.

266 THE NATURE OF MATHEMATICAL KNOWLEDGE

least two members which do not occur in B_1, then the number corresponding to (A_1, A_2) is greater than the number corresponding to (B_1, B_2); if A_1 has the same members as B_1 or if A_1 contains only one member which does not belong to B_1 then the numbers corresponding to the two cuts are equal. On the basis of this treatment, Dedekind is able to provide a careful argument for the trichotomy principle, proving that for any real numbers α,β, $\alpha>\beta$ or $\alpha=\beta$ or $\alpha<\beta$. It is worth quoting the passage in which he summarizes this discussion:

> As this exhausts the possible cases, it follows that of two different numbers one is necessarily the greater, the other the less, which gives two possibilities. A third case is impossible. This was indeed involved in the use of the *comparative* (greater, less) to designate the relation between α,β; but this use has only now been justified. In just such investigations one needs to exercise the greatest care so that even with the best intention to be honest he shall not, through a hasty choice of expressions borrowed from other notions already developed, allow himself to be led into the use of inadmissible transfers from one domain to the other.[59]

The final phrase echoes the Preface to Cauchy's *Cours d'Analyse* and makes clear Dedekind's own view of his enterprise. Historically, the real numbers were introduced geometrically without any explicit statement of their properties, so that, in proving theorems of analysis, mathematicians were forced to use geometrical representations because they had no arithmetical principles on which to rely. Dedekind has modified the language of analysis by offering (7) as a new way of fixing references to real numbers and by providing specifications of the conditions on order (these will be supplemented further by fixing addition, multiplication, and so forth). It is now incumbent on him to show that the familiar properties, previously ascribed to the reals on the basis of geometrical evidence, can be obtained from his definitions. In particular, Dedekind has to establish the continuity of the real numbers and to demonstrate the power of his approach to yield the troublesome theorems on limit existence.

The continuity of the real numbers is relatively straightforward. Consider any division of the real numbers into two classes A, B, such that, for any x, y, if x belongs to A and y belongs to B then $x<y$. Let A', B' be the subclasses of A, B consisting of all their rational members. Then (A',B') is a cut. By (7), there is a real number corresponding to this cut. Given the definition of the ordering relation on the reals, it is relatively easy to prove that this real number is either the greatest member of A or the least member of B. So Dedekind has shown that any division of the reals into two classes, such that all members of one class are less than any member of the other, must meet the condition that one class has a greatest member or the other has a least member. And, on his account, this is what we mean in asserting the continuity of the reals.[60]

59. *Ibid.*, pp. 17–18.
60. The proof is given at *ibid.*, pp. 20–1. To see how it goes, consider the following. Let α be the number "produced by" (A',B'). Then assume, for reductio, that α is neither the greatest

The real test of Dedekind's approach comes in the final section of the memoir. Can he establish, using his new language, the theorems that had baffled Cauchy? Consider the theorem that a monotonically increasing sequence, bounded above, is convergent. (This theorem would have enabled Cauchy to complete his "analytic proof" of the Intermediate Zero Theorem; it was also recognized explicitly by Bolzano, who struggled to obtain it.) Using contemporary notation, we can state Dedekind's theorem as follows:

(8) Let $\{s_n\}$ be a sequence of real numbers such that, for every n, $s_n < s_{n+1}$. Suppose that, for all n, there is a real number α, such that $s_n < \alpha$. Then there is a real number β such that for any $\epsilon > 0$ there is an N such that, for every $n > N$, $|\beta - s_n| < \epsilon$.

Dedekind's proof is simple and elegant. Let A_2 be the class of all those real numbers meeting the condition that, for every n, $s_n < \alpha$. Let A_1 be its complement in the class of real numbers. By the principle of continuity, either A_1 has a greatest member or A_2 has a least member. Let this real number be β. It is easy to show that β belongs to A_2 and that if β is *not* the limit of $\{s_n\}$ then A_2 contains a real number less than β (contradicting the definition of β).[61] Hence the theorem is proved.

We may summarize Dedekind's achievement as follows. Recognizing that the old methods of fixing the referents of expressions in the language of real numbers did not enable mathematicians to prove arithmetically fundamental theorems of analysis, Dedekind proposed a new method of fixing the referents of those expressions. This new method characterized the real numbers explicitly in what Dedekind took to be an arithmetical vocabulary, and the new characterization was justified by showing that it allowed for the derivation of familiar theorems about reals and for the tricky limit existence theorems. Yet, of course, Dedekind's work itself brought new problems to the fore. The vocabulary which Dedekind allowed himself not only contained purely arithmetical expressions but also terms from the language of classes (or sets). These references generated new questions for the mathematical community. Could

---

member of A nor the least member of B. Since A,B divides the reals, α belongs to A or to B. Suppose, without loss of generality, that α belongs to A. Then, by hypothesis, there is a real number, β, in A, such that $\alpha < \beta$. Let β be produced by the cut (C',D'). Since $\alpha < \beta$, C' must contain at least two members not belonging to A'. From this it follows that there is a rational c such that $\alpha < c < \beta$, and that c does not belong to A'. However, since β belongs to A and $c < \beta$, c must belong to the set of rationals included in A, i.e. A'. So we obtain a contradiction.

61. See *ibid.*, p. 25. If β does not belong to A_2, then there is an n such that $\beta < s_n$. Hence, since $\{s_n\}$ is monotonically increasing, $\beta < s_{n+1}$. But now we know that $\beta < s_{n+1} < s_{n+2}$, so that s_{n+1} belongs to A_1, contradicting the definition of β as the greatest member of A_1 or the least member of A_2. So β belongs to A_2. Suppose, then, that for some $\epsilon > 0$, there is *no* N such that for $n > N$ $|\beta - s_n| < \epsilon$. Since $\{s_n\}$ is monotonically increasing, it follows that, for every n, $\beta - s_n \geq \epsilon$. Hence $\beta - \epsilon > s_n$, from which we infer that $\beta - \epsilon$ is in A_2. This contradicts the specification of β as the least member of A_2. Hence, for any $\epsilon > 0$, there is an N such that, for $n > N$, $|\beta - s_n| < \epsilon$. Therefore $\lim_{n \to \infty} s_n = \beta$.

one state principles about the existence of sets which would make explicit the assumptions inherent in Dedekind's reasoning? The story of attempts to answer this question is a familiar one.

VIII

At this point I shall conclude my presentation of the history of analysis, using one final episode to underscore a point that I have made already. In treating the foundational work of the great nineteenth-century mathematicians, I have insisted that it was not inspired by any exalted epistemological aims, but was, instead, an attempt to respond to the needs of mathematical research. I believe that this is evident in the work of Cauchy, Abel, Weierstrass, and Dedekind. A different way to make my case is to describe the effects of an explicitly philosophical call for rigor. From 1884 to 1903, Frege campaigned for major modifications in the language of mathematics and for research into the foundations of arithmetic. The mathematicians did not listen. They were unmoved by the epistemological considerations which Frege advanced, and, as the years went by, Frege's judgments became progressively harsher. In *The Foundations of Arithmetic*, he is relatively optimistic. Although he concedes that mathematicians may consider him to have gone "further back into the foundations than [is] necessary," Frege claims that his ideas are "in the air" and will repay the attention of the mathematical community.[62] On the account of the interests which motivated research into the foundations of analysis that I have offered above, we can easily understand why Frege's investigations into the concept of number were ignored. None of the techniques of elementary arithmetic cause any trouble akin to the problems generated by the theory of series or results about the existence of limits. Instead of continuing a line of foundational research, Frege contended for a new program of rigor at a time when the chain of difficulties which had motivated the nineteenth-century tradition had, temporarily, come to an end.

There can be no doubt that the lack of response to his work changed Frege's attitude. In the Introduction to his *magnum opus*, *The Basic Laws of Arithmetic*, he is pessimistic about the possibility of reaching a wide audience.

> Perhaps the number of mathematicians who trouble themselves over the foundations of their science is not great, and even these frequently seem to be in a great hurry until they have got the fundamental principles behind them.[63]

Frege had come to understand that the interests of contemporary mathematicians might be different from his own—but he continued to insist that they

---

62. *The Foundations of Arithmetic*, pp. ix and x.
63. *The Basic Laws of Arithmetic*, pp. 9–10.

ought to be concerned with the problems which he investigated. He did so by offering fierce criticisms of the casual remarks about the natural numbers with which the allegedly rigorous analysts began their discussions, by attempting to uncover confusions in definitions, and, in general, by trying to show that the mathematicians were extremely unclear about the subject matter of their discipline. Frege's comments are always trenchant, frequently witty, and occasionally bitter. They failed to provoke a response because the unclarities to which Frege called attention—indicated in an inability to give satisfactory characterizations of numbers and functions—did not stand in the way of mathematical research. Those whom Frege attacked could shrug off his criticisms with the response that philosophical fussing about the nature of the numbers was none of their business. Frege's investigations paid no obvious dividends for mathematical research, and Frege himself offered the natural way for the mathematicians to respond to them: *metaphysica sunt, non leguntur.*[64]

The difference between Frege's philosophical ideals and the attitude of the mathematicians emerges clearly from a passage in an unpublished essay of 1914. Frege begins with an ironic characterization of the view he opposes.

> One might think: "The content of mathematical sciences is of no real concern to the mathematician, but is the province of the philosopher; and all philosophy is quite inexact, uncertain and really unscientific. A mathematician who attends to his scientific calling will have nothing to do with it. Perhaps, in a weak moment, even from the best mathematician, a definition will occasionally slip out, or, at least, something which appears as if it were a definition; but this is of no significance. It is as though he had merely sneezed. In fact, what is of real concern is only that everybody should agree in terminology and in formulas. A mathematician who has not caught the disease of philosophy will content himself with that."[65]

I claim that if we subtract the irony then Frege's characterization is perfectly accurate, and that the unwillingness of the mathematical community to respond to Frege's scolding is to be understood in this way. My claim is supported by Frege's attempt to delineate clearly the benefits which will result from answering the questions he has posed. These are epistemological benefits which embody the ideals of mathematical apriorism. When Frege emphasizes the possibility of complete clarity and certainty in mathematical knowledge, he is advancing a picture of mathematics that is almost irrelevant to the working mathematician.

When we disentangle the factors which led to the Weierstrassian rigorization of analysis, we find a sequence of local responses to mathematical problems. That sequence ends with a situation in which there were, temporarily, no fur-

64. *Ibid.*

65. *Nachgelassene Schriften,* pp. 233–34. Similar sentiments can be found in the final paragraph of "Logical Errors in Mathematics" (*ibid.,* pp. 180–81).

ther such problems to spur foundational work. Frege was wrong to portray himself as continuing the nineteenth-century tradition (and he learned that he was wrong), and some of his admirers have erred in accepting his early assessment of his inquiry.[66] By a curious twist, Frege's work began to be appreciated by the mathematicians because of the incident which seemed to him to bring his whole enterprise down in ruins. After the discovery of the set-theoretic paradoxes there was work for *some* of Frege's formal logic to do (Frege's advances in semantics were only recognized much later). From the attempts of Cantor and Dedekind to systematize the set-theoretic references needed for analysis, a new mathematical problem had emerged, and Frege's work showed to advantage in allowing for a precise and stark presentation of the problem. Yet Frege did not join in the mathematical work of reconstruction. His logical studies after 1903 continue to emphasize points on which he had harped earlier: the inadequacy of "formal arithmetic," troubles with the notion of function, and so forth. In comparison with the difficulties of set theory, such worries seemed minor, and, just as they had failed to listen to Frege's earlier critiques, so too the mathematicians ignored his reiteration of the old themes.

IX

In the preceding account, I have tried to display the growth of the calculus and its metamorphosis into late nineteenth-century analysis (the analysis of Weierstrass and Dedekind) as a rational process. If I have been successful then I have shown how, in sequence, the transitions identified in Chapter 9 can transform mathematical practice. For the episodes I have discussed involve rational modification of components of the prior practice in accordance with the patterns of change which Chapter 9 described. Question-answering and the systematization of previous problem solutions led to the warranted adoption of the Newton-Leibniz calculus. The resultant practice, particularly that of the Leibnizians, generated new questions, and, in the hands of Euler, the practice was modified as previous ideas were generalized and new language and new reasonings were introduced to answer questions hailed as important. A justified shift in the corpus of accepted questions, prompted largely by the demands of mathematical physics and Fourier's proposal for responding to them, lent new urgency to an old problem of rigor. So, for Cauchy and Abel, the question of clearing away the anomalies of the theory of infinite series became important. Cauchy's proposed rigorization was thus warranted because it satisfied the needs of research. Yet Cauchy's new language spawned new questions and introduced new prob-

66. See, for example, Michael Dummett's comment in his article on Frege in the *Encyclopedia of Philosophy* (Volume 3, p. 226). "Frege's primary object in devising this logical system was to achieve the ideal of that rigour to which all nineteenth century mathematics had been tending. . . ." There was no such ideal.

lems of rigor. These questions and problems made rational the rigorization and systematization achieved by Weierstrass and Dedekind. At each stage in the development of the calculus, precisely the transitions I have identified in Chapter 9 play a crucial role.

In my account, I have paid particular attention to the problem of understanding work in the foundations of mathematics. My reason for selecting the "foundational" episodes in the history of the calculus should be relatively obvious: such episodes seem to present the greatest challenge to an account of the growth of mathematical knowledge along the lines I favor, and they are also intimately connected with the acceptance of new axioms and definitions. Traditionally, philosophical accounts of mathematical knowledge have rested on the thesis that mathematicians have basic a priori knowledge of axioms and definitions. My attack on this claim needed to be supplemented with an answer to two questions: How do mathematicians know axioms and definitions? What is the point of foundational work if not to provide a priori knowledge of that part of mathematics for which foundations are sought? In the foregoing account, I have implicitly answered these questions. But it may help to offer, explicitly, a simplified précis. Axioms and definitions are accepted because they systematize previously accepted problem solutions. Foundational study is motivated by the need to fashion tools for continuing mathematical research. The discussions of Cauchy, Dedekind, and Weierstrass above articulate and qualify these stark responses. My concluding remarks on Frege are intended to underscore the difference between the pragmatic concerns of the mathematicians and the epistemological ideals (misguided epistemological ideals, I would suggest) of the philosophers.

At first glance, the remoteness of contemporary mathematics from perceptual experience is so striking that it appears to doom any empiricist account of mathematical knowledge. The empiricist must respond by explaining how "higher" mathematics could emerge from those rudimentary parts of the subject which can be perceptually warranted. The history of the calculus, as I have presented it, does not take us all the way from proto-mathematical knowledge to our present situation. Yet it should not be too difficult to see how the narrative could be completed, how the practice which Newton and Leibniz inherited can be traced back to the mathematics of the Greeks and of their predecessors, and how the practice which Weierstrass and Dedekind bequeathed can be unfolded into the mathematics of the Bourbaki and their descendants. Moreover, my history yields the general moral that the types of changes I have identified, occurring in succession, can lead to a practice remarkably different from that which initiated the sequence. To adapt Newton's famous figure, when giants continually stand on the shoulders of giants, someone lucky enough to sit at the top can see a remarkably long way.

Bibliography

Abel, Niels. *Oeuvres Complètes.* 2 vols. New York, 1965.

Appel, K., and W. Haken. "The Solution of the Four Color Map Problem." *Scientific American,* 137 (1977): 108–21.

Ayer, A. J. *Language, Truth and Logic.* London, 1946.

Beck, L. W. "Can Kant's Synthetic Judgments Be Made Analytic?" In *Kant,* edited by R. P. Wolff, pp. 3–22. New York: Doubleday, 1967.

Belnap, N., and T. Steel. *The Logic of Questions and Answers.* New Haven: Yale University Press, 1967.

Benacerraf, P. "What Numbers Could Not Be." *Philosophical Review,* 74 (1965): 47–73.

———. "Mathematical Truth." *Journal of Philosophy,* 70 (1973): 661–79.

Benacerraf, P., and H. Putnam, eds. *Philosophy of Mathematics: Selected Readings.* Englewood Cliffs: Prentice-Hall, 1964.

Berkeley, G. *The Analyst.* In *The Works of George Berkeley,* vol. 4, edited by A. A. Luce and T. Jessop. London: Nelson, 1950.

Bernays, P. "On Platonism in Mathematics." In Benacerraf and Putnam, *Philosophy of Mathematics.*

Birkhoff, G. *A Source Book in Classical Analysis.* Cambridge: Harvard University Press, 1973.

Bochner, S. *The Role of Mathematics in the Rise of Science.* Princeton: Princeton University Press, 1966.

Bolzano, B. *Rein Analytischer Beweis.* Leipzig: Ostwald, 1905.

———. *Paradoxes of the Infinite.* Translation by D. Steele of *Paradoxien des Unendlichen.* New Haven: Yale University Press, 1950.

———. *Theory of Science.* Translation by R. George of parts of *Wissenschaftslehre.* Berkeley: University of California Press, 1972.

Boolos, G. "The Iterative Conception of Set." *Journal of Philosophy,* 68 (1971): 215–31.

Bourbaki, N. *Elements of Mathematics: Theory of Sets.* Paris: Hermann, 1968.

Boyer, C. *The History of the Calculus and Its Conceptual Development.* New York: Dover, 1949. (Formerly entitled: *The Concepts of the Calculus.*)

Brouwer, L. E. J. *Collected Works.* Vol. 1. Amsterdam: North Holland, 1975.

————. "Consciousness, Philosophy and Mathematics." Printed in part in Benacerraf and Putnam, *Philosophy of Mathematics,* and completely in *Collected Works,* vol. 1.

Burian, R. "More than a Marriage of Convenience: On the Inextricability of History and Philosophy of Science." *Philosophy of Science,* 44 (1977): 1–42.

Cantor, G. *Gesammelte Abhandlungen.* Edited by E. Zermelo. Berlin: Springer, 1932.

————. *Briefwechsel Cantor-Dedekind.* Edited by E. Noether and J. Cavaillès. Paris: Hermann, 1937.

Cardano, G. *The Great Art.* Translation of parts of *Ars Magna* by T. R. Witmer. Cambridge: MIT Press, 1968.

Carnap, R. *Foundations of Logic and Mathematics.* Chicago, 1939.

————. *Meaning and Necessity.* Chicago, 1956.

————. *Logical Foundations of Probability.* Chicago, 1950.

————. "The Aim of Inductive Logic." In E. Nagel, P. Suppes and A. Tarski, editors, *Logic, Methodology, and Philosophy of Science.* Stanford: Stanford University Press, 1962.

Castañeda, H-N. "Indicators and Quasi-Indicators." *American Philosophical Quarterly,* 4 (1967): 85–100.

————. "On the Logic of Attributions of Self-Knowledge to Others." *Journal of Philosophy,* 65 (1968): 439–56.

Cauchy, A. *Oeuvres Complètes.* 25 vols. Paris, 1882–1932.

————. *Cours d'Analyse.* Series 2, vol. 3 of *Oeuvres Complètes.*

————. *Résumé de Leçons Données à L'Ecole Polytechnique.* Series 2, vol. 4 of *Oeuvres Complètes.*

Chihara, C. *Ontology and the Vicious-Circle Principle.* Ithaca: Cornell University Press, 1973.

Chisholm, R. *Theory of Knowledge.* Englewood Cliffs: Prentice-Hall, 1966.

Chomsky, N. *Language and Mind.* New York: Harcourt Brace Jovanovitch, 1972.

————. *Rules and Representations.* New York: Columbia University Press, 1980.

Crowe, M. "Ten 'Laws' Concerning Patterns of Change in the History of Mathematics." *Historia Mathematica,* 2 (1975): 161–66.

————. *A History of Vector Analysis.* Notre Dame: University of Notre Dame Press, 1967.

Curry, H. B. *Outline of a Formalist Philosophy of Mathematics.* Amsterdam: North-Holland, 1951.

d'Alembert, J. Articles "Limite," "Différentiel," in the *Encyclopédie.* Paris, 1751–65.

Dauben, J. *Georg Cantor.* Cambridge: Harvard University Press, 1979.

Davidson, D. "Truth and Meaning." *Synthèse,* 17 (1967): 304–23.

Dedekind, R. *Dedekind's Essays on the Theory of Numbers.* Edited by W. Beman. New York: Dover, 1963.

————. *Continuity and Irrational Numbers.* Translation by W. Beman of *Stetigkeit und irrationale Zahlen.* In *Dedekind's Essays on the Theory of Numbers.*

————. *The Nature and Meaning of Numbers.* Translation by W. Beman of *Was Sind und Was Sollen die Zahlen?* In *Dedekind's Essays on the Theory of Numbers.*

Descartes, R. *Philosophical Writings.* Edited by E. S. Haldane and G. R. T. Ross. Cambridge University Press, 1911–12.

——. *Geometry*. Translation and facsimile reproduction of *La Géométrie* by D. E. Smith and M. L. Latham. New York: Dover, 1954.

Detlefsen, M., and M. Luker. "The Four-Color Problem and Mathematical Proof." *Journal of Philosophy*, 77 (1980): 803–24.

Donnellan, K. "Proper Names and Identifying Descriptions." In *Semantics of Natural Language*, edited by D. Davidson and G. Harman. Dordrecht: Reidel, 1972.

——. "Speaking of Nothing." *Philosophical Review*, 83 (1974): 3–31.

——. "The Contingent A Priori and Rigid Designators." *Midwest Studies in Philosophy*, 2 (1977): 12–27,

Dugac, P. "Eléments d'analyse de Karl Weierstrass." *Archive for the History of the Exact Sciences*, 10: 41–176.

——. *Richard Dedekind et les Fondements des Mathématiques*. Paris: Vrin, 1976.

Dummett, M. A. E. *Frege: Philosophy of Language*. London: Duckworth, 1973.

——. "What Is a Theory of Meaning (II)?" In *Truth and Meaning*, edited by G. Evans and J. McDowell. Oxford: Oxford University Press, 1976.

——. *Elements of Intuitionism*. Oxford: Oxford University Press, 1978.

——. *Truth and Other Enigmas*. London: Duckworth, 1979.

——. "Frege." In *The Encyclopedia of Philosophy*, edited by P. Edwards. New York: Macmillan, 1967.

Euclid. *The Thirteen Books of Euclid's Elements*. Edited and translated by T. Heath. New York: Dover, 1956.

Euler, L. *Opera Omnia*. Leipzig: Teubner, 1911–36.

Feigl, H. "The 'Orthodox' View of Theories: Remarks in Defense as Well as Critique." In *Minnesota Studies in the Philosophy of Science*, vol. 4, edited by M. Radner and S. Winokur. Minneapolis: University of Minnesota Press, 1970.

Feyerabend, P. "Explanation, Reduction, and Empiricism." In *Minnesota Studies in the Philosophy of Science*, vol. 3, edited by H. Feigl and G. Maxwell. Minneapolis: University of Minnesota Press, 1962.

——. "Problems of Empiricism." In *Beyond the Edge of Certainty*, edited by R. Colodny. Englewood Cliffs: Prentice-Hall, 1965.

——. *Against Method*. London: Verso Books, 1975.

——. *Science in a Free Society*. London: New Left Books, 1978.

Field, H. "Quine and the Correspondence Theory." *Philosophical Review*, 83 (1974): 200–228.

Firth, R. "The Anatomy of Certainty." *Philosophical Review*, 76 (1967): 3–27.

Fodor, J. A. *The Language of Thought*. New York: Crowell, 1975.

Frege, G. *The Foundations of Arithmetic*. Translation by J. L. Austin of *Die Grundlagen der Arithmetik*. Oxford: Blackwell, 1950.

——. *Die Grundgesetze der Arithmetik*. Hildesheim: Olms, 1962.

——. *The Basic Laws of Arithmetic*. Translation by M. Furth of parts of *Die Grundgesetze der Arithmetik*. Berkeley: University of California Press, 1964.

——. *Kleine Schriften*. Hildesheim: Olms, 1967.

——. *Nachgelassene Schriften*. Hamburg: Felix Meiner Verlag, 1969.

——. *On the Foundations of Geometry and Formal Theories of Arithmetic*. Edited and translated by E.-H. W. Kluge. New Haven: Yale University Press, 1971.

Friedman, M. "Explanation and Scientific Understanding." *Journal of Philosophy*, 71 (1974): 5–19.

Galois, E. *Ecrits et Mémoires Mathématiques*. Paris: Gauthier-Villars, 1962.

Gibson, J. J. *The Ecological Approach to Visual Perception*. Boston: Houghton Mifflin, 1979.

Glymour, C. "The Epistemology of Geometry." *Noûs*, 11 (1977): 227–51.

Gödel, K. "What Is Cantor's Continuum Problem?" In Benacerraf and Putnam, *Philosophy of Mathematics*.

Goldman, A. I. "A Causal Theory of Knowing." *Journal of Philosophy*, 64 (1967): 357–72.

————. "Innate Knowledge." In *Innate Ideas*, edited by S. P. Stich. Berkeley: University of California Press, 1975.

————. "Discrimination and Perceptual Knowledge." *Journal of Philosophy*, 72 (1976): 771–91.

————. "What Is Justified Belief?" In *Justification and Knowledge*, edited by G. Pappas. Dordrecht: Reidel, 1980.

Goodman, N. *Fact, Fiction and Forecast*. Indianapolis: Bobbs-Merrill, 1965.

Gottlieb, D. "The Truth about Arithmetic." *American Philosophical Quarterly*, 15 (1978): 81–90.

Grattan-Guinness, I. *The Development of the Foundations of Analysis from Euler to Riemann*. Cambridge: MIT Press, 1970.

Grice, H. P., and P. F. Strawson. "In Defense of a Dogma." *Philosophical Review*, 65 (1956): 141–58.

Grosholz, E. "Descartes' Unification of Algebra and Geometry." In *Descartes, Mathematics and Physics*, edited by S. Gaukroger. Hassocks, Eng.: Harvester, 1980.

Hallett, M. "Towards a Theory of Mathematical Research Programmes." *British Journal for the Philosophy of Science*, 30 (1979): 1–25, 135–59.

Hamilton, W. R. *Lectures on Quaternions*. Dublin, 1853.

Hanson, N. R. *Patterns of Discovery*. Cambridge: Cambridge University Press, 1961.

Hardy, G. H. *A Mathematician's Apology*. Cambridge University Press, 1941.

Harman, G. *Thought*. Princeton: Princeton University Press, 1973.

Hart, W. D. "On an Argument for Formalism." *Journal of Philosophy*, 71 (1974): 29–46.

Hawkins, T. *Lebesgue's Theory of Integration*. Madison: University of Wisconsin Press, 1970.

Heath, T. *A History of Greek Mathematics*. Oxford: Clarendon Press, 1921.

Hempel, C. G. *Aspects of Scientific Explanation*. Glencoe: Free Press, 1965.

————. *Philosophy of Natural Science*. Englewood Cliffs: Prentice-Hall, 1966.

Heyting, A. *Intuitionism: An Introduction*. Amsterdam: North-Holland, 1956.

Hilbert, D. "On the Infinite." In Benacerraf and Putnam, *Philosophy of Mathematics*.

Hume, D. *A Treatise of Human Nature*. Edited by L. A. Selby-Bigge. Oxford: Oxford University Press, 1973.

Joseph, G. "The Many Sciences and the One World." *Journal of Philosophy*, 77 (1980): 773–91.

Jourdain, P. E. B. "The Origins of Cauchy's Conception of the Definite Integral and of the Continuity of a Function." *Isis*, 1 (1913): 661–713.

Jubien, M. "Ontology and Mathematical Truth." *Noûs*, 11 (1977): 133–50.

Kant, I. *Critique of Pure Reason*. Translated by N. Kemp Smith. London: Macmillan, 1965.

Katz, J. J. "Recent Criticisms of Intensionalism." In *Minnesota Studies in the Philosophy of Science*, vol. 7, edited by K. Gunderson. Minneapolis: University of Minnesota Press, 1975.

Kiernan, B. "The Development of Galois Theory from Lagrange to Artin." *Archive for the History of the Exact Sciences*, 8, no. 1/2 (1971): 40–154.

Kitcher, P. S. "Fluxions, Limits, and Infinite Littlenesse." *Isis*, 64 (1973): 33–49.

———. "Kant and the Foundations of Mathematics." *Philosophical Review*, 84 (1975): 23–50.

———. "Bolzano's Ideal of Algebraic Analysis." *Studies in the History and Philosophy of Science*, 6 (1975): 229–71.

———. "Hilbert's Epistemology." *Philosophy of Science*, 43 (1976): 99–115.

———. "On the Uses of Rigorous Proof." Review of *Proofs and Refutations* by I. Lakatos, *Science* (1977): 782–83.

———. "The Nativist's Dilemma." *Philosophical Quarterly*, 28 (1978): 1–16.

———. "The Plight of the Platonist." *Noûs*, 12 (1978): 119–36.

———. "Theories, Theorists and Theoretical Change." *Philosophical Review*, 87 (1978): 519–47.

———. "Frege's Epistemology." *Philosophical Review*, 88 (1979): 235–62.

———. "A Priori Knowledge." *Philosophical Review*, 89 (1980): 3–23.

———. "Apriority and Necessity." *Australasian Journal of Philosophy*, 58 (1980): 89–101.

———. "Arithmetic for the Millian." *Philosophical Studies*, 37 (1980): 215–36.

———. "Mathematical Rigor—Who Needs It?" *Noûs*, 15 (1981): 469–93.

———. "Explanatory Unification." *Philosophy of Science*, 48 (1981): 507–31.

Kline, M. *Mathematical Thought from Ancient to Modern Times*. New York: Oxford University Press, 1972.

Kordig, C. *The Justification of Scientific Change*, Dordrecht: Reidel, 1971.

Kornblith, H. "Beyond Foundationalism and the Coherence Theory." *Journal of Philosophy*, 77 (1980): 597–612.

Kreisel, G. "Hilbert's Program." In Benacerraf and Putnam, *Philosophy of Mathematics*.

———. "Mathematical Logic—What Has It Done for the Philosophy of Mathematics?" In *Lectures on Modern Mathematics*, edited by T. L. Saaty. New York, 1965.

Kripke, S. "Identity and Necessity." In *Identity and Individuation*, edited by M. K. Munitz. New York University Press, 1971.

———. *Naming and Necessity*. Cambridge: Harvard University Press, 1980.

Kronecker, L. *Werke*. 5 vols. Leipzig: Teubner, 1895–1931.

Kuhn, T. S. *The Structure of Scientific Revolutions*. Chicago: Chicago University Press, 1970.

———. *The Essential Tension*. Chicago: University of Chicago Press, 1977.

———. "Objectivity, Value Judgment and Theory Choice." In *The Essential Tension*.

———. "Mathematical versus Experimental Traditions in the Development of Physical Science." In *The Essential Tension*.

Lagrange, J. *Théorie des Fonctions Analytiques*. Paris, 1797.

Lakatos, I. *Proofs and Refutations*. Cambridge: Cambridge University Press, 1976.

———. *Philosophical Papers*. 2 vols. Cambridge: Cambridge University Press, 1978.

(Vol. 1: *The Methodology of Scientific Research Programmes;* vol. 2: *Mathematics, Science and Epistemology.*)

Laudan, L. *Progress and Its Problems.* Berkeley: University of California Press, 1977.

Lear, J. "Sets and Semantics." *Journal of Philosophy,* 74 (1977): 86–102.

Lehrer, K. *Knowledge.* Oxford: Oxford University Press, 1974.

Leibniz, G. *Mathematische Schriften,* edited by C. Gerhardt. Halle, 1849–63.

Lewis, C. I. *Mind and the World Order.* New York: Dover, 1956.

Lewis, D. K. "The Paradoxes of Time Travel." *American Philosophical Quarterly,* 13 (1976): 145–52.

———. *Convention.* Cambridge: Harvard University Press, 1969.

———. "Propositional Attitudes *De Dicto* and *De Se.*" *Philosophical Review,* 88 (1979): 513–43.

L'Hôpital, G. F. A. *Analyse des Infiniments Petits.* Paris, 1696.

L'Huilier, S. *Exposition Élémentaire des Principes des Calculs Supérieures.* Paris, 1786.

Locke, J. *Essay Concerning Human Understanding.* Edited by A. C. Fraser. New York: Dover, 1959.

Maclaurin, C. *Treatise on Fluxions.* Edinburgh, 1742.

Maddy, P. "Perception and Mathematical Intuition." *Philosophical Review,* 89 (1980): 163–96.

Mahoney, M. S. *The Mathematical Career of Pierre de Fermat.* Princeton: Princeton University Press, 1973.

Masterman, M. "The Nature of a Paradigm." In *Criticism and the Growth of Knowledge,* edited by I. Lakatos and A. Musgrave. Cambridge: Cambridge University Press, 1970.

Mendelson, E. *Introduction to Mathematical Logic.* Princeton: Van Nostrand, 1964.

Michaeis, C., and C. Carello. *Direct Perception.* Englewood Cliffs: Prentice-Hall, 1981.

Mill, J. S. *A System of Logic.* London: Longmans, 1970.

Mittag-Leffler, G. "Une page de la vie de Weierstrass." *Proceedings of the Second International Congress of Mathematicians.*

Moore, G. E. *Some Main Problems of Philosophy.* New York: Collier Books, 1962.

Nagel, E. *The Structure of Science.* London: Routledge and Kegan Paul, 1961.

———. "Impossible Numbers." In *Teleology Revisited.* New York: Columbia University Press, 1979.

———. "The Formation of Modern Conceptions of Formal Logic in the Development of Geometry." In *Teleology Revisited.* New York: Columbia University Press, 1979.

Neugebauer, O. *The Exact Sciences in Antiquity.* New York: Dover, 1969.

———. *Vorgriechische Mathematik.* Berlin: Springer, 1934.

Newton, I. *Mathematical Works.* 2 vols. Edited by D. T. Whiteside. New York: Johnson, 1964.

———. *Treatise on Quadrature.* In *Mathematical Works,* edited by D. T. Whiteside. New York: Johnson, 1964.

———. *The Mathematical Papers of Isaac Newton.* Volume 1. Edited by D. T. Whiteside. Cambridge: Cambridge University Press, 1967.

———. *The Mathematical Principles of Natural Philosophy.* Edited and translated by A. Motte and F. Cajori. Berkeley: University of California Press, 1960.

Nieuwentijdt, B. *Considerationes circa analyseos ad quantitates infinite parvas appli-*

catae principia, et calculi differentialis usum in resolvendis problematibus geo-metricis. Amsterdam, 1694.

O'Malley, M. T. *The Emergence of the Concept of an Abstract Group.* Unpublished dissertation. Columbia University, 1973.

Parsons, C. "Infinity and Kant's Conception of the 'Possibility of Experience'." *Philosophical Review,* 73 (1964): 183–97.

———. "Ontology and Mathematics." *Philosophical Review,* 80 (1971): 151–76.

———. "What Is the Iterative Conception of Set?" In *Logic, Foundations of Mathematics, and Computability Theory,* edited by R. E. Butts and J. Hintikka. Dordrecht: Reidel, 1977.

———. "Mathematical Intuition." *Proceedings of the Aristotelian Society* (1979–80): 145–68.

Perry, J. "Frege on Demonstratives." *Philosophical Review,* 88 (1977): 474–97.

———. "The Problem of the Essential Indexical." *Noûs,* 13 (1979): 3–21.

Poincaré, H. *Science and Method.* New York: Dover, 1952.

———. *Science and Hypothesis.* New York: Dover, 1952.

———. Obituary of Weierstrass. *Acta Mathematica,* 22: 1–18.

Polya, G. *Induction and Analogy in Mathematics.* Princeton: Princeton University Press, 1954.

———. *Patterns of Plausible Inference.* Princeton: Princeton University Press, 1954.

Popper, K. R. *The Logic of Scientific Discovery.* London: Hutchinson, 1959.

Putnam, H. *Philosophical Papers.* 2 vols. Cambridge: Cambridge University Press, 1975.

———. "What Is Mathematical Truth?" In *Philosophical Papers,* vol. 1. Cambridge: Cambridge University Press, 1975.

———. "The Thesis that Mathematics is Logic." In *Philosophical Papers,* vol. 1. Cambridge: Cambridge University Press, 1975.

———. "The Meaning of 'Meaning.' " In *Philosophical Papers,* vol. 2. Cambridge: Cambridge University Press, 1975.

———. "Explanation and Reference." In *Philosophical Papers,* vol. 2. Cambridge: Cambridge University Press, 1975.

———. "Meaning and Reference." *Journal of Philosophy,* 70 (1973): 699–711.

———. "Two Dogmas Revisited." In *Contemporary Philosophy,* edited by G. Ryle. Oxford: Oriel Press, 1977.

———. "There Is At Least One A Priori Truth." *Erkenntnis,* 13 (1978): 153–70.

Quine, W. V. "Two Dogmas of Empiricism." In *From a Logical Point of View.* New York: Harper and Row, 1963.

———. *The Ways of Paradox.* New York: Random House, 1966.

———. "Carnap and Logical Truth." In *The Ways of Paradox.* New York: Random House, 1966.

———. "Truth by Convention." In *The Ways of Paradox.* New York: Random House, 1966.

———. *Word and Object.* Cambridge: MIT Press, 1960.

———. *Philosophy of Logic.* Englewood Cliffs: Prentice-Hall, 1970.

Quine, W. V., and J. Ullian. *The Web of Belief.* New York: Random House, 1970.

Rawls, J. "Kantian Constructivism in Moral Theory." *Journal of Philosophy,* 77 (1980): 515–72.

Reichenbach, H. *The Philosophy of Space and Time*. New York: Dover, 1958.

Resnik, M. D. "Mathematical Knowledge and Pattern Recognition." *Canadian Journal of Philosophy*, 5 (1975): 25–39.

———. *Frege and the Philosophy of Mathematics*. Ithaca: Cornell University Press, 1980.

———. "Mathematics as a Science of Patterns: Ontology." *Noûs*, XV (1981): 529–50.

Robins, B. *The Mathematical Tracts of Benjamin Robins*. London, 1761.

Russell, B. *An Introduction to Mathematical Philosophy*. London: Allen and Unwin, 1960.

———. *An Inquiry into Meaning and Truth*. Baltimore: Penguin, 1962.

———. *Human Knowledge: Its Scope and Limits*. New York: Simon and Schuster, 1962.

Scheffler, I. *Science and Subjectivity*. Indianapolis: Bobbs-Merrill, 1967.

Schlick, M. "The Foundation of Knowledge." In *Logical Positivism*, edited by A. J. Ayer. New York: Free Press, 1966.

Sellars, W. *Science, Perception, and Reality*. London: Routledge and Kegan Paul, 1963.

———. "Empiricism and the Philosophy of Mind." In *Science, Perception, and Reality*. London: Routledge and Kegan Paul, 1963.

Shapere, D. "Meaning and Scientific Change." In *Mind and Cosmos*, edited by R. Colodny. Pittsburgh: University of Pittsburgh Press, 1966.

———. "Notes towards a Post-Positivistic Interpretation of Science." In *The Legacy of Logical Positivism*, edited by P. Achinstein and S. Barker. Baltimore: Johns Hopkins Press, 1969.

———. "Scientific Theories and Their Domains." In *The Sturucture of Scientific Theories*, edited by F. Suppe. Urbana: University of Illinois Press, 1977.

Sklar, L. *Space, Time, and Space-Time*. Berkeley: University of California Press, 1974.

Stalnaker, R. "Possible Worlds." *Noûs*, 10 (1976): 65–76.

Steiner, M. *Mathematical Knowledge*. Ithaca: Cornell University Press, 1975.

———. "Mathematical Explanation." *Philosophical Studies* (1978): 135–51.

Struik, D. J. *A Source Book in Mathematics: 1200–1800*. Cambridge: Harvard University Press, 1969.

Teller, P. "Computer Proof." *Journal of Philosophy*, 77 (1980): 797–803.

Tharp, L. *Myth and Math*, unpublished manuscript.

Toulmin, S. *Human Understanding*. Vol. 1. Princeton: Princeton University Press, 1972.

Troelstra, A. S. *Principles of Intuitionism*. Berlin: Springer, 1969.

Tymoczko, T. "The Four-Color Problem and Its Philosophical Significance." *Journal of Philosophy*, 76 (1979): 57–83.

Wang Hao. *From Mathematics to Philosophy*. London: Routledge and Kegan Paul, 1974.

Weierstrass, K. *Werke*. 7 vols. Berlin, 1894–1927.

Weyl, H. *Philosophy of Mathematics and Natural Science*. Princeton: Princeton University Press, 1949.

White, N. "What Numbers Are." *Synthese*, 27 (1974): 111–24.

Whittaker, E. T., and G. N. Watson. *A Course of Modern Analysis*. Cambridge: Cambridge University Press, 1915.

Wilder, R. *Evolution of Mathematical Concepts*. New York: Wiley, 1975.

Williams, M. *Groundless Belief*. New Haven: Yale University Press, 1977.

Wittgenstein, L. *Remarks on the Foundations of Mathematics*. Oxford: Blackwell, 1956.

Wussing, H. *Die Genesis des Abstrakten Gruppenbegriffs*. Berlin, 1969.

Index

Abel, N., 149, 185, 188, 205, 215, 217, 249, 250, 251, 254, 256, 257, 268, 270
Abstract objects, 102. *See also* Platonism
Addition, 112, 135, 136
Affordance, 11–12
   universal, 12, 108n
Analyticity, 68n, 80. *See also* Conceptualism
Apollonius, 197
Appel, K., 46n
A priori knowledge. *See also* Mathematical apriorism
   approximate, 89
   certainty and, 42–43
   indubitability and, 56
   innateness and, 21
   linguistic ability and, 73, 76ff
   necessity and, 29–30, 32–35
   positivist analysis of, 15–16
   psychologistic analysis of, 21–24, 88–89
   Quine's critique of, 80ff
   self-knowledge and,, 27–30
   unrevisability and, 80–81
Apriority-preserving rule of inference, 38
Apsychologistic account of knowledge
   analyticity and, 68n
   a priori knowledge and, 14–15
   difficulties of, 15–17
   influence on conceptualism, 65
   influence on philosophy of mathematics, 39, 47
   main features of, 14
   mathematical proof and, 37
Archimedes, 9

Argand, J., 176
Aristotle, 158
Arithmetic, reconstruction of, 112–19
Axiomatization. *See also* Systematization
   of Euclidean geometry, 219–20
   of group theory, 220
Ayer, A. J., 14n, 15n

Barrow, I., 232
Basic a priori statement, 38
Beck, L. W., 85n
Belnap, N., 186n
Benacerraf, P., 59, 60n, 68n, 102, 103, 104n, 140n
Berkeley, G., 200, 214, 239, 240, 241n, 252, 259
Bernays, P., 38n
Bernoulli, D., 245, 249, 254
Bernoulli, Jacques, 229, 236, 241, 245
Bernoulli, Jean, 229, 236, 237, 241, 245
Birkhoff, G., 247n
Bochner, S., 248n
Bolyai, J., 158, 159, 160, 224
Bolzano, B., 190n, 191, 192, 209, 210, 211, 215, 219, 260, 262, 263, 264, 267
Bolzano-Weierstrass Theorem, 135n
Bombelli, R., 176
Boolos, G., 131, 132, 133, 134, 145
Bourbaki, N., 127, 134, 138, 271
Boyer, C., 248n, 256n
Boyle-Charles Law, 116, 117, 138, 140